P9-ARB-290

ALLE ZEIT WACH
1842

Springer Series in Synergetics

Synergetics, an interdisciplinary field of research, is concerned with the cooperation of individual parts of a system that produces macroscopic spatial, temporal or functional structures. It deals with deterministic as well as stochastic processes.

Volume 1 **Synergetics** An Introduction 2nd Edition
By H. Haken

Volume 2 **Synergetics** A Workshop
Editor: H. Haken

Volume 3 **Synergetics** Far from Equilibrium
Editors: A. Pacault and C. Vidal

Volume 4 **Structural Stability in Physics**
Editors: W. Güttinger and H. Eikemeier

Structural Stability in Physics

Proceedings of Two International Symposia on Applications of Catastrophe Theory and Topological Concepts in Physics

Tübingen, Fed. Rep. of Germany, May 2–6 and December 11–14, 1978

Editors: W. Güttinger and H. Eikemeier

With 108 Figures

Springer-Verlag Berlin Heidelberg New York 1979

Professor Dr. *Werner Güttinger*

Institute for Information Sciences, University of Tübingen
D-7400 Tübingen, Fed. Rep. of Germany, and

Department of Physics and Astronomy, University of Wyoming,
Laramie, WY 82071, USA

Dr. *Horst Eikemeier*

Institute for Information Sciences, University of Tübingen
D-7400 Tübingen, Fed. Rep. of Germany

ISBN 3-540-09463-6 Springer-Verlag Berlin Heidelberg New York
ISBN 0-387-09463-6 Springer-Verlag New York Heidelberg Berlin

Offset printing: Beltz Offsetdruck, Hemsbach
Bookbinding: J. Schäffer OHG, Grünstadt
2153/3130-543210

Preface

This volume is the record and product of two International Symposia on the Application of Catastrophe Theory and Topological Concepts in Physics, held in May and December 1978 at the Institute for Information Sciences, University of Tübingen.

The May Symposium centered around the conferral of an honorary doctorate upon Professor René Thom, Paris, by the Faculty of Physics of the University of Tübingen in recognition of his discovery of universal structure principles and the new dimension he has added to scientific knowledge by his pioneering work on structural stability and morphogenesis. Owing to the broad scope and rapid development of the field, the May Symposium was followed in December by a second one on the same subjects. The symposia, attended by more than 50 scientists, brought together mathematicians, physicists, chemists and biologists to exchange ideas about the recent fascinating impact of topological concepts on the physical sciences, and also to introduce young scientists to the field. The contributions, covering a wide spectrum, are summarized in the subsequent Introduction.

The primary support of the Symposia was provided by the "Vereinigung der Freunde der Univertät Tübingen" (Association of the Benefactors of the University). We are particularly indebted to Dr. H. Doerner for his personal engagement and efficient help with the projects, both in his capacity as Secretary of the Association and as Administrative Director of the University. We also would like to thank the President of the University, A. Theis, for his interest and encouragement. The December Symposium received additional support from the British Council and the German Academic Exchange Service. We are deeply indebted to their respective officers, Miss L.J. Bennet and Dr. G. Schulz, for their kind cooperation.

It is a most pleasant task to express on behalf of the lecturers and participants our gratitude to Miss E. Kohn and Miss B. Santorini for their invaluable and efficient assistance in the organization of the Symposia and for their expert and cheerful help in preparing difficult manuscripts. Furthermore, we are grateful to Professor M. Dal Cin, and, in particular, to Dr. E. Dilger for their help and also to Dr. H. Lotsch of Springer-Verlag for his patient cooperation.

Tübingen, March 1979

W. Güttinger
H. Eikemeier

Contents

Part V *Statistical Mechanics and Phase Transitions*

Part VI *Solitons*

Part VII *Dynamical Systems*

Introduction

The Symposia on Structural Stability in Physics had the objective to discuss recent applications of topological concepts deriving from the notion of structural stability -- notably, catastrophe and singularity theory -- in disparate physical systems exhibiting analogous behavior on different scales. The flow of ideas of the central lecture topics of this rapidly expanding field, which are at the frontier of current research, reflects the organization of these Proceedings:

The first part of the book, GENERAL CONCEPTS, begins with R. Thom's reflections about the need for a revival of natural philosophy and the role catastrophe theory plays in it. E.C. Zeeman then summarizes the use of catastrophes for modeling in the sciences, and W. Güttinger provides a sketch of recent trends in applying catastrophe theory to physics. H. Haken closes out the chapter by linking synergetics with catastrophe and bifurcation theory.

The book then turns, in the chapter on TOPOLOGICAL ASPECTS OF WAVE MOTION, to recent developments in the theory of random waves, fractals and diffraction catastrophes. This part starts with M.V. Berry's surveys of the roles catastrophes and fractals play in the theory of intensity fluctuations in random waves and in the distribution of modes in fractal resonators. J.F. Nye concludes the chapter with an analysis of stable optical caustics generated by irregular, shape-changing water drop lenses and the emerging umbilic diffraction patterns.

The papers of the chapter on CATASTROPHES IN INFINITE DIMENSIONS deal with attempts to use catastrophe techniques in spaces with infinitely many dimensions, i.e., in continuum physics, and establish links with bifurcation theory. R. Magnus and T. Poston begin by showing in functional analytic terms how catastrophe theoretic methods can be used to gain insight into modeling infinite-dimensional systems. G. Dangelmayr analyzes bifurcation phenomena in variational problems by means of catastrophes. G. Dangelmayr, W. Güttinger and W. Veit then discuss semiclassical approximations of path integrals on and near caustics in terms of catastrophe theory. M.V. Berry concludes the chapter by analyzing the bifurcation of quantum bound states in the semiclassical limit using the notion of quantum maps.

The contributions of the chapter on DEFECTS AND DISLOCATIONS center around the topological classification of singularities in ordered media, flow fields and wavefronts. R. Thom begins by developing new ideas about combining the notion of pseudogroup with catastrophe theory to describe stable defects in ordered media. J.F. Nye then discusses the changing topology of evolving flow fields, describes their structurally stable singularities in terms of catastrophes and applies the results to geophysical phenomena. F.J. Wright concludes the chapter by showing how catastrophe theory can be used to analyze and classify dislocations of wavefronts encountered in various branches of physics.

The next chapter is devoted to topics in STATISTICAL MECHANICS AND PHASE TRANSITIONS. M. Rasetti begins with a thorough discussion of the fundamental role played by the notion of structural stability in statistical mechanics. G. Dukek then analyzes the validity of the 180° rule of phase diagrams of thermodynamics in terms of a butterfly catastrophe model. K. Keller, G. Dangelmayr and H. Eikemeier show how to describe qualitatively and quantitatively various critical phenomena in terms of higher order catastrophes. F. Schlögl then discusses generalized stability criteria for nonequilibrium phase transitions. W.T. Grandy concludes the chapter by developing ideas about how catastrophe theory may focus, via the structure of partition functions, on the ultimate physical origin of phase transition phenomena.

The papers of the chapter on SOLITONS deal with both the mathematical aspects and the physical significance of the soliton concept, whose topological background is still almost unexplored. R.K. Bullough and R.K. Dodd begin with a comprehensive survey of the role solitons play in various branches of physics and then move to their mathematical description. P.W. Kitchenside, R.K. Bullough and P.J. Caudrey close out the chapter with a discussion of spin wave breathers in ^{3}He B in terms of the double sine-Gordon equation.

The chapter on DYNAMICAL SYSTEMS crosses the borders of physics. R. Brause and M. Dal Cin report on applications of catastrophe theory to discontinuous phenomena in pattern recognition, learning algorithms and decision making. F.J. Seif then models experimental results of the complex system of pituitary thyrotropin secretion in terms of catastrophes at the macroscopic and microscopic level. O.E. Rössler concludes the chapter with an overall view on chaos with its many fascinating problems.

In summing up, the papers of these Proceedings describe topological textures of physics as well as diverse points of view towards the physical forces inducing topologies. This reflects the continuing objective of the Tübingen symposia: encouragement of new approaches to the deep interplay between mathematics and physics.

The Editors

Part I

General Concepts

Towards a Revival of Natural Philosophy *

René Thom

Institut des Hautes Etudes Scientifiques, 35, route de Chartres
F-91440 Bures-sur-Yvette, France

I would like to discuss the synthetic, global point of view of Natural Philosophy by sketching how I see the role played by catastrophe theory in terms of it. Of course, one may say that in doing so I am describing a kind of program, and it is well-known that sometimes programs are realized and sometimes they are not. I am perfectly aware of this fact, but at least it is perhaps good to try to know, where one wants to go, and in that respect, describing a program is perhaps not so useless as it seems. The starting point of my consideration is basically a reflection about the role of mathematics in the sciences and, eventually, in philosophy. It is well-known, of course, that mathematics has to do with science, say, with scientific explanation, but the precise reason why we use mathematics in science is still rather mysterious and one may offer several possible explanations for that. Let me start by trying to define what is science: It is, of course, the well-known task of epistemology to define what is science and you know that epistemologists, in general, do not agree about the definition of science. Well, the definition I want to propose to you is rather simple, but of course it is also rather vague. But one cannot escape some vagueness in that matter, so I would put as a principle that a proposition, if it is to be called scientific, first has to be true, and second that the proof, the veracity, the proof of the truth of the proposition should be understandable, in principle, by anybody. This I would take as a basic definition of science. If you take that as a definition, then, of course, some consequences immediately follow: First, the rejection of any kind of authority arguments because there are no authorities in science. In principle, a proof, the testing of the validity of some proposition should be accessible to anybody. So there should be no experts, no authority at all in science. The second consequence is the cumulative process of science. Namely, if a proposition has been recognized as scientific at a given time and at a given place, then it has to be recognized as scientific for all later times. So the quantity of scientific truth cannot but increase. In this respect, the progress of scientific knowledge is by no means a mystery, it is a tautology. And when people in science are always glorifying themselves with the progress of their discipline, as you see it in the press, then you should not be mistaken by this kind of talk: they are just saying tautologies. Any science, in some sense, progresses by its very definition, contrary to other human disciplines, like philosophy or art, of which one cannot say that they are progressing. Now, if we want to know what makes a scientific proposition accessible and amenable to proof, then one is practically obliged to make the following distinction. Among scientific propositions, there are first those propositions which refer to a specific spatio-temporal event which describes a part of reality and this part of reality has always to be localized in space and time in order that one can make sure the thing is like it is. And in this respect, for instance, I think that history is a science like all other sciences. As to history, many people say that only phenomena which are reproducible can be considered as scientific and that events which are not reproducible should not be considered as scientific. I myself do not share this point of view: I think, for instance, that

---

* Verbatim lecture delivered by R. Thom on the occasion of his being awarded an honorary doctorate by the Faculty of Physics, University of Tübingen, Germany, in 1978.

a historical fact inasmuch its existence has been proved by strict documentation and testimony, could be considered as true on the same solid grounds as any other kind of scientific fact. So, in this respect, the source of all truth in science is fundamentally the reference to spatio-temporal localization. To this extent, I agree with the standard neo-positivist philosophy, which claims, more or less, that the source of all truth is the reference to a specific spatio-temporal event. If we follow this line of reasoning, then we see that there are fundamentally two kinds of proposition in science, the descriptive propositions which refer to a specific event in space-time and the organizational propositions which have, so to speak, a theoretical value and try to organize the data of spatio-temporal experience.
So, I think this distinction between descriptive propositions and spatio-temporal, organizational, theoretical propositions is of fundamental importance and it is from this point of view that I want to start with. Among the organizational propositions, of course, you have first the propositions of mathematics. Mathematical objects are not, of course, directly connected to the real world. The mathematical entities are abstract entities, whether they are constructed or given in a sort of Platonic universe, no matter for the time being. But, in principle, the mathematical propositions do not need any reference to spatio-temporal events to be true. So, in that respect, one could say, the epistemological status of mathematics is somewhat ambiguous and some epistemologists, essentially the neo-positivists, have said that mathematics is nothing but a huge tautology and the information content in mathematics is nil. This is a point of view which I think is very acceptable, to some extent, and I think this makes up precisely the peculiarity of the position of the mathematician because it uses a world which has, in principle, nothing to do with any specific reality. So, let us forget mathematics for the moment and let us consider the problem of theorization of science. Why do we want to make theories in science? Well, I think there is a very good reason for that. Namely, if we accept the idea that science has to do with the description of reality, then, obviously, reality is a rather vague concept and in principle, if one would like to describe the totality of all reality, one would have, so to speak, to make a tremendous catalog of all events in all parts of space-time, which would, of course, be absolutely impossible. So, among all the facts which occur in space-time, we have to make a selection. First, to recognize among the facts which we can see, perceive or detect, to single out among these facts those which we consider more interesting than others. And second, given the totality of all these data, which is so huge that no computer could store all of them, then we need to find a way of condensing or reducing these data in a manageable way, and for this, of course, theorization is a fundamental tool. By theorization I mean, first, to recognize the regularities among the spatio-temporal appearances, patterns or structures. Among the data we are perceiving, we must be able to recognize the underlying regularities, and then we can express these either in terms of the reproducibility of phenomena, like in physics, or by more vague concepts, like the concepts given by Gestalt theory, some sort of vague equivalence class between appearances. And this vague equivalence class, of course, may give rise to a conceptualization which may be used later to systematize the data and to organize the structure of the data of our experience. If you look at things that way, then you are led to the definition that theorization can be considered as the reduction of arbitrariness in the description. Given any kind of experimental field, you have a lot of data whose total description, in general, is impossible because of the immensity of the details which are to be stored. Then by theorization, you are able to single out of these data the meaningful elements and out of these elements to reconstruct a fairly coherent theory. So, in this respect, one could say that the basic object of theory in science is the reduction of the arbitrariness which is needed in the description. Now, how could this reduction of arbitrariness be realized? I think, one of the basic methods to be used for this end could be stated by remembering the old maxim of the French physicist Jean Perrin, who said, "Expliquer, c'est remplacer du visible compliqué par de l'invisible simple". Now, this maxim can be understood in several ways. When Perrin said that, he had, of course, in mind the explanation of natural phenomena by atoms. Atoms were, of course, invisible because of their smallness and this led, of course, to the standard reductionist explanation, which in some sense introduces new beings, atoms, for instance, and forces between atoms out of which you can reconstruct the visible morphology of phenomena. But there is, I think, another approach to theorizing natural phenomena, one that I would call the Platonist approach. You know, the Platonist approach uses the well-known myth of

the cave in front of any morphology. One could try to explain this morphology by introducing unknown parameters or hidden variables, and in the new space obtained by adjunction of these parameters introduce conceptually simpler objects, whose projection on the space of observables would yield the given data. So the Platonist approach takes essentially the following point of view: Here is the space of observables, the space U which is, so to speak, the support or carrier of the experimental morphology, and we get a lot of complicated forms in this space, which we don't know how to explain. Then what we do is to introduce a space of unknown or hidden parameters. So let me call U the space of observables and let S be the space of hidden parameters. Then I will construct in the product space U x S simpler objects which, by projection, will reconstruct the complex morphology I have in the observable space. This is a general procedure for scientific explanation from the Platonist's approach. And, you might perhaps be surprised, but this Platonist approach is followed even in standard physics, it is not just philosophy! To take a very simple example, consider the morphology of a linear oscillator. That is in q-space you see a point moving on an interval according to a sine law, $q=\sin(\omega t)$, e.g., a mass moved by a spring. The q-space is the space U of observables. Then you may introduce a space S of hidden parameters which in that case will be the momentum or p-space, and in the (p,q)-space you will find a nice geometric object, namely the circle $p^2+q^2=1$, and this circle describes a uniform motion with uniform angular velocity ω. Projecting down this motion, you obtain in the observable space the given oscillatory movement. The standard interpretation in science is always following these lines of introducing hidden parameters in order to simplify the description. Now, it happens, of course, that physicists are becoming so used to these new p-coordinates, the momenta, that they now consider momentum coordinates p far more important and far more intrinsic than position coordinates q which for them are a kind of anthropocentric vestigium. But nevertheless, I think it remains still true that the momentum coordinates are created, essentially, for explaining of what happens in the support space of q-coordinates. Now, if we want to proceed a bit more along this way of looking at the scientific theorization, then we are led to the problem of somehow pushing forward this Platonist approach and trying to find out essentially what we mean by something being conceptually simple. And here, of course, one enters a fairly complicated situation because the problem of simplicity is something which is not easily to be defined. The notion of simplicity is perhaps relatively easy in itself, but the contrary notion of complexity is something much more difficult to define. I have often to argue with biologists because they are always claiming that man is much more complicated than a bacterium, and I am always holding the opposite point of view, namely, that the bacterium is much more complicated than man because we understand a lot of things in man which we don't understand in a bacterium. But that point of view is, of course, not accepted by biologists in general, who stick to the strict reductionist point of view. Now, one of the aims of catastrophe theory is precisely to introduce a way of measuring complexity. In some sense, we can say that when the form here is more complex than another, then essentially we try to explain its form as projection of simple objects in the product space, the more complicated is the algebra of the object upstairs, the less complicated one is the form downstairs. So by this procedure, we are, so to speak, able to create a hierarchy of complexity between forms, and with this tool we are able, at least theoretically, to recognize when a form -- a given structure or an experimental morphology -- is exceptional or when, on the contrary, it is rather natural or ordinary. And that's where I see one of the basic interests and consequences of catastrophe theory: it is of help in recognizing when a phenomenon is surprising. It should not be so difficult to recognize when a phenomenon is surprising, but it is a fact that there is no decisive criterion for that. In the absense of a theory you are simply unable to recognize when a phenomenon is surprising. And this is, in a sense, the big difficulty of present-day biologists: they have a lot, indeed an immensity of data on chemistry, biochemistry, molecular biology and so on. Given all these data, but missing any unifying principles, it is very difficult for the biologists to say where the most interesting and significant phenomena are taking place. Now, the complexity we find in these models -- for instance in the elementary catastrophe theory models Christopher Zeeman was describing in his lectures, you are dealing with singularities. So you are led to considering the singularity of a degenerate map, a map having a singular point at a specific critical point of space and then using the theory of singularities, one is able to recognize the complexity of the given map, one is able to classify the complexities of these singular points and to measure, so to speak, their intrinsic complexity. Thus, if one uses this theory, then one is able to solve the problem of ordering the complexity of a given morphology in terms of specific standard forms. Now, let us have a glance at the sciences, as they are existing presently, from the point of view of their theorization. Let me list some of them: physics, chemistry, biology, social sciences, psychology, sociology etc. Mathematics, as I said before, has its own world and, in principle, has nothing to do with the natural world. But mathematics applies directly to physics, and it does so -- essentially -- for the following reason: In physics we have laws, quantitative laws, and these physical laws are extremely precise. I think it was E. Wigner, who once worried about the unreasonable accuracy of physical laws. Now, we can set up exact quantitative models for some physical situations, for instance, in mechanics. Planetary motions, e.g., form a typical example, in which we can formulate exact quantitative models even if sometimes these models cannot be solved exactly because of their intrinsic difficulty or complexity but, nevertheless, the model is there and with sufficiently powerful tools or computers, we could, in principle, solve these models and find the corresponding laws of evolution and then make quantitative predictions. And as you know, this works out fairly well for a good part of physics, namely for that part of physics which deals precisely with fundamental laws, like Newton's law of gravitation, Maxwell's law of electromagnetism, and so on. Now, what is the main reason for, first, the existence of these laws and, second, for the fact that we can use them for predictions? I think it is because the two things are conceptually separated. You might well have a situation in which you have fairly exact laws, but if you are unable to express them in terms of standard mathematical tools, then it might be very difficult for you to use them in a very predictive way. I will come back later to this problem. I think that the physical laws exist, essentially, because of the fact that physicists can and have to communicate between themselves. Two observers should be able to compare their observations or visions of the universe and, as you know, in physics, the comparison of the observations of two observers is always realized through a geometric object called a change of transformations -- a change of frames -- so that the basic physical entity is always some vector space on which you have some representations acting, representations dealing with the change And now, fundamentally, all laws of physics are expressed in terms of the invariance with respect to these representations, and it is a kind of miracle which states that these representations are analytic. If you consider, let's say, an element of the matrix of such a change of frame, then each element, each entry of the matrix, has to be an analytic function of the spatio-temporal coordinate of the observer. And this analyticity requirement, I think, is something rather mysterious. I myself have only one explanation for it, namely, the fact that for compact groups--compactly acting in a space--it is a fact that each representation is analytic. So, I believe that this analyticity one finds there is a result of some sort of compactification of the group action which means that one does not take into account what happens at infinity, or one believes that everything smooths down to zero at infinity, so that one is practically left with a compact situation, and because of that one gets this analyticity property. And now this analyticity property, of course, has the very important consequence, that if you know the experimental data in any domain of your space, then, knowing the data to be analytic, you can extend them by analytic continuation and this extension, of course, allows you to make quantitative predictions which can be compared with reality and tested by experiment.

So the analyticity of representations, I think, is something fundamental in explaining the fecundity of physics, of physical laws, and the possibility of their providing quantitative predictions. Now, if we pass from elementary fundamental physical laws to macroscopic physics, e.g. solid state physics, equations of states of real fluids, and so on, then the situation degrades rapidly. That's because here one finds, of course, empirical laws. For instance, if you have a real fluid, you have an empirical equation of state relating pressure, temperature, and density, but it is impossible to find a complete analytic expression for the real equations of state, and because of this, all extrapolation procedures to some extent break down. Of course, you can make approximations--it's always possible to make approximations and to use approximating procedures--but one is no longer sure of the theoretical validity of such procedures. That's the main point. When we have elementary underlying physical laws, we are capable of obtaining good estimates for the validity of interpolating or extrapolating procedures. But given empirical laws, the situation is already quite bad, and if we pass to chemistry and biology, etc., we realize immediately that the use of quantitative mathematics is no longer really so important. In chemistry, I think, the use of mathematics is related, first, to the local study of interacting molecules, for which you do in fact need a lot of approximating procedures, the validity of which may be viewed with some suspicion, and you also have to discuss chemical kinetics, which generally leaves a lot of coefficients in the equation which you don't know how to estimate. Because like the law of mass action, all these laws are rather approximate and, in general, one does not have an accurate model. So using mathematics in chemistry already becomes quite cumbersome. When we pass to biology, it is fair to say that mathematics is used in biology in basically two fields. First, to elaborate local models, e.g., the circulation of blood in the aorta or propagation of nerve impulses or things of that sort, and second, to elaborate the theory of population, what I would call mathematical ecology. The local models, in my opinion, do not, in general, have great importance from the general theoretical point of view. They represent more of less accurately a local phenomenon. About the theory of population which, of course, is a nice mathematical theory, I think, it is fair to say that its biological relevance is rather dubious. It is a thing most people were quite enthusiastic about, including those in mathematical genetics, but now one is slowly starting to realize that while this work is fairly nice from a theoretical point of view, its biological relevance is rather dubious, essentially because the relation between genetics, genotype and phenotype cannot be formalized, so that all these constructions are more or less in the air. Thus, you see that when you pass from physics to chemistry to biology--and from there to psychology and sociology--you find here too only the use of statistics, with one single exception, viz., mathematical economics, which is a nice theory, but whose relevance for real economics also seems very dubious at this time. There are very much the same objections as to population theory in biology. But nevertheless, I may say that with mathematical economics, e.g., the exchange model of Walras-Pareto, we enter a field in which a first beginning of mathematical conceptualization has been made, which seems to be a bit more promising than previous considerations. And in the so-called "humanity sciences" as contrasted with the "inhumanity sciences," I could just speak about linguistics and ethnology in which one has several algebraic models of very definite structuralistic type. They are, I think, of more theoretical than practical interest. But they may nevertheless have some interest by showing that when you go further up to the domain of man's activities, in a sense you recover some mathematical character. That is a point I want to stress. I would say that the minimum of existing standard mathematics lies somewhere between biology and psychology.

Now, the fact that the mathematical tool loses its strength and its operational character as we go down the scale of sciences is something which, of course, is well-known among scientists, but it is a thing that nobody likes to speak very clearly about and this for essentially three reasons. Three reasons which are very easy to understand: The first is that those sciences which do not have as efficient tools at hand as physical laws would like to be like physics, and therefore they tend to build a methodology similar to that of physics and try to appear in the eyes of other people as precise as physics. Every science wants to become mathematized because it believes that way it would be put on the same footing as fundamental physics. That is, so to speak, the external reason. The second, internal reason now works in the reverse sense: Inasmuch as a given science does not allow for precise mathematization it opens practically indefinite working possibilities to scientists in that field, because they can make models of all kinds, with approximations, statistical hypotheses, and so on, and there is practically no limit to the possibility of building models in situations which actually do not allow for specific, exact quantitative models. So from the internal point of view, the imprecision of a science is a good thing for scientists working in that field because it allows them to work. And to work, I would say, almost indefinitely. And the third reason, of course, is the computer industry's lobby: Every laboratory wants to have its own computer, and as soon as the lab has got it, it wants to have its computer working even in situations where a priori there is no reason to believe that you can extract any kind of useful information out of the things you have put into the computer. So far, this degradation of the mathematical tool is something which is well-known but also rather well hidden from our external point of view.

Now, I think, what catastrophe theory can offer in this respect may be stated as follows. I was speaking earlier about science and said that science deals with propositions which are true. Now, the truth of a proposition is something which is not, perhaps, as important as it seems. I think that truth has to be compared with interest. That is, one should not only look at the truth of a proposition but one should also look at its interest. And interest divides, of course, into two parts, practical interest and theoretical interest. Practical interest means essentially to predict and to act; theoretical interest means essentially to understand the phenomenon. Of course, this scission might perhaps seem strange to you, and many people believe that one cannot separate practical interest from theoretical interest. I'm not so sure about that. It seems to me that there are situations which one understands perfectly well but where, nevertheless, one finds oneself unable to act. For instance, if you are in a house which is being threatened by a flood, and the flood is rising higher and higher, you climb on the roof of the house and continue to see the flood getting higher, then you understand the situation perfectly, but you are unable do anything about it. Conversely, there are situations in which one may be able to act quite efficiently without understanding: I think present-day medicine provides proof of that almost entirely--inasmuch as it acts effectively, of course, which remains to be seen. But fundamentally, there are quite a number of situations which one can handle by means of experimentally known tricks, while their deep reasons are still not understood. So, what I wanted to say is that the truth of a scientific proposition, as I explained earlier, has to do, of course, with spatio-temporal reference, so if I would push my position to the extreme, I would say that the only really true propositions in science are descriptive propositions. All organizational or theoretical propostions are to some extent doubtful, or even suspicious. I think the main exception to this statement are the fundamental physical laws which are organizational propositions, but nevertheless true. But this, I think, is a sort of exception in the whole situation of science, and in general, any kind of organizational sentence could be considered as not completely true. This is well-known since, for instance, the Popperian analysis: "No hypothesis can ever be verified by experiment because whatever the number of experiments you make to verify an hypothesis, you are never sure that a further experiment would not falsify your hypothesis." So, in this regard, I think it is fair to say that all theoretical propositions in science are to some extent conjectural and in that respect they do not share the status of truth of the purely descriptive propositions. This in a sense amounts to saying that the distinction between science and philosophy--if you put it that way--is not as clear as one ordinarily believes. There are disciplines or theories like psychoanalysis, for instance, which are not or cannot be termed scientific because, as for instance Karl Popper has shown, any kind of failure of the theory can be explained in terms of the theory itself; so such theories cannot be falsified. Thus, strictly speaking, they cannot be said to be scientific. But nevertheless, I would claim that psychoanalysis from the general point of view of the history of ideas is a theory which is far more important in the history of mankind than many present-day scientific theories. Thus, in point of view of interest, there is no doubt that some theories which are not scientific are nevertheless quite important. And the importance of a theory, then, has to be judged from these two directions, either from the practical importance or from the theoretical importance to help understanding. Now I think that catastrophe theory is essentially an attempt to construct a general method for scientific explanation and I may basically localize it as follows. There are three ways for scientific explanation. First, models of the physical type, exact quantitative models. Then you have, let us say, the biological type, which is essentially description plus, I would say, low-level theory. And then you have the social sciences, where the problems are different: If you read papers in social science, then generally you are astonished by the fact that their authors are extremely intelligent. I am not joking: it is true that people writing on sociology, ethnology, or anthropology are in general extremely clever. They are highly cultured and skilled in rhetoric and are able to see a lot of things and compare them, so what we are dealing with here is verbal conceptualization which is essentially of extreme finesse and intelligence. This cannot be said of present-day theorization in biology. Now, what I claim is that catastrophe theory in a sense may be able to fill the gap, this hole between these three types of theorization in science. And it can do so in the following way. There is a connection here with exact quantitative models in catastrophe theory and this connection, of course, is made when one makes the usual approximations in these models. One naturally obtains models which are halfway between quantitative models and qualitative models. The strict qualitative model is, of course, a model which is described not by an equation but by an equivalence class of equations. One has a lot of equations and there is no reason for choosing a specific class from among them. But all these equivalence classes of equations have the same qualitative structure so, in that respect, this describes the phenomena up to qualitative analogical classification. But there is a sort of obscure domain here: One does not pass from strict quantitative models to purely qualitative models in a sharp way. There are intermediate domains when one uses empirical approximation, statistical hypotheses, approximate models, and so on. This is the domain about which Rutherford was quite true in saying that qualitative is nothing but poor quantitative. It is this obscure but fascinating intermediate domain. And it is this reproach of using unfounded but controlled approximations, I guess, which one can level against some catastrophe theory models but the same reproach can be levelled at many models applying standard applied mathematical methods.

Now, catastrophe theory also may help in connecting quantitative models with verbal conceptualization and this, I think, relates to the attempt I have made to modelize geometrically the syntactic structure of sentences: Inasmusch as we will be able to geometrize their meaning, we can put verbal conceptualization on a universal basis. One of the basic objections against verbal conceptualization, of course, is the impossibility of making sure that a given word in a language has a unique translation in another language. Science, as I said, has to be universal. Hence, any kind of scientific concept should be translatable into all the languages of the world. And this defines that vague, almost invisible but important boundary between science and philosophy: A scientific concept has a unique meaning all over the world, in all languages, whereas a philosophical concept does not. If you want to translate the French word "raison," then you have the choice between "Verstand" and "Vernunft" in German, and it is difficult to understand which word to choose. That is just one example to demonstrate that if we want to make the theorization of social sciences into a genuine scientific discipline, we have to make these people less intelligent than they actually are and to oblige them to use concepts which are so poor and so primitive that they can be visualized in a geometric way, in a way which can be grasped by everybody. In fact, one of the basic difficulties of the social sciences is that they use concepts which cannot be expressed in terms of morphologies. A concept such as "power" of "social class" or things of this sort are concepts which roughly speaking cannot be expressed in a morphological way. And this, certainly, is a big drawback for the social sciences. Now, in biology, of course, you have the morphological level just in front of you, and I would say that in general you have no more. Then, the main interest of catastrophe theory is to create a space of hidden variables above the given space here and to introduce objects of algebraic nature, like the one I was describing earlier, which would be able, so to speak, to unify the morphological events and reduce the arbitrariness of the description. This is how I see, from the point of view of natural philosophy, the place of catastrophe theory in the sciences: at the very center, I would say, of the above-mentioned three types of theorization. Of course, only its future development can tell us whether it will fulfill this role and its promises, but at least that is the program I want to outline here.

Catastrophe Theory*

E.C. Zeeman

Mathematics Institute, University of Warwick
Coventry, Warwickshire CV4 7AL, U.K.

Ladies and gentlemen, it is privilege and pleasure to speak in honour of René Thom. It is also a very European occasion that you invite an Englishman to deliver the laudatio for a Frenchman receiving an honorary degree at a German university.

As Professor Güttinger has observed, the unifying factor behind Thom's work, behind his mathematics, his science and his philosophy, is that he is a geometer. Now a geometer needs an overall picture, a "Gestalt" point of view; he needs to classify his forms so that he can turn them over in his hand, and develop an affection for them until they are old friends. This is exactly what Thom does both in his early great work on manifolds prior to his Fields Medal [6] *in 1958, and also in his later work on singularities and applications, which has become known as catastrophe theory, and which (in my opinion) will prove to be even more far-reaching.*

But before I get on to catastrophe theory, let me first describe briefly René Thom's early work on the classification of manifolds up to cobordism. Two n-manifolds X and Y are said to cobound if together they bound an (n+1)-manifold.

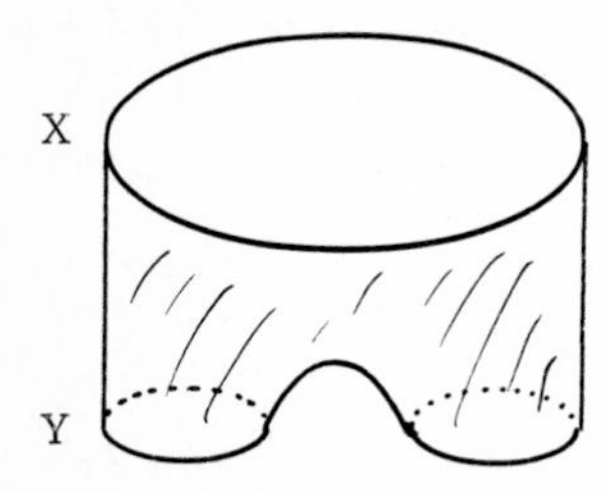

Figure 1: Cobounding manifolds

For example in the picture n=1, X is the circle at the top, Y consists of the two circles at the bottom, and together they bound the pair of trousers. This is an equivalence relation; let Ω^n denote the set of equivalence classes, and $\Omega=\Sigma\Omega^n$. Then Ω can be given a ring structure, defining addition by disjoint union, subtraction by reversal of orientation, and multiplication by the cartesian product. Of course, what made this definition significant was the fact that Thom was able to compute the ring in terms of the homotopy groups of an auxilary space manufactured out of the orthogonal group.

---

* Verbatim laudatio delivered by E.C. Zeeman in honour of R. Thom on the occasion of the latter being awarded an honorary doctorate by the Faculty of Physics, University of Tübingen, in 1978.

To appreciate the significance of this achievement, you should be aware of the geometric despair of topology around the 1950's, because at that time the subject had been all but gobbled up by algebra. Surfaces (= 2-manifolds) had been classified in the nineteenth century, and when Poincaré began his attempt to classify 3-manifolds in 1899 he ran up against the famous Poincaré Conjecture: is a simply connected 3-manifold a sphere? The failure to solve this conjecture - indeed it is still unsolved today - blocked all further attempts at classification. Meanwhile topologists began to bypass the problem: they invented the tools of algebraic topology, the homology and homotopy groups, which proved so rich and splendid that topologists were seduced away from the real subject matter of topology and in effect all became algebraists. I remember as a student in 1951 the first time I heard Henry Whitehead lecture he confessed rather sadly that topologists had given up studying the homeomorphism problem because it was too difficult. Moreover this tendency was exaggerated in France, where algebra reigns supreme. So you must imagine Thom's predicament as a geometer, starting out as a lone figure swimming against the stream and in struggling to express what homology really meant in geometric terms, he had the temerity to cut right through this despair, by classifying the cobordism classes of manifolds. This was the first step towards the general classification of manifolds, which is still very much at the centre of research in topology today. It was not that he avoided the algebra; on the contrary he used it with mastery, but as a servant to the geometry, which is its proper place in topology.

After his Fields Medal, Thom retired into his shell for a while, for nearly a decade, to think about the next step. What was the next step? Whenever one has a category in mathematics, the first step is to study the objects, the next step is to study the maps, then the automaps, and so on. Therefore in the category of smooth manifolds the first step was the classification problem, the next step was to study smooth maps between manifolds, then the diffeomorphism problem, and so on. The study of smooth maps resolved itself into a local problem, singularity theory, and a global problem of how these singularities fit together. During the last two decades singularity theory has flourished, starting with Whitney, followed by Thom, who was responsible for several of the fundamental concepts concerning stability, transversality, unfolding, etc., and who in turn stimulated Arnold, Malgrange, Mather and many others. Meanwhile the global problem has hardly been touched upon as yet; as Thom puts it, we need to develop a language, in which the phonemes are singularities, the words are configurations of singularities determined by sections of higher dimensional singularities, and in which the grammar is the constraint imposed by the global structure, including the algebraic topology (illustrated for instance by the Thom polynomials).

At the same time as studying singularities, Thom's established position in mathematics gave him the leisure, authority and opportunity to think about the creation and evolution of forms in nature, which had always fascinated him and been close to his heart. Gradually he put these two ideas together, on the one hand the mathematics of stability, singularities and dynamical systems, and on the other hand the applications to morphogenesis in several branches of science. Each idea fed upon and stimulated the other, and the result was his remarkable book "Structural Stability and Morphogenesis", [11]. This is primarily a work in the philosophy of science, but, unlike most works in that field, is oriented towards the future rather than the past: it not only offers novel approaches, but reassesses the objectives of applied mathematics, and proposes broad programmes for future research in both mathematics and the physical and biological sciences. It is this mixture of philosophy of science, applications, programmes of research, theorems and conjectures drawn from several related branches of mathematics, that has become loosely known as catastrophe theory in the popular press. Now, the main subject of my talk is to explain a little of catastrophe theory to those of you who have not heard it before, and so, at the risk of boring those who already know it, I will start with an elementary introduction.

The name "catastrophe" refers to the unexpectedness of discontinuous effects when they are produced by continuous causes, thus violating our intuition, which would

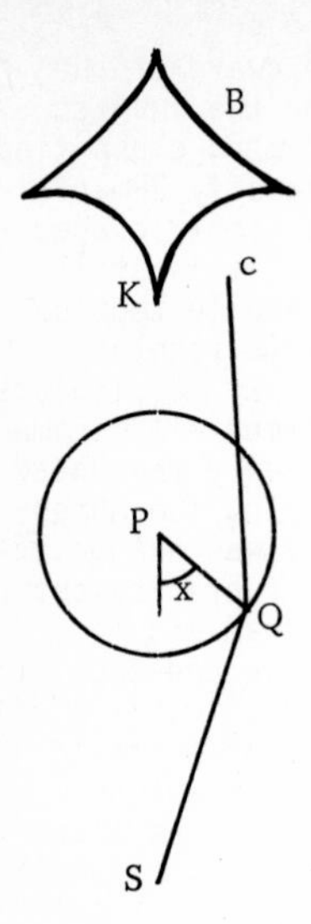

Figure 2: Catastrophe Machine

normally lead us to expect continuous effects. The little machine shown in Figure 2 is designed to illustrate this phenomenon. It consists of a disk, freely pivoted at P, with two elastic bands attached at Q, one of which has its other end fixed at S. The other end of the other elastic band is held at a point c, which we call the control point. The disk seeks a stable equilibrium position, which is a local minimum of the energy V_c in the elastic bands; we measure this position by the angle x = QPS. Here x lies in a circle, which we denote by X. Let C denote the plane. If we move c smoothly around in C, then most of the time the disk responds by moving smoothly, but sometimes it jumps. The jumps occur when c crosses a diamond-shaped curve B, which has four cusps; moreover the jumps only occur upon exit after entering from the other side. The reason for the jumps is that when c lies inside B then V_c has two local minima, but when c lies outside B then V_c has only one minimum. As c grosses B going outwards, the one minimum disappears (as in Figure 3) and so a stable equilibrium breaks down; if the disk happened to be sitting in that equilibrium, then it will jump into the other surviving stable equilibrium. B is called the bifurcation set, because that is the set above which the equivalence class of V_c changes, and hence the qualitative nature of the statics and dynamics change.

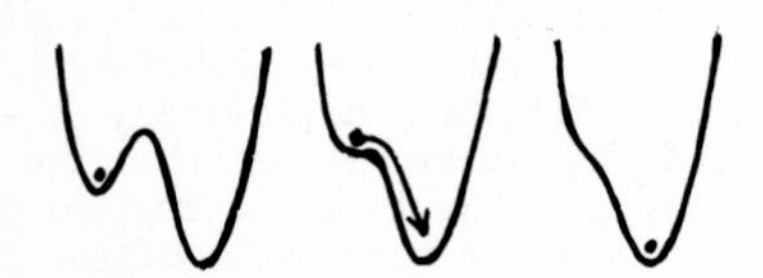

Figure 3: Energy Graphs

If we plot the stationary values of V in XxC, for all cϵC, we obtain the graph M of cause and effect, where c = cause and x = effect. Restricting attention to a neighbourhood of the cusp point K in C, and a neighbourhood of 0 in X, the resulting subset of the graph is shown in Figure 4. The equilibria form a smooth surface M, single-sheeted over the outside of B, triple-sheeted over the inside of B, with the upper and lower sheets corresponding to stable equilibria (local minima of V) and the middle shaded sheet corresponding to unstable equilibria (local maxima of V); the projection M $\to$ C has fold singularities over B-K and a cusp singularity

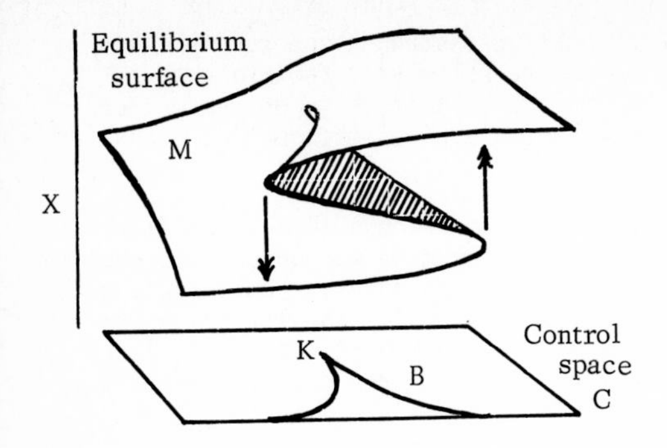

Figure 4: Cusp Catastrophe

over K. The two vertical arrows represent the catastrophic jumps exhibited by the disk when the control point is moved to the left and right across the front of the picture. The significance of the surface in Figure 4 is contained in the following theorem.

Elementary Catastrophe Theorem (Thom, Mather).

Given an n-dimensional behaviour space X, a 2-dimensional control space C, and a smooth locally-stable function V on X parametrised by C, then the stationary points of V form a smooth 2-dimensional surface $M \subset X \times C$, and the only singularities of the projection $M \to C$ are folds and cusps. Moreover, most functions are locally stable (more precisely, there is an open dense set of them of measure one in the space of all smooth functions).

In other words Figure 4 illustrates the most complicated possible local structure of M; of course globally there may be many cusps - for instance there are four in Figure 2. Since the hypothesis is so general, one would expect the surface illustrated in Figure 4 to crop up frequently, as indeed it does, in many branches of science. The theorem also extends to higher dimensional control spaces.

Elementary Catastrophe Theorem (continued)

If C is k-dimensional then M is a smooth k-manifold, and the number of types of singularities is given by:

k	1	2	3	4	5	6	...
number of new singularities	1	1	3	2	4	4	...

When k = 1 only the fold appears; when k = 2 the cusp appears. When k = 3 three new singularities appear, whose bifurcation sets are shown in Figure 5.

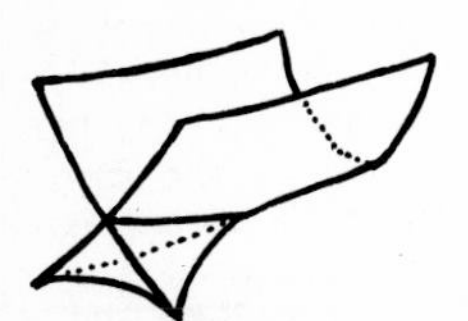

Swallowtail

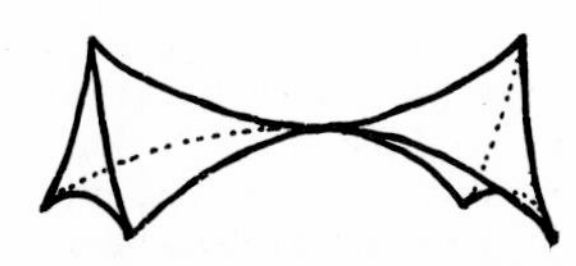

Elliptic umbilic

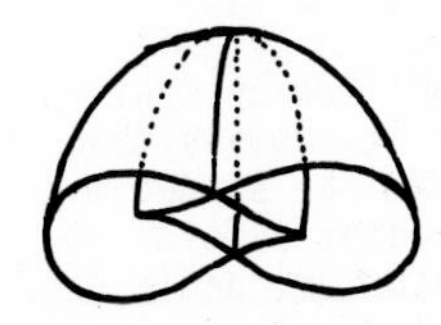

Hyperbolic umbilic

Figure 5: Bifurcation sets, k=3

The first surprising thing about the elementary catastrophe theorem is that classification should be possible at all; there are relatively few branches of mathematics that are sufficiently clear for the stable objects to be dense. But it is exactly this stability that is so desirable in applied mathematics for scientific modelling.

The next surprise is the fewness of singularities on Thom's list. With a little imagination one can visualise all sorts of the ways for M to map onto C, so why don't they occur? For example why can't we have a winding point, such as occurs in the map $z \to z^2$ of the complex plane? The answer is that this singularity is unstable, as Whitney [13] showed, because the perturbation $z \to z^2 + \varepsilon\bar{z}$ converts the singular point at the origin into a bifurcation set with three cusps, in fact a triangular hypocycloid.

Whitney's work in the 1950's on the classification of stable singularities was the inspiration for Thom's theory.

During the 1960's Thom introduced the essential concepts and laid down the main structure for the proof of the elementary catastrophe theorem. He persuaded others to provide the necessary subsidiary results, and fill in details of proofs: Malgrange proved the preparation theorem, and Mather proved the uniqueness of unfoldings. A complete exposition of the proof for $k \leq 5$ can be found in [14]. Thom emphasised the first seven singularities corresponding to $k \leq 4$ because he was particularly interested in applications to embryology, in which C represented 4-dimensional space-time. In dimension $k \geq 5$ the classification has been considerably extended by Arnold [1] and the Russian school.

At first sight it seemed as though Thom's classification was the same as Whitney's, because Whitney had already proved that the only stable singularities of maps between surfaces are folds and cups; it turns out that the two theories are subtly different, because the surface M that appears in the elementary catastrophe theorem above is not an arbitrary surface, but is constrained to be the critical set (= set of stationary points) of a function V. In other words Thom classifies singularities that are stable under this constraint. Although the constraint makes no difference when $k = 2$, it causes the two theories to diverge when $k \geq 3$. Most of Whitney's singularities do not appear on Thom's list since they cannot satisfy the constraint, and conversely most of Thom's singularities do not appear on Whitney's list because if the contraint is removed they become unstable. For example if $k = 3$, the three singularities on Thom's list are shown in Figure 5, but when the constraint is removed then the two umbilics become unstable, leaving only the swallowtail on Whitney's list. In order to distinguish between the two lists, and since Whitney had already pre-empted the term stable singularities, Thom called his list the elementary catastrophes. Arnold prefers to call them Lagrangian singularities because the same singularities also appear in the more specialised context of Lagrangian manifolds in symplectic geometry.

In fact there are good reasons for both points of view, which we now explain. In many applications there is, in addition to the function V, an associated dynamic that locally minimises V (in other words a dynamic for which V is a Lyapunov function), making the critical set M into a set of equilibria. For example in the machine described above, V is the energy in the elastic bands, given by Hooke's law, and the associated dynamic is given by Newton's law and friction, in other words by the damped Hamiltonian system (see below). The presence of a dynamic has several consequences: it implies that some points of M will be attractors, and other points repellors, like the shaded middle sheet in Figure 4. It implies that when the control point crosses a fold, going in the right direction, then there will be a breakdown of equilibrium, and a catastrophic jump to another equilibrium. It implies delay and hysteresis. To emphasise all these additional qualities Thom coined the term "elementary catastrophe".

The word elementary refers to the fact that the associated dynamic is gradient-like. Non-elementary catastrophes can occur in non-gradient dynamics; the classical example is the Hopf bifurcation, where an attractor point bifurcates into a repellor point inside a small attracting cycle, as occurs for instance at the onset of aircraft flutter, or a bicycle speed wobble. Recent work of Takens [10] on bifurcations of differential equations in the plane shows that the Hopf bifurcation is so inextricably intermingled with the elementary catastrophes that they must be considered as belonging to the same theory. Thus in this context (non-elementary) catastrophe theory is the same as multiparameter bifurcation theory, although it would perhaps be more accurate to call them both a programme of research rather than a theory, because in higher dimensions knowledge is still at the level of exploring examples, and as yet far from any general classification.

Returning to elementary catastrophe theory, in some applications there is a global minimising principle of the function V (rather than a local minimising principle like a dynamic) causing the system to seek the absolute minimum of V rather than a local minimum - for example in the maximisation of entropy in phase transition, or in the minimisation of risk in Bayesian decision making [9]. In these applications the equilibrium surface is again decomposed into attractors and repellors, and also meta-attractors, corresponding to meta-stable states where the minimum is local but not global. Again there are catastrophic jumps (except that these now occur at Maxwell points rather than at bifurcation points). Therefore it is again appropriate to use Thom's term "elementary catastrophes" in these contexts.

However, there are other applications in which there is no minimising principle. Therefore there are no catastrophic jumps. Nor is there any meaningful distinction between the maxima, minima and saddles of V, and therefore no meaningful decomposition of M into attractors and repellors (the middle sheet in Figure 4 should no longer be shaded). In fact M should no longer be called an equilibrium set, but more appropriately a critical set. If, furthermore, M has arisen from Hamiltonian-Jacobi equations, then it is more appropriate to follow Arnold's terminology and call the singularities Lagrangian. The classical example is light caustics; Figure 5 shows the three possible types of singular point that a generic 3-dimensional light caustic can have. Incidentally it was while experimenting with light caustics in the early 1960's that Thom first discovered catastrophe theory. According to Whitney's list he expected to find only the swallowtail, and in addition he found the two umbilics appearing stably, to his surprise, until he realised the nature of the mathematical constraint that was in effect being imposed by Fermat's principle (or equivalently the method of stationary phase). Thus light caustics gave birth to catastrophe theory; and today in return, catastrophe theory is breathing new life into optics. On the theoretical side the knowledge of the Lagrangian singularities has improved the understanding of oscillatory integrals [4]. Berry and Nye have extended the classical interference patterns of Airy (for the fold) and Pearcey (for the cusp) to higher singularities. Nye [8] has done beautiful experiments revealing the geometry of important 6- and 8-dimensional singularities through their sections, by shining laser beams through water droplets. Berry [3] had used caustics to explain the twinkling of star-light, and to predict how the n-th moment of the intensity depends upon some exponent of the wave number. This is perhaps the most sophisticated application of catastrophe theory to date, since to calculate the exponents he needed to use Arnold's complete list.

Going back to the elementary catastrophe theorem above, one of the surprising features is that the result is independent of n. This enables it to be used in contexts where n is large and k is small. For example there might be implicitly too many behaviour variables (n large) to measure or compute, but nevertheless the behaviour might be modellable by an explicit low dimensional manifold M^k (k small) using only a few of those variables that happened to be convenient to measure. For example in an economic model n might be 10^3 to represent the important variables in an economy; in an embryological model n might be 10^4 to represent protein concentrations in a cell; in a brain model n might be 10^{10} to represent neuronal activity; in a physical model n might be infinite to represent possible eigenvalues.

Another point in the elementary catastrophe theorem that is worth drawing attention to is the little word "are" in the phrase "the only singularities are folds and cusps". What does it mean to say that a singularity "is" a cusp? If we look up the formula for the cusp catastrophe we find some nice little canonical polynomial such as $V = ax + bx^2 + x^4$, where $c = (a,b)$ is the control point and x the behaviour variable. On the other hand if we actually calculate V in some specific application, for example, if we work out the energy in the elastic bands in the machine described above, we obtain a Taylor series $V = \Sigma\lambda_i x^i$, where the coefficients λ_i depend upon c. I well remember as a student struggling, and failing, to acquire the proper applied mathematician's approach towards approximating a Taylor series by its first few terms; but even on those rare occasions when I could prove convergence, and satisfactorily estimate that boundedness of the error, I confess I could never cheerfully forget that small truncated tail of the Taylor series. Perhaps this intolerance of imperfection springs from too great an affection for the perfection of geometric forms, a penalty that any geometrically minded mathematician may have to pay. One of the attractions of theorems like the elementary catastrophe theorem is that it resolves this difficulty. In effect it says that if at a certain control point, the first non-vanishing Taylor coefficient is λ_4, then there is a smooth change of coordinates with respect to which V will have the canonical form; there is no need to approximate because the Taylor series is a polynomial of degree 4, all higher terms being zero. In other words if the problem satisfies a computable algebraic test, then there are coordinates with respect to which it has a perfect geometric form. This may explain why certain physical laws can take polynomial form, even in the absence of symmetry, when at first sight polynomials would appear to be much too special. No doubt, the aesthetic appeal of this approach will attract more geometrically minded mathematicians towards applied mathematics, and I would venture to suggest that this is an ingredient of Thom's own mathematical taste.

Of course, to make quantitative estimates and quantitative predictions in terms of the original coordinates one may have to return to the original Taylor series, but to obtain qualitative understanding and make qualitative predictions one can utilise the known geometry of the canonical models. Also different problems may be "solved" at different levels of detail. For instance, it may be possible to solve

I. the full dynamics
II. only the statics (and not the dynamics)
III. only the bifurcation set (and not even the statics)
IV. only statistical properties.

We illustrate the four levels by examples.

I. Consider a parametrised damped Hamiltonian system; for each value of the parameter (or control) the damping causes the system to seek a local minimum of the Hamiltonian. A simple example is given by the catastrophe machine described above. Before we do the mathematics, let me tell a story about a physicist friend of mine, who made a machine for himself; only instead of using a cardboard disk and drawing pins like a mathematician would, he got his departmental workshop to make it, and they used heavy steel disk pivoted on high quality ball bearings. Of course, it didn't work, because when the control point crossed the bifurcation set, instead of jumping into the other equilibrium position, the disk went into large steady oscillations. When I laughingly pointed out that it didn't work, he replied that it was my mathematics that didn't work, because it didn't describe his physical model. Of course, the correct mathematics does describe both his machine and my machine; they have the same statics, and qualitatively similar dynamics, but differ quantitatively in the size of the friction term, as follows.

The configuration space of the disk is a circle, Q say. The state space of the disk, $X = T^*Q$, is the cotangent bundle of Q, which is topologically a cylinder. The state of state of the disk is given by a point $x \epsilon X$. We can write x in coordinates, $x = (p,q)$, where $q \epsilon Q$ denotes the position of the disk and $p \epsilon T^*_q Q$ denotes the angular momentum. The Hamiltonian $H: X x C \to R$ is given by

$$H(x,c) = V_c(q) + \frac{p^2}{2I}$$

where $V_c(q)$ is the potential energy in the elastic when the control point is at c and the disk at q, and $p^2/2I$ is the kinetic energy, I being the moment of inertia of the disk. The dynamic is given by the damped Hamiltonian equations

$$\dot{q} = \frac{\partial H}{\partial p} = \frac{p}{I}$$

$$\dot{p} = -\frac{\partial H}{\partial q} - \varepsilon p = \frac{\partial V}{\partial q} - \varepsilon p \quad ,$$

where $\varepsilon > 0$ denotes friction. Therefore the dynamic locally minimises H, because $\dot{H} = -\varepsilon p^2/I < 0$, if $p \neq 0$.

If the damping, ε, is large then the behaviour is dominated by the statics: the system homes rapidly towards the equilibrium surface M, which is given by

$$\frac{\partial H}{\partial p} = \frac{\partial H}{\partial q} = 0, \quad \text{in other words by } p = 0, \ \frac{\partial V}{\partial q} = 0.$$

Now M is contained in the 4-dimensional space $T^*Q{\times}C$, but since $p = 0$ it is actually sitting inside the 3-dimensional subspace QxD, where Q is identified with the zero section of the bundle T^*Q. Therefore if we are only interested in the statics, in other words in the equilibria and the catastrophic jumps (rather than the full dynamics), then it suffices to consider only the embedding $M \subset Q{\times}C$, given by $\partial V/\partial q=0$, as illustrated in Figure 4. The advantage of Figure 4 is that it gives good geometric intuition of the statics, but the disadvantage is that it does not contain the dynamics. For instance, a catastrophic jump really looks like Figure 6, the damped oscillations being the image, under the projection $T^*Q \to Q$, of an orbit spiralling in towards the attractor, as in my physicist friend's machine.

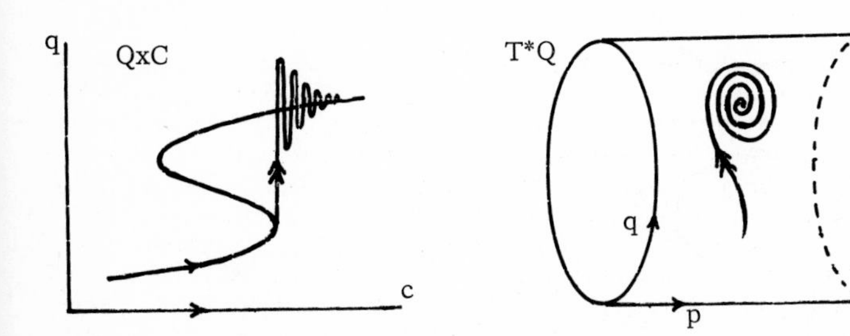

Figure 6: Damped Hamiltonian Catastrophe

Exactly the same equations and geometry occur in the rolling of a ship in still water [14]. Here the parameter c represents the position of the centre of gravity of the ship, and the diamond-shaped bifurcation set represents the evolute of the buoyancy locus determined by the shape of the hull; the catastrophic jump corresponds to capsizing. Modern ships have a more rectangular hull, which produces a butterfly-catastrophe modification of the bifurcation set, implying greater stability. More complicated singularities occur in the 3-dimensional problem, for instance in the study of the stability of icebergs. To represent the effect of waves a forcing term must be added to the equations, and in this case the dynamics is as yet unsolved; moreover, this is a serious problem because even today the number of ships that capsize in heavy weather is still high. Whether or not our improved understanding of the singularities of the statics in still water can lead to a better qualitative understanding of the dynamics in the seaway remains to be seen.

II. In some problems one is more interested in the statics than the dynamics, and here the classification of higher dimensional singularities is proving to be useful. For example, when looking at the von Karman equations for a buckling plate, Magnus and Poston [7] identified a double-eigenvalue as a double-cusp catastrophe, and knowing that the latter was 8-dimensional, were able to select 8 parameters that gave a full unfolding of the equations. By imposing symmetries on the problem Golubitsky and Schaeffer [5] have given a 1-dimensional unfolding.

III. In some problems both the statics and dynamics are insoluble, but nevertheless the bifurcation set B in the parameter space C may be not only identifiable, but measurable and usable for quantitative prediction. For example Benjamin [2] observed the formation of Taylor cells in viscous fluid flow inside a cylinder with D-shaped cross section, the curved part of whose boundary was being rotated with constant velocity. He used two parameters

$$a = \text{rotation of boundary } (\sim \text{Reynolds number})$$
$$b = \text{length of cylinder.}$$

The state space X is the infinite dimensional space of all possible flows, and the dynamic on X is the evolution determined by the Navier-Stokes equations, which of course depend upon the parameters. The equilibrium states were observed to be stable steady flows, and the set of equilibria formed a cusp catastrophe surface M over the parameter space, as shown in Figure 7.

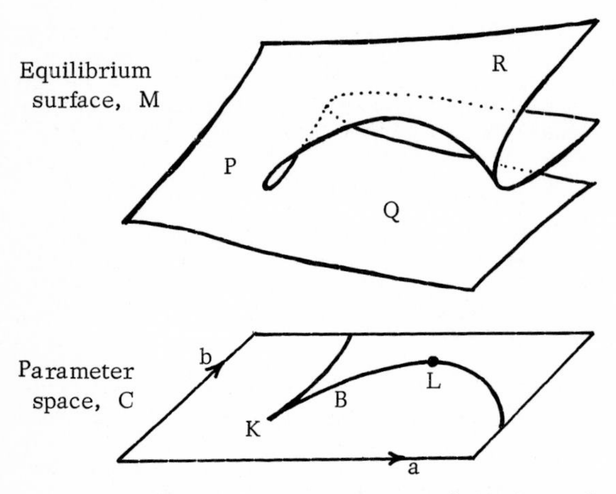

Figure 7: Fluid flow and embryological determination

The regions P, Q, R on M correspond to 0, 2, 4 Taylor cells respectively. When the rotation of the boundary was altered then the flow adapted smoothly, unless the bifurcation set was crossed in the appropriate direction, in which case a catastrophic jump occured, consisting of the cell pattern breaking up and the flow resetling relatively quickly into the other stable pattern. In particular, the behaviour exhibited hysteresis near the cusp point, since, as to be expected from generic considerations, the cusp axis was not parallel to the a-axis. Thus although neither the dynamics (the evolution) nor the statics (the steady state flow in a bounded cylinder) were soluble - the Navier-Stokes equations being as yet too difficult - nevertheless the bifurcation set was observable, measurable and predictable.

Exactly the same picture occurs in a model for differentiation in embryology [14]. Here the parameters a, b represent time, space. X represents the possible states of a biological cell, and the dynamics on X represents the homeostasis of the cell, which of course depends upon the parameters of its position, a, in the embryo and the stage, b, of development of the embryo. Neither the dynamics nor the statics is measurable, but the three qualitatively different types of cell P, Q, R were observable. For example, if the model is applied to gastrulation, P represents blastoderm before differentiation while Q and R represent mesoderm and ectoderm after differentiation. This application is due to Thom [11], and my modification was to turn the cusp axis so that it was generically not parallel to the time-axis (the a-axis in Figure 7). Wasserman [12] confirmed that the modified model was locally stable in the group of diffeomorphisms of state-space-time, XxSxT, preserving the projections XxSxT $\rightarrow$ SxT $\rightarrow$ S, in other words the group leaving invariant the basic concept of the developmental path of a cell. Furthermore, under a suitable simplicity hypothesis, it can be shown that Figure 7 is unique up to global diffeomorphism in the group.

The interesting consequence of this result is that the arc KL of the bifurcation set B represents catastrophic switches of developing cells from R to Q, in other words a wave of Q-determination sweeping through the embryo. In the case of gastrulation this can be identified with the wave of mesoderm determination, which can be measured and predicted in space-time. Moreover it appeared to be related to subsequent morphogenetic movements. When combined with a clock, it gave a simple model for somite formation, that explained regulation, and led to predictions that were subsequently confirmed by experiment [14]. Again we emphasise, here is an application of catastrophe theory in which neither the statics nor the dynamics is measurable, but the bifurcation set is.

IV. The most striking result concerning the statistics of singularities is Berry's application to twinkling of stars [3], mentioned above. The intensity of starlight is affected by many caustics arising from many waves in the turbulent atmosphere, and so it is necessary to average or integrate their effects. By analysing the oscillatory integral associated with a singularity Berry isolated two indices, α derived from the germ of the singularity, measuring the intensity of light at the caustic, and β derived from the unfolding of the singularity measuring the width of the caustic, or in other words the duration of its effect. Thus the contribution of that type of singularity to the average intensity is proportional to $\alpha\beta$, and the contribution to the n-th moment is proportional to $\alpha^n\beta$. Since the indices are proportional to the wave number raised to some exponent, each moment will be dominated by the random appearances of that singularity with the highest exponent. Thus statistical considerations led, surprisingly, to a particular singularity for each n, and the precise computation of rational exponents.

I have tried to illustrate the breadth of the impact of Thom's work by giving a brief selection from amongst the hundreds of different applications of catastrophe theory. Naturally, in each field of application the specialist has gone further than Thom himself, but I am sure that Thom's own breadth of vision enabled him to see from the beginning both the central position that his work would occupy in mathematics, and its potential use in all branches of science.

References

1. V.I. Arnold, Critical points of smooth functions, Proc. Int. Cong. Math., Vancouver (1974) 19-39.
2. T.B. Benjamin, Bifurcation phenomena in steady flows of a viscous fluid, Proc. Roy. Soc. Lond. A 359 (1978), 1-43.
3. M.V. Berry, Focusing and twinkling: critical exponents from catastrophes in non-Gaussian random short waves, J. Phys. A; Math. Gen., 10, 12 (1977) 2061-2081.

4. J.J. Duistermaat, Oscillatory integrals, Langrangian immersions, and unfoldings of singularities, Comm. Pure Appl. Math., 27 (1974) 207-281.

5. M. Golubitsky & D. Schaeffer, A theory for imperfect bifurcation via singularity theory, M.I.T. preprint (1978).

6. H. Hopf, The work of R. Thom, Proc. Int. Cong. Math., Edinburgh (1958), lx-lxiv.

7. R. Magnus & T. Poston, On the full unfolding of the von Karman equations at a double eigenvalue, Battelle Report, 109 (1977) Geneva.

8. J.F. Nye, Optical caustics in the near field from liquid drops, Proc. Roy. Soc. Lond. A361 (1978) 21-41.
Optical caustics from liquid drops under gravity: observations of the parabolic and symbolic umbilics, Phil. Trans. (in press).

9. J.Q. Smith, P.J. Harrison & E.C. Zeeman, The analysis of some discontinuous decision processes, (to appear).

10. F. Takens, Unfoldings of certain singularities of vector fields: generalised Hopf bifurcations, J. Diff. Equations, 14 (1973) 476-493.

11. R. Thom, Stabilité structurelle et morphogénèse, Benjamin, 1972.

12. G. Wassermann, Stability of unfoldings in space and time, Acta. Math. 135 (1975) 57-128.

13. H. Whitney, Mappings of the plane into the plane, Ann. Math. 62 (1955) 374-470;
Singularities of mappings of Euclidean spaces, Symp. Int. Top. Alg., Mexico City (1958) 285-301.

14. E.C. Zeeman, Catastrophe theory, Selected papers 1972-1977, Addison-Wesley, 1977.

Catastrophe Geometry in Physics: A Perspective

W. Güttinger

Institute for Information Sciences, University of Tübingen
D-7400 Tübingen, Fed. Rep. of Germany

and

Department of Physics and Astronomy, University of Wyoming
Laramie, WY 82071, USA

1. Introduction

The mutual penetration of mathematical, physical and philosophical thinking has been in the past of vital importance to the development of the exact sciences. It suffices to recall Newton's mechanics and the differential calculus, relativity theory and Riemannian geometry, or quantum mechanics and the Hilbert space. Within the last decades, however, the gap between mathematics and physics has steadily widened and the impetus mathematics has gained from physics has not prevented it from becoming ever more abstract: The increasing refinement and precization of mathematical notions and methods inevitably results in a loss of intuition and in a decreasing capability of grasping empirical facts in a qualitative way. This also happened to topology which, after Poincaré conceived it toward the end of the nineteenth century, soon turned away from its intuitive geometric roots and rapidly followed quite entangled algebraic paths. However, differential topology right now provides a rare opportunity to reunite physics and mathematics since it determines a new and more appropriate language in which to express the qualitative mathematical content of any physical theory.

Physics is the most fundamental and all-inclusive of the sciences. Based on observation, reasoning and experiment, it is surprising that its tremendous amount of data and results can be condensed into simple laws which summarize our knowledge. These laws are essentially of qualitative nature. In contradistinction, mathematics is not a natural science. The test of the validity of its statements is not experiment. With the exception of geometry, the only thing certain about the remarkable relationship between mathematics and physics is that they impregnate themselves in the dark but, face to face, readily fail to appreciate and even disavow one another. In the physical sciences an intuitive, phenomenological approach to complex problems generally is superior to, and more effective than, precise quantitative methods and, indeed, is always at the root of new fundamental conceptions. Relying on a qualitative understanding of processes that have proved themselves too complex for rational quantitative analysis is a basic human instinct. It provides us with conceptual guidance to single out the most significant phenomena and to simplify matters to the point where we can intuitively come to terms with them. That is why the basic physical phenomena and the laws governing them are inherently qualitative. Furthermore, the increasing diversification of the physical sciences during recent years has made it more and more imperative to search for unifying principles, by focusing on the wealth of profound, fascinating analogies discovered in the critical behavior of systems of various genesis, which, when passing through instabilities, suddenly exhibit new spatio-temporal patterns or modes of behavior. Their common characteristic is that one or more significant behavior or order parameters undergo sudden, discontinuous changes (or cascades of these) if slow, competing but continuously driving control parameters or forces cross a bifurcation set and enter into conflicting regimes causing instabilities in the behavior variables or order parameters. Common to all these phenomena, besides their qualitative similarity, is their universality expressed by the fact that the details (e.g., the interaction) of the system undergoing sudden transitions are almost irrelevant. While this hints at thermodynamic principles, the dimensional analysis underlying scaling laws points to a qualitatively invariant description of the phenomena under consideration. This means that we have to disregard algebraic structures for a while (we cannot add two phases to yield a third one) and confine ourselves to precisely those two concepts which alone appear sufficient to allow qualitative conclusions, viz. order structures and topology. It amounts to saying that the qualitative laws of nature are written in the language of thermodynamics and geometry.

The first one who realized again that physical and nature's necessities are always within the mathematical possibilities was R. Thom [1] when he returned to topology's intuitive geometric roots and, relying on the notion of structural stability, started to classify the local singularities of smooth stable maps between manifolds, or objects for that matter, thus conceiving a geometric theory of analogies. The resulting mixture of deep theorems on unfoldings, transversality, singularity and general bifurcation theory became popular as catastrophe theory since, interpreted in its broadest terms, it is designed to provide a method for modeling sudden qualitative structural changes in complex systems in geometric terms.

Truly enough, ever since Pythagorean times the Gods have made geometry. Thus, catastrophe theory and its extension into singularity theory by Arnold [2], Zeeman [3] and others can only explain and describe the local geometry and not the forces that are shaping it. On the other hand, looking at local forces alone cannot explain the global geometry unless structure formation can be attributed to local minima of entropy production, e.g., in the sense of Prigogine's [4] dissipative structures, or in the spirit of Haken's synergetics [5]. Thus, right now, catastrophe and singularity theory have already become too narrow to satisfy all the needs of the physicists and the story starts anew: Continuum physics requires generalizations to infinite-dimensional spaces, to variational and optimization problems [6] and to establishing interrelationships with the bifurcation theory of operator equations (Hale [7], Sattinger [8] and the references quoted there). Symmetries and their breaking lead us into the world of infinite codimension. Hopf bifurcations [9], [10], strange attractors and singularity theory are now becoming so intermingled that one can hardly discuss the one without the other in setting up a comprehensive theory of dynamical systems. Since in applying singularity theory to the physical sciences we have hardly scratched the surface (Poston and Stewart [11]), exploring Thom's grand program remains a challenge for years to come.

2. Singular Points of Smooth Functions

Catastrophe and singularity theory deals with the classification of critical or singular points of smooth (C^∞) real-valued, parametrized families of functions of n real variables. The singular points of a smooth function $f:R^n \to R$ are the points x where the differential vanishes: $(df)(x)=\text{grad}\, f(x)=0$. The function f has a nondegenerate (or Morse) singularity at x if the second differential $(d^2f)(x)=(\partial^2 f(x)/\partial x_i \partial x_k)$ is a nondegenerate quadratic form, i.e., if the Hessian $H(x)=\det(d^2f)$ does not vanish. Using suitable local coordinates, the function f can in some neighborhood of a nondegenerate singular point be represented in the Morse normal form $f=\text{const} + \Sigma(\pm x_i^2)$. Morse singularities are stable in the sense that a small perturbation g of f (with g, dg, d^2g near f, df, d^2f) also has a Morse singularity. On the other hand, every degenerate singular point x of f, for which H=0, bifurcates into some nondegenerate points after an arbitrarily small deformation. The function $f(x)=x^3/3$ $(x\epsilon R)$ and its perturbation $f_t(x)=x^3/3+tx$ furnish a familiar example. We see that degenerate singular points appear naturally if the function depends upon parameters, i.e., if one considers not an individual function but a parametrized family of functions. Then it is possible for a non-Morse function to appear as a member of a "stable family". In the above example, $f=x^3/3$ is not stable at x=0 because a small perturbation of it, generated by adding tx, radically changes its shape when t is varied: for t<0, $f_t(x)$ has two extrema while for t>0 it has none. Thus, the family $f_t(x)=x^3/3+tx$ is stable as a (t-)parametrized family near x=0. We call $f_t(x)$ the unfolding of f, t an unfolding parameter, and t=0 a catastrophe point of the family because it separates the stable regions $t \lessgtr 0$. Formally, one may continue playing by considering the function $f=x^4/4$, $x\epsilon R$, and its unfolding (or perturbation) $f_{uv}(x)=x^4/4+ux^2/2+vx$ near its singularity $x=0$. The singular points are given by an overhanging cliff S, determined by $f'_{uv}(x)=x^3+ux+v=0$, in (x,u,v)-space, and projecting the tangents (parallel to x) of its two edges, given by $f''_{uv}=3x^2+u=0$, vertical onto the (u,v)-plane by eliminating x from the two equations, one obtains the familiar cusp equation $\Sigma : 4u^3+27v^2 = 0$, viz., the bifurcation set on which two stationary points (a minimum and a maximum) of f_{uv} coalesce.

We have embedded f into the unfolding f_{uv}. While f is a very fragile object because its singularity is degenerate, the shapes of f_{uv} remain basically the same inside or outside the cusp, respectively (two minima of f_{uv} if u and v vary inside Σ, one minimum of f_{uv} if u and v stay outside of Σ): f_{uv} remains stable under small variations of (u,v) not crossing Σ. Σ is a singularity of the map χ of S onto the plane C, made up by (u,v), $\chi:S\to C$ induced by the projection $R\times C\to C$. Intuitively, modeling S by an elastic fiber, deforming S smoothly into a surface S' (by a diffeomorphism) and deforming C smoothly into another surface C' by a change of coordinates, does not change the qualities of the "cusp catastrophe": The ensuing map $\chi':S'\to C'$ remains equivalent to χ (i.e., equal up to smooth transformations). More generally, take a two-dimensional piece of elastic fiber (a piece cut from a lady's stocking does the job!), viz. a two-dimensional manifold $M\subset R^3$. Lift it into R^3, deform it as you please and map it vertically down onto R^2. You will discover four sorts of points on the fiber M: (a) regular points which lie smoothly over R^2, singular points,i.e., some on the edges of folds (b) and some marking the origin of an overhanging cliff (c), and, finally, (d) points which resemble none of the former. Deform the fiber slightly near any of these points: After a lot of experimentation the following picture emerges. The location of points of type (a), (b), and (c) will be shifted a bit but their "quality" remains the same and they cannot be made to disappear under such small deformations of the fiber. But all the messy points of type (d) or worse, turn, under small deformations, into points of type (a), (b) or (c). This is an experimental "proof" of the Thom-Whitney theorem: Consider the space F of all 2-parameter families of functions $f=f_{uv}(x)$ and their singularities $df=0$. Call $f\in F$ structurally stable if it has a neighborhood of equivalents, i.e., if χ_f is equivalent to $\chi_{f'}$ for all $f'=f+\delta f\in F$. Then the only singularities of the projection $S\to C$ are folds and cusps. This means that the most complicated behavior that can happen locally is the cusp. While this is a geometric fact, its physical meaning is obvious: If a system with one behavior variable or order parameter x is slowly driven from one phase to another by a control variable v, and if an orthogonal drive u sets in to split the quality of the phases (different states of order or symmetry, etc.) and if a phase may persist for a while with the transition to the other delayed, then the cusp catastrophe is intuitively the simplest model and, as we have seen, the least fragile.

The generalization of these -- admittedly oversimplified -- considerations to higher dimensions, due to Thom, Siersma, Mather and others [1], [2], [3] result in the following classification theorems, here stated in a qualitative way:

(1) For $f\in C^\infty(R^n\times R^k)$ let $M\subset R^n\times R^k$ be given by $\mathrm{grad}_x f(x,u) = 0$, where x and u are coordinates on R^n and R^k, respectively. Let $\chi:M\to R^k$ be induced by the projection $R^n\times R^k\to R^k$. Then, for $k\leq 5$, there is an open dense set of functions f in $C^\infty(R^n\times R^k)$ -- with the Whitney C^∞-topology -- such that (a) M is an r-dimensional manifold, (b) any singularity of χ is equivalent to one of a small number of types, called the elementary catastrophes, (c) χ is locally stable at all points of M under small perturbations of f. The number N of elementary catastrophes depends only on k : N(1) = 1, N(2) = 2, N(3) = 5, N(4) = 7, N(5) = 11; M does not depend on the number n of variables x. In fact, in appropriate coordinates, the elementary catastrophes -- or normal forms -- depend only upon one or two variables in an essential way.

The simplest example is provided by a gradient system $\dot{x}=X(x,\mu)=-\mathrm{grad}_x f(x,\mu)$ with $x\in R^n$, $\mu\in R^k$, governed by a "potential function" $f(x,\mu)$. $df=0$ defines a surface in $R^n\times R^k$ with stable equilibrium points x_0, defined by $\dot{x}=0$, corresponding to the extrema of f. "Critical" equilibrium states, where a loss of stability can be expected, occur if $H=\det(d^2f)=0$. If $n=2$, the equilibrium point x_0 is given by the intersection of the two solution curves C_1, C_2 of the equations $X_1(x_1,x_2,\mu) = 0$ and $X_2(x_1,x_2,\mu) = 0$, respectively. For a given μ, C_1 and C_2 either may intersect transversally (i.e., "without contact"), in which case x_0 is nondegenerate ($\det J \neq 0$, J=Jacobian matrix), or with a common tangent (i.e., "with contact") in which case x_0 is degenerate ($\det J=0$). Changing μ results in a deformation of C_1, C_2 which gives rise to new types of stationary points if $\det J=0$ (i.e. $H=0$), and, therefore to new topological situations and modes of behavior.

When the parameters are few, only the simplest degeneracies appear, one can list them explicitly and give normal forms for functions and families. When the number of parameters increases beyond $k=5$ more complicate degeneracies spring up, but Arnold [2] has discovered that even then one can extend the hierarchy of singularities and relate their classification to the Lie, Coxeter, Weyl and braid groups and to the classification of the platonics in 3-dimensional Euclidean space:

(2) By introducing the notion of modality, Arnold has shown that unimodular germs reduce to one of a series of one-parameter families of normal forms [2].

One of them, the double-cusp, also discussed by Zeeman and Callahan, seems to play a basic ingredient in describing buckling phenomena, ternary fluid mixtures and cuspoid geometries in ion optics and plasma physics.

3. Catastrophes in Physics

Many systems and structures encountered in nature enjoy an inherent stability property: They preserve their quality under slight distortions, otherwise we could hardly describe or think about them, and today's experiment would not reproduce yesterday's result. When one accepts the idea that, because of their universality, critical or analogous phenomena in physics have a common topological basis, one may expect them to be governed by unfolding singularities in a structurally stable way and be classifiable by means of singularity theory.

However, the stability concept of systems or objects found in physics cannot easily be transformed into the mathematical concept of structural stability because the relation between physical variables (or observables) and mathematical ones is not necessarily differentiable. Furthermore, singularity theory deals with the classification of local singularities, and, therefore, is a rather static affair without any physical dynamics (i.e., forces shaping the geometry) whatsoever. Last but not least, catastrophe theory (in its present setting [11], [13]) is restricted to spaces with finite dimensions and can only be applied rigorously to problems which come naturally equipped with a C^∞ parametrized family of functions. It is because of these reasons that in many applications of the theory one restricts oneself to the determination of bifurcation sets from $df(x,\mu)=0$, $d^2f(x,\mu)=0$, obtaining nice qualitative results but no quantitative ones. This amounts to looking for catastrophes in physics while the real issues ought to be formulating physics in terms of catastrophes and singularities and arriving at quantitative predictions. Quite slowly, research in the field turns in the latter direction, as I shall briefly sketch.

Perhaps the most exciting applications are found in geometrical optics, scattering and S-matrix theory [14]. This is to be expected because here one is concerned with families of trajectories of wavefields. In contrast with an individual trajectory, a family exhibits the property of focusing, giving rise to caustics, and it is now well-known that stable caustics fall into the classification scheme of singularity theory. The mathematical reason for this is that caustics and wavefronts belong to the class of Lagrange and Legendre singularities [2],[15]. Since geometrical optics is one-to-one to Hamilton-Jacobi theory, it is not surprising that basic statements of the latter can be formulated via singularity theory, especially developed by Arnold in terms of symplectic geometry. The topological origin of the optical-mechanical analogy may, indeed, be traced to the parallelism between the theory of Lagrange and Legendre singularities.

While the classification of singularities in geometrical optics and mechanics possesses its own flavor, it gains particular importance when one considers these fields as semiclassical approximations of linear and nonlinear wave optics, quantum mechanics and, in particular, quantum field theory, which is presently revisited by sewing quantum flesh on classical bones [16]. The central role is played here by oscillatory and path integrals because their formulation incorporates global properties of the system and they can give answers which cannot be given by partial differential equations. The starting points are WKB-type approximations to wave equations in the large frequency (τ) limit ($\tau = 1/\hbar$ in quantum theory). In case of the reduced wave equation $(\Delta + \tau^2)\, u = 0$, for example, one assumes that $u(x,\tau) = A(x,\tau) \exp(i\,\tau\,\phi(x))$ with an asymptotic expansion $A = \Sigma\, a_n(x)/\tau^n$. Equalizing coefficients of τ^{-n}, the first term gives the eiconal equation $|\Delta\phi|^2 = 1$, while the transport equations for a_n can be solved recursively once ϕ is known. Determining ϕ falls into the domain of the Hamilton-Jacobi theory, ϕ plays the role of the action S and thus is multivalued around caustics. Intuitively, allowing for all rays or particle paths in whatever wave equation considered, one is led to replace the above series by a continuously infinite superposition

$$u(x,\tau) = \int_{R^n} A(x,\alpha,\tau)\, e^{i\tau\phi(x,\alpha)}\, d\alpha$$

where ϕ is an n-parameter phase, and A an amplitude with asymptotic expansion $A(x,\alpha,\tau) = \tau^\mu\, \Sigma\, a_n(x,\alpha)/\tau^n$. In general, the oscillatory integral $u(x,\tau)$ plays the role of a diffraction or scattering amplitude, or that of an S-matrix element. For large τ the stationary phase method tells us that the main contribution to u comes from those points (x,α) for which $\partial\phi(x,\alpha)/\partial\alpha = 0$. This indicates that catastrophe theory may be applied (with α as "state" or "behavior" variables, x as "control" variables and ϕ playing the role of V of Sec. 2). If $\alpha_i(x)$ is a solution of the equation $\partial\phi/\partial\alpha = 0$, then, if the singularity is Morse, we obtain for large τ a contribution u_i to u,

$$u_i(x,\tau) \approx \frac{\tau^{\mu-n/2}}{H^{1/2}}\, a_o(x,\alpha_i)\, e^{i\tau\phi(x,\alpha_i) + i\pi\sigma/4}$$

where H is the Hessian of $\phi(x,\alpha)$ at $\alpha_i(x)$ and σ its signature. The Hessian vanishes on the caustics, in which neighboring trajectories meet, and u_i blows up. This is the degenerate or catastrophe case, corresponding to maximal intensity, and the unfolding techniques of catastrophe theory can be applied by transforming ϕ into one of the standard forms, $\phi \to \psi(y,\alpha)$, whence, near the caustics, the problem is reduced to considering

$$\tilde{u}(y,\tau) = \int_{R^n} e^{i\tau\psi(y,\alpha)}\, d\alpha,$$

where ψ is a catastrophe polynomial, giving rise to an asymptotic expansion of the form $\Sigma\, C_{\mu\nu}\tau^\mu(\log\tau)^\nu$. This provides a unified treatment of a variety of phenomena in optics, ion collisions, etc. [14]. For example, inserting for ψ the cusp polynomial $\psi = \alpha^4 + y_1\alpha^2 + y_2\alpha$ gives for $|\tilde{u}|^2$, as a function of y_1, y_2 the wave-like patterns of intensity around the caustics which cannot be obtained from geometrical optics. Experimentally, however, intensities do not become infinite on caustics (as H = 0 seems to imply).

In a more general setting, therefore, the next step is to turn to infinite-dimensional spaces, in particular, to Feynman path integrals for the propagator K(p,p'),

$$K(p,p') = \int D_{pp'}\,[x(t)]\, \exp\,[iS_{pp'}(x(t))/\hbar]$$

where p,p' are space-time points, $D_{pp'}[x(t)]$ is the path differential and $S_{pp'}$ is the action functional defined on paths $x(t)$ connecting p with p', and to develop semiclassical approximations which remain finite in conjugate (focal) points and on caustics, exhibit the experimentally observed behavior near these and reduce to familiar approximations far away. How this can be done will be explained later at this conference [6]. After all that has been said above, it is not surprising that instead of the conventional approximation $K \simeq |\det \partial^2 S^c/\partial q \partial q'|^{-1/2} \exp(iS^c/\hbar)$, which diverges on caustics, one obtains a finite approximation of K, determined by what we call a "generalized Airy integral"

$$\phi(d) = \int_{R^n} d\alpha \; e^{ig(\alpha,d(\sigma))/\hbar\Delta t},$$

with a catastrophe polynomial g whose bifurcation properties correspond to those of the solutions of the Euler-Lagrange equations. This has, of course, far-reaching implications not only for optics and scattering problems of all sorts but also for quantum corrections to classical fields [16], tunneling phenomena (e.g., in "instanton" physics) and for statistical mechanics.

Extending singularity and catastrophe theory to infinite-dimensional spaces and establishing interrelationships between bifurcation theory of operator equations $F(u,\lambda) = 0$ (where $F: B_1 \times R^k \to B_2$, B_i being Banach spaces and R^k the space of parameters λ) carries enormous potential in geometrizing physical phenomena in a unifying way [6], [7]. In fact, fundamental classification theorems of singularity theory can be carried over to infinite-dimensional spaces and, for example, the bifurcation behavior of Euler-Lagrange equations deriving from variational problems can be characterized by catastrophe polynomials. Potential applications range from elasticity theory to nonequilibrium thermodynamics. Thus, for example, the buckling modes of a column can be described in terms of the now famous double-cusp, while the classification statements of singularity theory can be carried over (via Glansdorff-Prigogine's stability criteria) to phase transitions of irreversible thermodynamics [6]. By applying these techniques to reaction-diffusion equations [19], new light may be shed on problems of pattern formation in systems with activator-inhibitor dynamics, thus coming close to Thom's expectation that systems with infinitely many degrees of freedom also should exhibit standard topological forms at the physical level because, ultimately, there should prevail a very deep structural relationship between geometric terms and the general physical forces giving rise to them (a program akin to general relativity).

Singularity theory may also be expected to play a considerable part in our understanding of nonlinear hyperbolic or dispersive wave motions. The simplest case of a conservation law is provided by the equation $\phi_t + c(\phi)\phi_x = 0$ with initial condition $\phi(x,0) = f(x)$ $(x,t \in R)$ with the implicit solution ϕ given by $G \equiv \phi - f(x - c(\phi)t) = 0$. If $dG/d\phi = 1 + tf'c' = 0$, $\phi(x,t)$ develops a cusp surface for some $t > 0$ if $c'(\phi) > 0$ and $f' < 0$. Using the Maxwell convention, or entropy condition, a shock wave is seen to develop and it is not difficult to analyze shock collisions in terms of elementary catastrophes. The theory can be extended to one single quasilinear conservation law $\phi_t + q(\phi)_x = 0$ with $t \in R$, $x \in R^n$ [17] and the resulting singularities can be classified in terms of catastrophes. Physical examples range from the Clausius-Clapeyron equation, to binary fluid mixtures, Maxwell equations and even to classical mechanics in which, e.g., the Hamilton function appears as a singular solution (i.e., as the envelope) of a Clairaut equation. It is not difficult to see [20] that singular solution surfaces occur if Legendre transformations are not uniquely invertible. Singular surfaces are developable, regular ones are not. And to a developable surface there belongs a quasilinear conservation law. The problem is to extend the classification scheme of singularity theory to systems of nonlinear partial differential equations, notably to those of hydrodynamics, combustion and plasma physics, to gain a deeper insight into their structure and behavior. This has not yet been achieved, although simple dissipative-dispersive wave equations (such as the Korteweg-de Vries-Burgers equations) are amenable to a discussion in terms of singularity theory. Dispersive waves dominating nonlinear optics and modulation theory [18], on the other hand, seem to provide a proper laboratory for applications of catastrophe theory because here, especially in discussing waves of envelopes, the application of geometrical optics is particularly inviting. A detailed discussion of this subject is beyond the scope of this outline, but it may well lead us to an understanding of soliton physics in terms of singularity theory. This would, of course, be of great significance for various physical topics.

Beside these applications of catastrophe theory and the many others covering a vast area from laser radiation, pattern recognition to dislocations (not to talk about fields like nuclear physics which have not yet gotten in touch with that theory), there are branches of physics in which the underlying principle of structural stability confronts us with very difficult problems. One of them is that of incorporating symmetries into the theory. Thus, e.g., if ψ in $\tilde{u}$ or g in ϕ possess symmetries, the theory (as developed so far) rapidly breaks down. Recent work in this direction by Bierstone, Poênaru and Wassermann [21] might help to classify stable unfoldings in the presence of symmetries. Another problem is posed by the fact that in statistical mechanics the concept of structural stability implies a global requirement though phase transitions appear in a local way. And, finally, in combining singularity theory with stochastics, we have thus far made only a few small steps.

Returning to what I said at the beginning, although singularity theory has told us a lot about topological structures in physics, it is physical forces which induce geometric structures and singularities. One may conjecture that combining nonequilibrium thermodynamics with catastrophe theory will provide us with a real understanding of morphogenesis in physics and nature for that matter. At the end of his lectures on the variety of structures appearing in hydrodynamics, R.P. Feynman expressed his belief that the next great era of awakening of human intellect may well produce a method of understanding the qualitative content of complex structures and phenomena. R. Thom has undoubtedly opened the door to this era.

References

1. R. Thom, Structural Stability and Morphogenesis, Benjamin 1975
2. V.I. Arnold, Critical Points of Smooth Functions, in Proceedings of the International Congress of Mathematicians, Vancouver 1974
3. E.C. Zeeman, Catastrophe Theory, Addison-Wesley 1977
4. P. Glansdorff and I. Prigogine, Thermodynamic Theory of Structure, Stability and Fluctuations, Wiley-Interscience 1971
5. H. Haken, Synergetics, Springer 1978 and the following lecture in this volume
6. R. Magnus and T. Poston, this volume;
   G. Dangelmayr, W. Güttinger and W. Veit, this volume
7. J.K. Hale, Restricted Generic Bifurcations, in: Nonlinear Analysis, Academic Press 1978
8. D. Sattinger, Topics in Stability and Bifurcation Theory, Springer 1973, and preprints; M. Golubitsky and D.G. Schaeffer, Queens College preprints 1978
9. J.E. Marsden and M. McCracken, The Hopf Bifurcation and its Applications, Springer 1976
10. R. Abraham and J.E. Marsden, Foundations of Machanics (2nd edition), Benjamin 1978
11. T. Poston and I. Stewart, Catastrophe Theory and its Applications, Pitman 1978
12. W. Güttinger, Catastrophe Geometry in Physics and Biology, in: M. Conrad et al. Eds., The Physics and Mathematics of the Nervous System, Springer 1974
13. M. Golubitsky, An Introduction to Catastrophe Theory and its Applications, SIAM Review 20, 352, 1978. Here, as well as in [11], further references may be found.
14. M.V. Berry, Waves and Thom's Theorem, Adv. in Phys. 25, 1 (1976) and the papers in this volume; J.F. Nye, the papers in this volume
15. V.I. Arnold, Mathematical Methods of Classical Mechanics, Springer 1978
16. F.L. Gervais and A. Neveu, Extended Systems in Field Theory, Physics Reports 23 C, 237, 1976; F.L. Gervais, Relevance of Classical Solutions to Quantum Theories, in: Springer Lecture Notes in Physics Vol. 73, 1977; J. Zinn-Justin, Perturbation Series at Large Order and Vacuum Instability in: Springer Lecture Notes in Physics Vol. 77, 1978; M. Polyakov, Phase Transitions and Quark Confinement, ICTP preprint 1978
17. J. Guckenheimer, Solving a Single Conservation Law in: Springer Lecture Notes in Mathematics Vol. 468, 1974; D.G. Schaeffer, A Regularity Theorem for Conservation Laws, Adv. in Math. 11, 368 (1973); C. Ushiki, preprints 1977

18. G.B. Whitham, Linear and Nonlinear Waves, Wiley 1974; V.I. Karpman, Nonlinear Waves in Dispersive Media, Pergamon 1975; W. Güttinger, Lectures on Nonlinear Waves, Tübingen 1976
19. J.P. Keener, Studies in Appl. Math. 59, 1 (1973)
20. E. Obermayer, Thesis, University of Tübingen, 1977
21. E. Bierstone, Generic Equivariant Maps, Toronto preprint 1976; V. Poénaru, Singularités C^{∞} en Présence de Symétrie, Springer 1976; G. Wassermann, preprint, University of Regensburg 1976

Synergetics and a New Approach to Bifurcation Theory

H. Haken

Institute for Theoretical Physics, University of Stuttgart, Pfaffenwaldring 57
D-7000 Stuttgart, Fed. Rep. of Germany

1. Introduction

While science is progressing at an enormous pace, it splits up into more and more different disciplines. Furthermore, the different disciplines seem to use entirely different concepts and to speak entirely different languages. It thus becomes harder and harder for a scientist to overlook science. This situation leads to the provoking question if, in spite of all these diversifications, there are unifying principles or aspects. I think René Thom's theory of catastrophes can be considered as a widely known approach of this kind. Since his approach has been described in a number of monographs [1], [2] and papers [3], I need not repeat it here.

Some years ago I was also led to the question whether there are any unifying principles at work [4]. In pursuing this idea I found striking analogies between the behavior of systems in quite different disciplines ranging from physics over chemistry and biology to sociology [5]. Furthermore it turned out that there seem to exist principles common to seemingly quite different systems in certain situations. These results convinced me that it is worthwhile to study these phenomena within a new interdisciplinary field of research which I called synergetics. In the course of this enterprise links could be established to phase transition theory in physics on the one hand and to bifurcation theory in mathematics on the other. In particular, it turned out that there are interesting links between catastrophe theory and synergetics. This is quite remarkable because the origins of these two approaches were quite independent and different from each other.

I personally was led into these problems when treating the behavior of the laser [6], [7], a new kind of light source whose emitted light has unique properties. The laser is a comparatively simple but non-trivial example for self-organization and it allows establishing links with bifurcation and phase transition theory [8]. Therefore I describe its features in some detail. Let us consider first a lamp, for instance, a gas discharge tube. In it the excited atoms spontaneously and entirely independently emit short wave tracks of light. In the laser, however, after a certain excitation rate is reached, the atoms emit a nearly infinitely long coherent wave in an entirely self-organized manner. If we could hear light, light from thermal sources would sound like noise. On the other hand, laser light would sound like a pure tone. When in the laser device the excitation rate is changed a sharp transition between one state ("lamp") and another one ("laser") occurs. In mathematical terms the old state becomes unstable and is replaced by a new stable state.

2. Examples

There are numerous further cases known both in the natural sciences and other disciplines where pronounced changes of one state of a system into another one occur and I give only a few examples [5].

In the laser depending on physical parameters different states can be reached. Action as a lamp, conventional laser action with a constant amplitude wave, ultrashort pulses, laser light turbulence. In fluids we have the Bénard instability with the onset of well defined macroscopic patterns of the velocity field and eventually turbulence. In chemistry under certain nonequilibrium conditions spatial, temporal or spatio-temporal patterns can be formed. In biology by the interaction and diffusion of certain agents premorphogenetic patterns can be formed which lead to switching of genes and thus to morphogenetic processes. In sociology there exist models for the abrupt changes of social behavior, for instance, change of public opinion. It has turned out that most of these phenomena, if not all, can be traced back to a few principles.

In all these problems we have to deal with systems composed of very many subsystems. Their cooperation may produce patterns on a macroscopic scale in an entirely self-organized way. Since I have described the underlying principles on various occasions I shall not go into the details. I just mention the following points.

When external control parameters are changed, a system may change its macroscopic features dramatically. In the words of mathematics, also stressed by Thom, structural stability gets lost at such points. When I make a linear stability analysis close to an instability point I find modes of the system, some of which are stable, some others unstable. In practical cases, i.e. in dissipative systems, there are in general only very few unstable modes and still very many stable modes. The stable modes can be uniquely determined by the unstable mode amplitudes so that the equations of motion can be reduced to the unstable modes alone in an entirely exact way. The unstable modes serve as order parameters which slave the stable modes and thus determine the macroscopic state of the system. In many cases of practical interest the order parameter equations can be grouped into classes analogous to universality classes in phase transition theory. In contrast to bifurcation theory which neglects fluctuations, in practical cases fluctuations play a crucial role in the transition region and must not be neglected. I have elaborated the reduction scheme, which I just described, including fluctuating forces. In the first steps I started from a situation where the formerly stable state of the system referred to a spatially homogeneous and time independent state [9]. The approach can be extended, however, also to spatially inhomogeneous but still time independent cases. Here an interesting link with catastrophe theory has evolved. Catastrophe theory in its present state rests on the assumption that the equations of motion can be derived from a potential, V. It can easily be shown that all the above mentioned examples do not obey the potential condition. This must even be so because in systems far from equilibrium the detailed balance condition, which would guarantee the potential condition, is not fulfilled. However, to my surprise, after application of my reduction scheme, quite a number of order parameter equations fulfill the potential condition and thus allow us to apply a classification using catastrophe theory. This is the above mentioned link which could be established in synergetics with catastrophe theory and thus widens the applicability of catastrophe theory.

In this case, at least close to instability points, the changes between the steady states of a complex system can be traced back to catastrophes and thus to topology. Or, turning the argument around, topological relations prescribe the behavior of the individual modes (and parts) of a complex system (Table II). It should be noted that in a number of cases of practical importance the purely topological approach must be supplemented by a consideration of the impact of fluctuations.

I now come to another aspect. In following up the reduction scheme it turns out that there are cases in which even the order parameter equations do not fall into the class of potential equations. This is particularly so when we deal with situations in which the formerly stable state of the system has been time dependent. Recently two cases have become particularly interesting, namely, a bifurcation starting from limit cycles and, more generally, from quasi-periodic flows, i.e. motion on a torus. The concepts of order parameters and of slaving have been a useful guideline for me to develop a method, which seems novel and which allows me to treat the just mentioned bifurcation problems.

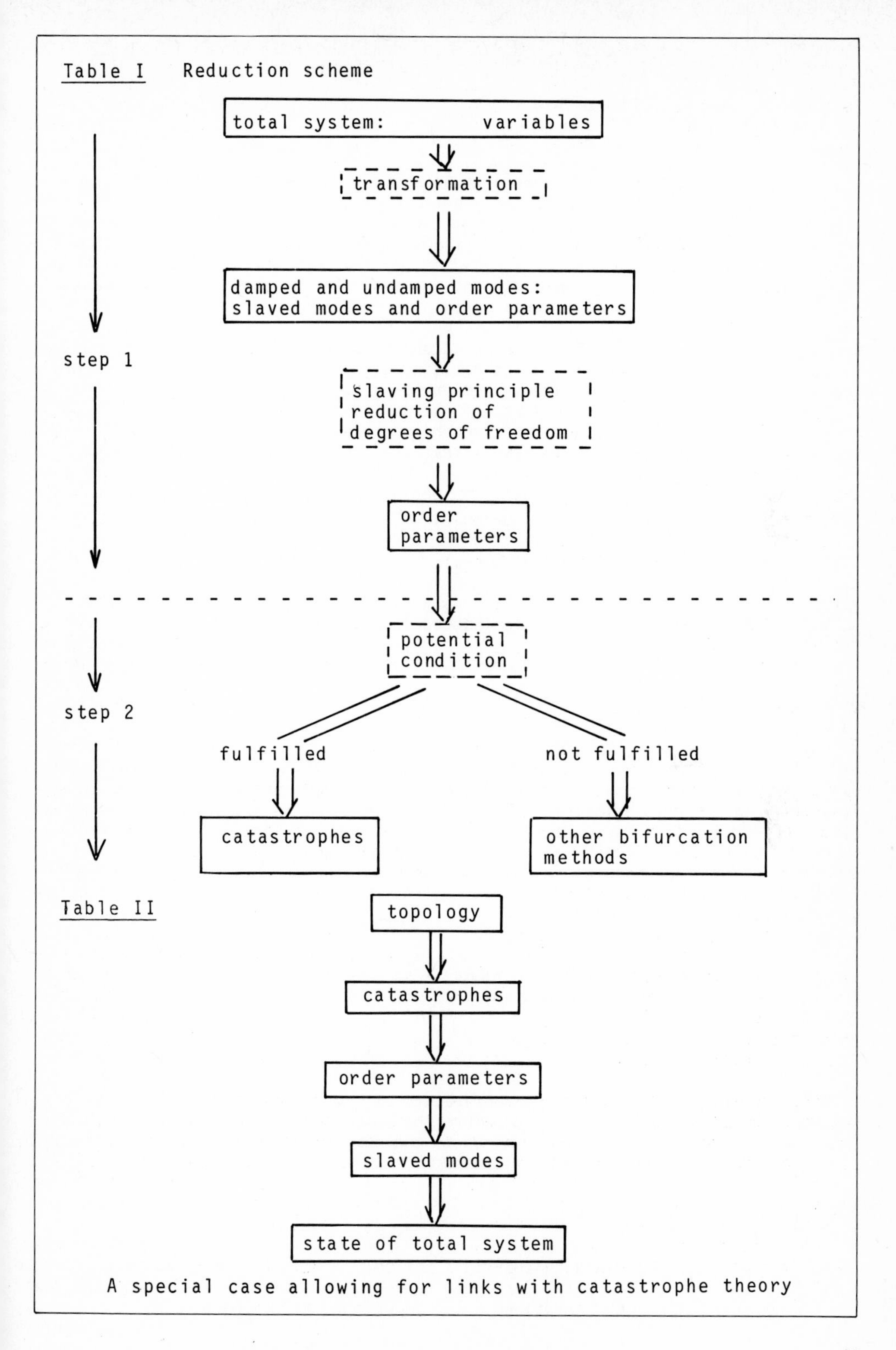
Table I Reduction scheme
total system: variables
transformation
damped and undamped modes:
slaved modes and order parameters
step 1
slaving principle
reduction of
degrees of freedom
order
parameters
potential
condition
step 2
fulfilled
not fulfilled
catastrophes
other bifurcation
methods
Table II
topology
catastrophes
order parameters
slaved modes
state of total system
A special case allowing for links with catastrophe theory

3. Nonequilibrium phase transitions and bifurcation of limit cycles and multiperiodic flows in continuous media

We consider equations of the form

$$\dot{q} = N(q, x, \nabla, \sigma) + F(t). \tag{3.1}$$

N is a nonlinear function of q depending on space points x and containing derivatives ∇. σ is a control parameter (or a set), $F(t)$ are fluctuating forces. We assume that for a certain range of control parameters σ there exists an attracting manifold on which the trajectories are dense. We assume that the manifolds are multidimensional tori. Other manifolds may also be admitted, for instance, multidimensional cylinders. To exhibit the main objective of our paper let us consider as an example a torus which becomes unstable for a certain control parameter $\sigma = \sigma_0$ and which is replaced by an outer and an inner torus. While we shall assume that the new tori are still close to the old one the point q on one of the new tori can move arbitrarily far away from the original point q. This suggests to introduce two kinds of variables describing a coordinate system on the original torus and vectors connecting the old and the new tori in a way of shortest distance. We assume that the original torus is described by a set of vectors r where we shall admit (but not require) that r may still depend on x. We assume that the original torus (or more general manifold) can be parametrized by $\Phi_1, \Phi_2, \ldots, \Phi_M$ and x. It is assumed that the Φ's are independent of x. We assume that r is periodic in the Φ's with period 2π.

We furthermore introduce trajectories by

$$q^0 = q^0(x, t, \sigma);\ q^0(x, 0, \sigma) = r(x), \tag{3.2}$$

which are defined so that they coincide with $r(x)$ at time $t = 0$. Since $r(x)$ is parametrized by the Φ's we may write instead of (3.1)

$$q^0 = q^0(x, t, \sigma; \{\Phi\}) \tag{3.3}$$

with

$$\{\Phi\} = (\Phi_1, \Phi_2, \ldots, \Phi_M). \tag{3.4}$$

We assume that the originally attracting manifold becomes unstable when σ approaches σ_0 from below. Now in principle two cases can happen. The original solution (3.3) can be extended into a region above σ_0 so that this new trajectory remains a solution of the (merely deformed) old manifold. We will treat this case in the next paragraph. The other possibility is that the meaning as a solution of q^0 is lost beyond $\sigma = \sigma_0$. This case is treated elsewhere.

4. Introduction of bifurcation variables

Since we want to treat nonlinear problems it is natural to assume that linear problems are solved. Therefore we assume that the following auxiliary linear problem is solved

$$\partial u_{j,m} / \partial t = (L u_j)_m. \tag{4.1}$$

In it the operator L is the Fréchet derivative of N with respect to q taken at $q = q^0$. In particular, L depends on $\{\Phi\}$. Therefore the solutions of (4.1) depend on x, t and $\{\Phi\}$. The solutions of (4.1) are subject to the same boundary conditions as the solutions of (3.1). We shall distinguish the set of solutions (4.1) by an index j

$$u_j(x, t;\ t;\ \{\Phi\}). \tag{4.2}$$

We now use the assumptions that q^0 is a periodic or multiperiodic function of time t and is dense on the bifurcating manifold (ergodic). Some analysis shows [10] that for a wide class of L's the solutions u can be written in the form

$$\underset{\sim}{u}_j(\underset{\sim}{x},t) = e^{\lambda_j t} \underset{\sim}{v}_j(\underset{\sim}{x},t) \tag{4.3}$$

In it $\underset{\sim}{v}_j$ is again a multiperiodic function which in certain degenerate cases may also depend on a finite number of powers of t. Here and in the following we adopt finite boundary conditions so that we may treat j as a discrete index. In the present paper I confine my treatment to the case where v does not depend on powers of t. The Floquet exponents λ_j can be shown to be independent of $\{\Phi\}$. One may readily show using (4.1) that the $\underset{\sim}{v}$'s obey the equations

$$\sum_{\ell} \partial N_m/\partial q_\ell^o \cdot v_{j,\ell} + (L\underset{\sim}{v}_j)_m = \lambda_j v_{j,m} + \partial v_{j,m}/\partial t \tag{4.4}$$

A certain number of $\underset{\sim}{v}$'s is already known, namely

$$\partial q_m^o \;/\partial \Phi_j \tag{4.5}$$

is a solution of (4.4) with Floquet exponent $\lambda = 0$. We now come to our basic hypothesis. We introduce the variables (trajectories) on the new bifurcating manifolds in the form

$$\underset{\sim}{q}(\underset{\sim}{x},t) = \underset{\sim}{q}^o(\underset{\sim}{x},t,\sigma;\{\Phi\ (t)\}\ ) + \sum_j{}' \xi_j(t)\ \underset{\sim}{v}_j(\underset{\sim}{x},t;\{\Phi\}\ ) \tag{4.6}$$

The basic idea is this: The Φ's take care of shifts on the old manifold, and the sum contains the vectors connecting the old and the new manifold. In (4.6) the prime at the sum indicates that we sum only over a set of $\underset{\sim}{v}$'s from which the $\underset{\sim}{v}$'s identical with (4.5) are excluded. Inserting the hypothesis (4.6) into (3.1) yields rather lengthy equations which we do not exhibit here.

We now introduce a set of functions $\bar{\underset{\sim}{v}}_j$ which obey an equation adjoint to (4.4) in the sense of a "backward equation". We introduce a scalar product by the definition

$$\langle \bar{\underset{\sim}{v}}_k | \underset{\sim}{v}_j \rangle = \int d^3x \sum_m \bar{v}_{k,m}\ v_{j,m} = \delta_{kj} \tag{4.7}$$

where the possible choice of the Kronecker symbol δ_{kj} can be proved in the same way as elsewhere [11]. We assume that the boundary conditions imposed on the solutions (4.2) are such that we obtain a discrete set of indices k. We arrange them in such a way that the first indices

$$k = 1,\ldots,M \tag{4.8}$$

cover the solutions (4.5), and all the other indices then start from M + 1

$$k = M + 1,\ \ldots \tag{4.9}$$

Taking the scalar product from the equations with $\underset{\sim}{v}_k$ for $k = 1,\ldots,M$ we immediately obtain

$$\dot{\Phi}_k + \sum_{j\ell}{}' \xi_j(t) \langle \bar{\underset{\sim}{v}}_k | \partial \bar{\underset{\sim}{v}}_j/\partial \Phi_\ell \rangle \cdot \dot{\Phi}_\ell = N^{(k)} + F^{(k)}(t) \tag{4.10}$$

with the abbreviations

$$N^{(k)} = \int d^3x \sum_m \bar{v}_{k,m} (N)_m \tag{4.11}$$

$$F^{(k)}(t) = \langle \underset{\sim}{\bar{v}}_k | \underset{\sim}{F} \rangle \tag{4.12}$$

$(N)_m$ means m-component of the nonlinear part of N (after subtracting the linear part containing L). Taking the corresponding scalar product for k = M + 1,... we obtain

$$\dot{\xi}_k(t) + \sum_{j\ell}{}' \xi_j(t) \langle \underset{\sim}{\bar{v}}_k | \partial \underset{\sim}{v}_j / \partial \Phi_\ell \rangle \dot{\Phi}_\ell = \lambda_k \xi_k + N^{(k)} + F^{(k)} \tag{4.13}$$

The new equations (4.10) and (4.13) are equivalent to the original set of equations (3.1). They represent a starting point to eliminate all stable modes. "Stability" refers to the negative sign of the Floquet exponents λ_k. We shall come back to the elimination procedure in section 6.

5. Elimination of slaved variables

The preceding chapters allow us to eliminate a good deal of the variables for the following reason well known from numerous explicit examples. When a system changes its behavior qualitatively usually only a few modes k become unstable (being characterized by non-negative real parts of their Floquet exponents), whereas all other modes are still stable (being characterized by negative real parts of their Floquet exponents)+. We divide the ξ's into "unstable" and "stable" ξ's replacing the indices k by indices u and s, respectively: $\xi_k \to \xi_u$ or ξ_s. For further simplification we introduce the abbreviation $\xi_u \to u$, $\xi_s \to s$. We further note that the phases Φ must be treated as unstable variables, ξ_u.

An important note to avoid misunderstandings must be added here: The distinction between stable and unstable modes is based on linear equations (4.1) and serves only to classify the modes. In our theory the unstable modes play a similar role as the null space of the conventional Lyapunov-Schmidt bifurcation theory. In the following we shall show how the stable modes can be uniquely and exactly expressed by the "unstable" modes. But the corresponding mode amplitudes ξ_u (or u) are no longer solutions of linear equations. They are the exact solutions of the fully nonlinear equations which result from the elimination of the "stable" mode amplitudes ξ_s (or s). Thus our procedure is based on a self-consistency approach according to the scheme: assume solution ξ_u → determine ξ_s as function of ξ_u → derive equations for ξ_u alone. Our method works, because we need to know only few general properties of the ξ_u's to derive their equations. The solutions ξ_u of these final equations are in general no more unstable. Choosing from the set of equations those which refer to stable modes we may write these equations in the form [11]

$$\Lambda \cdot \underset{\sim}{s} = \hat{\underset{\sim}{f}}(\underset{\sim}{u}, \underset{\sim}{s}) \tag{5.1}$$

where $\underset{\sim}{s}$ is a vector and Λ the matrix

$$\Lambda = d/dt - \begin{pmatrix} \lambda_{s_1} & & & \\ & \ddots & & \\ & & \lambda_{s_n} & \\ & & & \ddots \end{pmatrix} \tag{5.2}$$

The set of differential equations (5.1) must be supplemented by initial conditions. They depend on the physical situation under consideration. To secure the conver-

+ This asymmetry seems to be typical for dissipative systems. Under time reversal, the stable and unstable modes exchange their role and few become stable and many unstable. In non-dissipative systems (time-invariant systems) the "stable" and "unstable" modes play a symmetric role.

gence of the iteration procedure described below, the initial values of $\underset{\sim}{s}$ must be taken sufficiently small. When one is interested in the motion on the bifurcating attracting manifolds it suffices to require $\xi_s(t = -\infty) = 0$. This initial condition is assumed for what follows.

In general the right hand side of (5.1) contains a term independent of the variables $\underset{\sim}{u}$ and $\underset{\sim}{s}$ and further terms depending on $\underset{\sim}{u}$ and $\underset{\sim}{s}$

$$\hat{\underset{\sim}{f}}(\underset{\sim}{u},\underset{\sim}{s}) = \underset{\sim}{W}(t) + \underset{\sim}{f}(\underset{\sim}{u},\underset{\sim}{s}) \tag{5.3}$$

We assume that f is an analytic function of the variables. In a first step we expand $\underset{\sim}{f}$ into a series with respect to $\underset{\sim}{s}$ exhibiting explicitly the constant and linear term and a residual term which starts with the second power of $\underset{\sim}{s}$

$$\underset{\sim}{f}(\underset{\sim}{u},\underset{\sim}{s}) = \underset{\sim}{f}_0^{(o)}(\underset{\sim}{u}) + f_1^{(o)}(\underset{\sim}{u})\underset{\sim}{s} + \underset{\sim}{f}_2^{(o)}(\underset{\sim}{u},\underset{\sim}{s}) \tag{5.4}$$

We now describe an elimination procedure which generalizes one proposed previously [11] and which will allow us to express consecutively $\underset{\sim}{s}$ as a functional of $\underset{\sim}{u}$. We first define a new matrix Λ_o by

$$\Lambda_o = \Lambda - f_1^{(o)}(\underset{\sim}{u}) \tag{5.5}$$

and introduce a variable $\underset{\sim}{a}_o$ obeying

$$\Lambda_o \cdot \underset{\sim}{a}_o = \underset{\sim}{f}_o^{(o)}(\underset{\sim}{u}) \tag{5.6}$$

The formal solution of (5.6) reads

$$\underset{\sim}{a}_o = \Lambda_o^{-1} \cdot \underset{\sim}{f}_o^{(o)}(\underset{\sim}{u}) \tag{5.7}$$

This procedure can be explicitly continued up to infinite order as will be shown elsewhere. In this way our approach provides a new type of center manifold theorem. More precisely speaking, I determine in a new way the unstable manifold, which is continued into the newly evolving (i.e. bifurcating) attracting manifold.

6. A special case

As mentioned, it is possible to express all stable modes by the so-called unstable modes. When we insert the corresponding expressions into the equations (4.10) and (4.13) we find a closed set of equations for the unstable modes i.e. for the order parameters alone. Thus we see that the order parameter concept can be pushed rather far into regions where higher instabilities occur. The resulting order parameter equations can describe a variety of phenomena.

Due to lack of space I cannot go into details here and especially must skip all proofs. However, I want to illustrate the general procedure by a few typical examples.

(1) If there is only one real order parameter ξ_u we may find the bifurcation of an M-torus to two M-tori. To illustrate the procedure I consider a case with large enough $|\mathrm{Re}\lambda_s|$. The order parameter equations then assume the form

$$\dot{\xi}_u = \lambda_u \xi_u - c\,(\tilde{\Phi})\xi_u^3 + \ldots \tag{6.1}$$

and

$$\dot{\tilde{\Phi}}_k = \omega_k + \alpha_k(\tilde{\Phi})\xi_u^2 + \beta_k(\tilde{\Phi})\xi_u^3 + \ldots \tag{6.2}$$

Under suitable assumptions about the original equations and $\underset{\sim}{\omega}$ and thus about the coefficients of the equations (6.1), (6.2) one may show that a bifurcation of the original M-torus into two M-tori is established.

(2) As a second example I take a complex order parameter ξ_u with complex eigenvalue λ_u. A typical equation for it under additional simplifications, which can be shown as unessential for the type of solution, we obtain an equation of the form

$$\dot{\xi}_u = \lambda_u\xi_u + C(\tilde{\Phi})\xi_u|\xi_u|^2 + \ldots \tag{6.3}$$

$$\lambda_u = \gamma + i\omega_o$$

and one similar to (6.2). ξ_u introduces an additional frequency, so that when ξ_u is inserted into (4.6) we obtain now a quasi-periodic motion with M + 1 frequencies, i.e. motion on an M + 1 dimensional torus. Lack of space does not allow me here to discuss implications of these results and to discuss their connection with results by Ruelle and Takens [12] and Kolmogorov, Arnold and Moser. I just mention that in addition new phenomena occur when fluctuations are taken into account. They can cause a diffusion of the systems variables all over the tori. I have also found the possibility of resonances, when ω_o is a subharmonic of a fundamental frequency or of certain linear combinations of them. Some resonances had been studied by Joseph [13] for the case of bifurcating limit cycles.

7. A comment on "universal approaches"

In my talk I have outlined some results and ideas of the synergetics approach. I have chosen an aspect which particularly might be of interest to mathematicians. A rather simple but general concept, namely the slaving principle, has led synergetics in a straightforward manner to the forefront of a field of great current interest - bifurcation theory and here especially the bifurcation of motions on tori. On the other hand I have left aside many other interesting aspects of synergetics, for instance its relation to the theory of stochastic processes. In conclusion I should like to use this occasion to make a few general comments on "universal approaches". In view of the vast variety of different disciplines the possibility to develop some universal approach is certainly not self-evident. Nevertheless I think it is worthwhile to search for and further develop universal approaches. They seem to be the only way to understand or at least to describe our increasingly complex world. Needless to say that once the universal approach exists we can go from one field to another and use the results of one field to promote another field. However, we should never forget limitations of "universal approaches". It is highly dangerous to apply such an approach, if it has worked in a certain domain, to other domains as a dogma. Using any universal approach you must again and again check whether the prepositions made are fulfilled by the objects to which these approaches are applied.

In my opinion progress of "universal approaches" can be only achieved by preceding to more and more abstractions where we must heavily rely on mathematics which, after all, is the Queen of science. There might be a drawback of higher and higher abstractions. Fewer and fewer scientists might be able to follow up and to incorporate abstract theories into their own science. In my opinion it is up to those who develop (more or less) universal approaches to demonstrate explicitly how they apply to a given special problem. Having these and other limitations in mind it still remains a fascinating challenge to find unifying points of view on the laws of nature. I hope, synergetics is a useful step towards that goal.

I wish to thank the Volkswagen Foundation for financial support of the project "Synergetics".

References

1. R. Thom, Structural Stability and Morphogenesis, W.A. Benjamin 1975
2. E.C. Zeeman, Catastrophe Theory, Addison-Wesley 1977
3. cf. the contributions to this volume
4. H. Haken in lectures at Stuttgart University 1970
   H. Haken ed., Synergetics, Proceedings of a Symposium on Synergetics Elmau 1972. Teubner-Verlag, Stuttgart 1973
   H. Haken ed., Cooperative Effects, Progress in Synergetics, North-Holland, Amsterdam 1974
   H. Haken ed., Synergetics, A Workshop, Springer-Verlag 1977
5. H. Haken, Synergetics, An Introduction, Springer-Verlag 1977
6. H. Haken, Z. Physik 181, 96 (1964)
7. H. Haken, Laser Theory, Encyclopedia of Physics, Vol. XXV/2c, ed. S. Flügge, Springer-Verlag 1970
8. H. Haken, The Laser - Trailblazer of Synergetics, in Proceedings 4th Rochester Conference on Coherence and Quantum Optics, ed. L. Mandel and E. Wolf, 1978, Plenum
9. H. Haken, Z. Physik B 21, 105 (1975); B 22, 69 (1975)
10. H. Haken, to be published
11. H. Haken, Z. Physik B 29, 61 (1978)
12. D. Ruelle, F. Takens, Comm.Math.Phys. 20, 167 (1971); 23, 343 (1971)
13. D. Joseph in Synergetics, A Workshop, [4]

Part II

Topological Aspects of Wave Motion

Catastrophe and Fractal Regimes in Random Waves

M.V. Berry

H.H. Wills Physics Laboratory, University of Bristol, Tyndall Avenue
Bristol BS8 1TL, U.K.

1. Introduction

This is an account of two recent contributions to the theory of intensity fluctuations in random waves. Detailed treatments have been published elsewhere [1,2]; my purpose here is to give simplified outlines of the rather subtle arguments involved, and bring out the sharp contrasts between the two régimes considered.

Waves often acquire randomness in their wave functions ψ and intensities $I(\equiv|\psi|^2)$ by encounter with a random structure S. Familiar examples are starlight passing through turbulent atmosphere, sunlight reflected and refracted by water waves, and sound, radio and radar reflected by landscapes. The randomness of ψ is not related to that of S in any simple manner. In particular, if S is a spatially fluctuating refractive index with Gaussian randomness, ψ will not usually be a Gaussian random function of position. The most interesting statistics are those describing the intensity fluctuations. These are the moments I_n of the probability distribution of $|\psi|^2$, namely

$$I_n \equiv \langle I^n \rangle = \langle |\psi|^{2n} \rangle, \qquad (1)$$

where <> denotes averaging over the ensemble of S. (If ψ were a Gaussian random wave, I_n would be given by $I_1 n!$ for all problems to be considered here.)

In this paper I shall consider two random wave régimes where the effect of S on ψ cannot be approximated by perturbation theory. The régimes are distinguished by the absence or presence in S of detail on length scales close to the wavelength λ of ψ.

If S does not possess such detail, i.e. if it appears smooth on the wavelength scale, then methods based on geometrical optics can be employed. It is well known [3,4,5] that these lead to waves dominated by caustics (envelopes of the rays) which generically take the form of the catastrophes classified by Professor THOM [6]. In this 'diffraction catastrophe' régime, to be discussed in Section 2, ψ is characterised by violent fluctuations whose statistics are highly non-Gaussian and for which the moments I_n scale with λ according to 'critical exponents' that depend on the hierarchy of catastrophes.

If S does possess detail over a wide range of scales that includes λ, methods based on geometrical optics cannot be used. Instead, S can be modelled by a 'fractal' [7], that is, by a hierarchical structure with no length scale at all, whose Hausdorff-Besicovitch measure dimension D is not an integer. I call the corresponding waves 'diffractals' and discuss this régime in Section 3. The statistics of diffractals obey scaling laws very different from those for diffraction catastrophes, and involve integrals and asymptotic behaviour unfamiliar in wave theory.

Mathematically, it will be useful to think of the difference between these régimes in terms of the shortwave limit $\lambda \to 0$. If S is smooth on fine scales, then geometrical optics becomes valid as $\lambda \to 0$. But if S is a fractal, then as λ gets smaller, ever-finer levels of structure are exposed and the geometrical optics limit is never attained.

2. Random Diffraction Catastrophes

When λ is small enough, waves ψ diffracted by a smooth structure S can be described in terms of the rays of geometrical optics. The rays envelop caustics (focal manifolds) on which the intensity I rises to large values. On wavelength scales, the caustics are decorated with 'diffraction catastrophes' [8] characteristic of their topological type. In the language of catastrophe theory, the rays, waves, and caustics exist in the 'control space' $\underline{C}$. Now, it is crucial to the argument that when S is random $\underline{C}$ has many dimensions, corresponding not just to the spacetime point $\underline{r},t$ where ψ is measured but to the random variables $\underline{V}$ specifying the members of the ensemble of S. Therefore we can write the wave as $\psi(\underline{r},t;\, \underline{V})$, and the ensemble averages in (1) as

$$I_n = \int\dots\int d\underline{V}\, P(\underline{V})|\psi(\underline{r},t;\, \underline{V})|^{2n}, \tag{2}$$

where $\underline{r},t$ is fixed and where $P(\underline{V})$ is the density of realisations of S over its ensemble. The high dimensionality of $\underline{V}$ means that generically the caustics contain catastrophes of high codimension.

Consider first the case $\lambda = 0$. Then there is no diffraction at all, and I is infinity on the caustics. It follows from the conservation of energy, and can also be shown directly, that the infinities of I are integrable, so that the first moment $I_1 = \langle I \rangle$ exists. But higher powers of I cause the integrals in (2), and hence the moments I_n, to diverge. Therefore this simple argument based on geometrical optics confirms what has long been known [9,10], that the non-Gaussian strong fluctuations in ψ originate in focusing. However, it is too crude to account for the finite moments actually measured [11].

In practice several factors prevent I_n being infinite. Two examples are spatial and temporal incoherence of the source, as with the finite angular size and polychromaticity of the sun, which blur caustics refracted onto the sea bed [12,13]. Most fundamental, however, is the finite value of λ, which causes the divergences of I_n (as in, say, twinkling starlight) to be softened by diffraction.

A measure of this effect is the set of 'critical exponents' ν_n, defined by

$$\nu_n = \lim_{\lambda\to 0} \frac{d(\log I_n)}{d(\log 1/\lambda)}, \text{ i.e. } I_n \propto \lambda^{-\nu_n}, \tag{3}$$

which shows just how the I_n diverge as $\lambda \to 0$. Now I shall determine ν_n in terms of the catastrophes, using a simplified version of the argument in [1].

The main result will be that as n increases the exponents depend on catastrophes of ever higher codimension. Here is a simple physical argument to show why this must be so in the most familiar case where the randomness of S is stationary and averages are measured experimentally by time-averaging the intensity at fixed position $\underline{r}$: over long times, diffraction catstrophes of arbitrarily high codimension can pass arbitrarily close to $\underline{r}$. The higher-order catastrophes are rare but give rise to large localised fluctuations in the intensity. However, large rare fluctuations in I are precisely what dominate high moments I_n, which therefore depend on high-order catastrophes, as stated.

To calculate the exponents ν_n a scaling argument is employed. Consider the integral (2) for I_n. For small λ it will be dominated by those values of $\underline{V}$ lying on caustics. The integral will split into contributions $I_n(j)$ from the different catastrophes, here labelled j. Each contribution scales differently with λ and can be written

$$I_n(j) \propto \lambda^{-\nu_{nj}}, \tag{4}$$

where ν_{nj} is thus the exponent governing the contribution of the j'th catastrophe to the n'th moment.

What determines ν_{nj} are two aspects of the architecture of the j'th diffraction catastrophe. Firstly, there is the 'singularity index' β_j which describes how the wave amplitude $|\psi|$ on the caustic singularity diverges as $\lambda \to 0$:

$$|\psi|_{\text{on the caustic}} \propto \lambda^{-\beta_j}. \tag{5}$$

β_j is a measure of the strength of the diffraction catastrophe. And secondly, there is an index γ_j which describes how the diffraction pattern shrinks onto the caustic as $\lambda \to 0$:

$$\begin{array}{l}\text{hypervolume of maximum of}\\ \text{the diffraction catastrophe}\\ \text{in control space}\end{array} \propto \lambda^{\gamma_j} \tag{6}$$

γ_j is a measure of the extent of the diffraction catastrophe. The index β_j has been known for some time [14,15], but γ_j is a new quantity, introduced in [1] and recently shown [16] to be an invariant of the j'th catastrophe (related to the Jacobian of a diffeomorphism of control space).

In [1] the indices γ_j and β_j were determined from the diffraction integral representing ψ for the j'th catastrophe. Here the general procedure is illustrated in the simplest case of the 'fold' diffraction catstrophe where ψ is an Airy function:

$$\psi(V) = \frac{\text{constant}}{\lambda^{1/2}} \int_{-\infty}^{\infty} dS\ e^{2\pi i(S^3/3 + VS)/\lambda}. \tag{7}$$

The factor in brackets in the exponent is Thom's standard 'potential function' for the fold, where V is the single control parameter and S the 'state variable' (which usually represents postion on some initial wavefront from which the wave diffracts to $\underline{r}$, t). An obvious change of variables gives

$$\psi(V) = \frac{\text{constant}}{\lambda^{1/6}} \int_{-\infty}^{\infty} dS'\ e^{i(S'^3/3 + [V(\frac{2\pi}{\lambda})^{2/3}]S')}. \tag{8}$$

This shows that in terms of the standard Airy diffraction function, namely

$$\psi_{\text{standard}}(V') = \frac{1}{2\pi} \int_{-\infty}^{\infty} dS'\ e^{i(S'^3/3 + V'S')} \equiv Ai(V'), \tag{9}$$

$|\psi|$ rises to $O(\lambda^{-1/6})$ on the caustic itself (V = 0) and the width of the first bright fringe is $O(\lambda^{2/3})$. Therefore $\beta = 1/6$ and $\gamma = 2/3$ for the fold diffraction catastrophe.

Knowing β_j and γ_j, the corresponding contribution $I_n^{(j)}$ is estimated from the integral (2) as follows:

$$\begin{aligned} I_n^{(j)} &\propto [\text{maximum value of } |\psi|]^{2n} \times \begin{array}{l}\text{control space extent of}\\ \text{diffraction maximum}\end{array} \\ &\propto \lambda^{-2n\beta_j} \times \lambda^{\gamma_j}, \end{aligned} \tag{10}$$

whence (4) gives

$$\nu_{nj} = 2n\beta_j - \gamma_j. \tag{11}$$

It is thus established how each catastrophe contributes to each moment. Summing the contributions gives, for the n'th moment:

$$I_n \longrightarrow \sum_j C_{nj}\ \lambda^{-\nu_{nj}} \quad \text{as } \lambda \to 0, \tag{12}$$

where the sum is over all catastrophes j. The dominant term is clearly the one with the largest ν_{nj}, so that from (3) the critical exponents are given by

$$\nu_n = \max(\nu_{nj}). \tag{13}$$

This is the main result. To actually work out the ν_{ni}, and hence the ν_n, is a tedious exercise which I claim to be the most elaborate dimensional analysis (scaling) ever performed. What emerges are sets of rational numbers ν_n that depend only on the dimensionality d of the physical space in which ψ propagates. The reason for this is that although the codimension of the contributing catastrophes can be arbitrarily large the 'corank,' i.e. the number of state variables, cannot exceed d-1 since this is the dimensionality of wavefronts. Therefore for waves in two space dimensions only the corank-one 'cuspoid' catastrophes can contribute, and in three space dimensions only corank-one and corank-two catastrophes contribute; the extra singularities make ν_n larger in the latter case. Apart from this dimension-dependence the ν_n are universal: they do not depend on the details of S, only on the fact that it is smooth and its randomness is described by many variables $\underline{V}$.

Table 1 shows the first few ν_n for d = 2 and d = 3, together with symbols representing the dominant catastrophes; in more familiar terms [3], A_2 is the fold, A_3 the cusp, D_4 the elliptic and hyperbolic umbilics, and E_6 the symbolic umbilic. The value $\nu_2 = 0$ does not mean that the second moment I_2 is not singular as $\lambda \to 0$, only that its divergence is slower than any power of λ^{-1}. In fact, $I_n \propto \log(\lambda^{-1})$, as explicit (and elaborate) analysis [17,18,19] reveals.

Table 1 Critical exponents ν_n, and contributing catastrophes, for $2 \leq n \leq 5$

n	2	3	4	5
ν_n(2 space dimensions)	0	1/3	3/4	5/4
dominant catastrophe	A_2	A_2	A_3	A_3
ν_n(3 space dimensions)	0	1/3	1	5/3
dominant catastrophe	A_2	A_2 and D_4	D_4	D_4 and E_6

These values of ν_n constitute testable predictions about the wavelength-dependence of the moments I_n of such random waves. Usually I_n is measured as a function of other parameters, such as distance from a turbulent medium or strength of turbulence (see Section 3), but I am trying to arrange for direct measurements of ν_n to be made on the basis of the definition (3).

In this exposition I have not mentioned the serious problems [1] that arise in three space dimensions when n > 5, from the appearance of singularities with 'modality.' These singularities, which lie beyond Thom's classification, are discussed by ARNOL'D [15]. Making plausible assumptions it was possible to calculate the critical exponents up to ν_{13}.

This theory has tantalising analogies with the study of critical phenomena in statistical mechanics [20]. Here, critical behaviour ("$T \to T_c$") emerges as $\lambda \to 0$. Our analogue of the incorrect 'mean field theory' is geometrical optics. Like mean field theory, geometrical optics can be generated by a quadratic approximation in the exponent of an integrand (of a diffraction integral rather than a functional integral). And, just as in statistical mechanics, the correct behaviour is embodied in a series of 'universal' exponents. However, there is a serious difference: in the random waves problem the real work of calculating the exponents is made possible not by the 'renormalisation group' technique but by the Thom-Arnol'd classification of stable singularities of gradient maps.

Nevertheless, the analogies are sufficiently close to prompt the following question: is there a 'critical space dimensionality' d_c, analogous to (4) in statistical mechanics, beyond which all (or perhaps only some) ν_n are infinite, so that geometrical optics ('mean field theory') is valid? This would mean that catastrophes j of corank > d_c-1 and very high codimension would give contributions ν_{nj} that increase indefinitely with j (instead of reaching a maximum and then decreasing, as in cases so far studied), so that the dominant catastrophes are those of infinite codimension and I_n increases faster than any power of λ^{-1} as $\lambda \to 0$. At present singularity theory is not sufficiently developed to enable this question to be answered.

3. Random Diffractals

If the diffracting structure S is a fractal, ψ is a diffractal. Virtually nothing is known about diffractals. They constitute a new régime in wave physics, with potential to describe a wide range of phenomena from the sighing of the forest through the reflection of radio waves by landscapes to the propagation of light in fluids near their critical points.

In [2] I set up and solve what must be the simplest diffractal problem: free-space propagation of an initially-plane wave on which S has imposed a random fractal deformation of the wavefront at z = 0. Only propagation in two space dimensions x,z is considered. The initial wavefront is the fractal curve z = h(x), so that to a good approximation the diffraction problem has boundary condition

$$\psi(x\,,0) = e^{-2\pi i h(x)/\lambda} \tag{14}$$

Using diffraction theory, the propagation of ψ in the z direction can be studied, and the development of intensity fluctuations as a function of z can be followed by averaging over the ensemble of random wavefronts h(x).

To carry out this programme, h must be specified more precisely. It is here taken to be a Gaussian random function whose graph has fractal dimension D lying between 1 and 2 and whose correlations are described by the r.m.s. increment of h over distance X by

$$\sqrt{\langle [h(x+X) - h(x)]^2 \rangle} = L^{D-1}\,|X|^{2-D}. \tag{15}$$

The distance L (called the 'topothesy' of the wavefront) is the separation of points on the graph of h whose connecting chord has r.m.s. slope unity; L is a measure of the strength of the wavefront deformation. Both the variances $\langle h^2 \rangle$ and $\langle (\partial h/\partial x)^2 \rangle$ are infinite, but the existence of the average [15] is all that is required for diffracted statistics to be well defined. Fig. 1 shows a computed random function with D = 1.5. These graphs have the property of being self-similar under magnification [7] provided x and h are scaled in suitable ratio.

Diffraction theory shows that the intensity moments I_n (Eq. (1)) depend on D and also on one other parameter ζ which incorporates z, λ and L as follows:

$$\zeta = \frac{2\pi z}{\lambda}\left(\frac{2\pi L}{\lambda}\right)^{(D-1)/(2-D)} \Big/ 2^{1/(4-2D)} . \tag{16}$$

This is the most important of several diffractal scaling laws derived in [2]. It turns out that even for this apparently simple problem it is prohibitively difficult to calculate moments higher than the second (the first moment I_1 is unity for all ζ, as follows easily from (14)). I_2 is given by the following double integral:

$$I_2(\zeta) = \frac{4}{\pi\zeta}\int_0^\infty du \int_u^\infty dv\, \cos\frac{uv}{\zeta}\, e^{-[2u^{4-2D} + 2v^{4-2D} - (v+u)^{4-2D} - (v-u)^{4-2D}]} . \tag{17}$$

This integral has been studied by other authors (although not with diffractal interpretation) in connection with the propagation of laser beams [21] and radio [22] through turbulence; my results in [2] complement and extend theirs. The behaviour of the second-moment curves $I_2(\zeta)$ as D varies from 2 (extreme fractal with the graph of h just area-filling) to 1 (marginal fractal with the graph of h almost smooth) is summarised on Fig. 2.

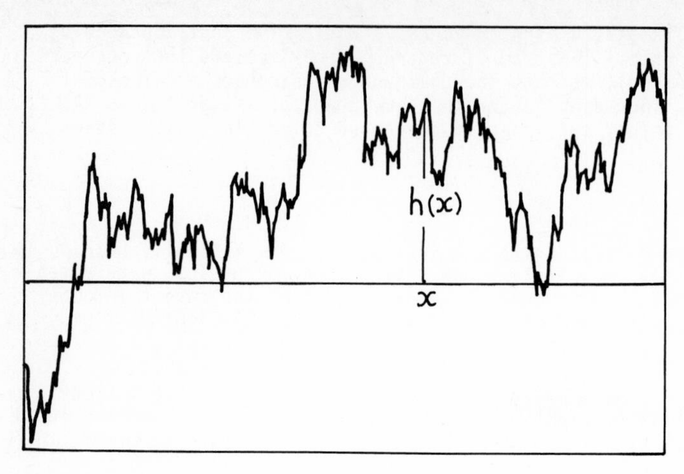

Fig. 1 Random fractal wavefront h(x) with D = 1.5 (computed by Z. V. Lewis)

In all cases $I_2(0) = 1$ (no intensity fluctuations near the initial wavefront where (14) shows that ψ is purely phase-modulated), and $I_2(\infty) = 2$ (Gaussian intensity fluctuations far from the initial wavefront). For intermediate values of ζ there may or may not be a maximum where $I_2 > 2$. In the case of the 'rougher' fractals $D \geq 1.5$ there is no maximum. For the 'less rough' fractals $D < 1.5$, however, there is a weak maximum in $I_n(\zeta)$. This can be regarded as an anticipation of the very strong maximum ($I_2 \propto \log \lambda^{-1}$) that occurs when the initial wavefront is smooth and which arises from diffraction catastrophes as explained in Section 2; this highly non-Gaussian second moment is illustrated in the bottom curve on Fig. 2.

The fractal and ordinary régimes are separated by the marginal case $D \to 1$. As shown on Fig. 2 the decay to Gaussian fluctuations is extremely weak for this case and takes the form of a term $(\log \zeta)^{-1}$. This asymptotic behaviour emerges from the analysis [2] as the result of an accumulation of power-law decays which is unprecedented in wave theory as far as I am aware.

4. Discussion

In this work I have tried to extend the boundaries of conventional random wave theory by studying two extreme régimes. In both cases the form of the intensity probability distribution is unknown and certainly not Gaussian. All we have is some information about the moments I_n. In the 'diffraction catastrophe' case of waves encountering smooth random structures S it was shown in Sec. 2 that the I_n obey universal scaling laws (3) as $\lambda \to 0$. However, this is very far from being a complete description of the statistics. For a start, each power $\lambda^{-\nu_n}$ is multiplied by a coefficient that depends on the measure of the dominant catastrophe in the space $\underline{V}$ of random variables of the ensemble of S, and this in turn depends on the nature of S--it is not universal. And then there is the question of the intensity correlations between different points rather than the fluctuations at a single point; it seems [17 - 19] that these correlations are characterised by several length scales, but there has been no analysis of the limit $\lambda \to 0$.

For diffractals the situation is just as bad. We know nothing about higher moments $I_{n>2}$. And we do not know whether there is any 'universality' about the behaviour of I_2 summarised on Fig. 2. It is probably a reasonable approximation to consider a random fractal S as imposing a random fractal deformation on a wavefront, but the randomness need not be Gaussian and the deformation need not be of

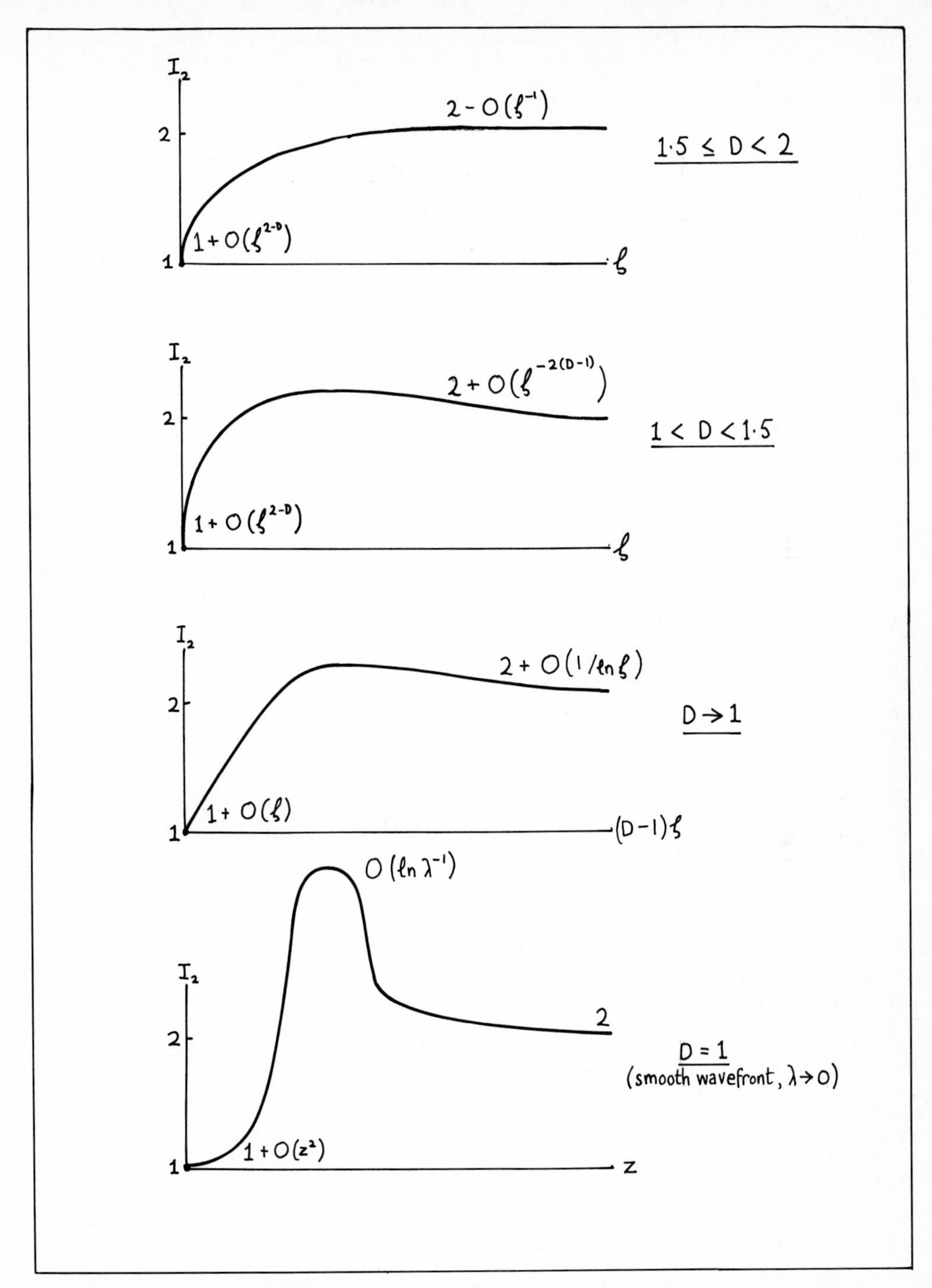

Fig. 2 Development of intensity fluctuations I_2 as a function of propagation distance for three diffractals (top three curves) from initial wavefronts with different fractal dimensions D, and one wave (bottom curve) that develops from a smooth initial wavefront (cf. Sec. 2)

a plane or of a straight line as assumed in Sec. 2. Moreover, we know little about waves within fractals (sound inside a tree, radio waves in the midst of turbulence, high-order modes of oscillation of an inland lake with random fractal boundary, etc.), although a conjecture on this subject is presented in [23].

In conclusion, it appears that the geometrical concepts of catastrophes and fractals can be fruitfully applied to wave physics. This is particularly the case in statistical problems where averaging over an ensemble of different realisations of a system implies that its properties will dominated by those morphologies that are structurally stable, as emphasised by Professor Thom.

References

1. M.V. Berry: J. Phys. A. 10, 2061-2081 (1977)
2. M.V. Berry: J. Phys. A.: to be published
3. T. Poston, I. Stewart: Catastrophe theory and its applications (Pitman, London 1978)
4. M.V. Berry: Adv. in Phys. 25, 1-26 (1976)
5. M.V. Berry: La Recherche 92, 760-768 (1978)
6. R. Thom: Structural stability and morphogenesis (Benjamin, U.S.A. 1975) (original French edition 1972)
7. B.B. Mandelbrot: Fractals (Freeman, San Francisco 1977)
8. M.V. Berry, J.F. Nye, F.J. Wright: Phil. Trans. Roy. Soc. Lond. In press (1978)
9. R.P. Mercier: Proc. Camb. Phil. Soc. 58, 382-400 (1962)
10. E.E. Salpeter: Astrophys. J. 147, 433-448 (1967)
11. E. Jakeman, E.R. Pike, P.N. Pusey: Nature 263, 215-217 (1976)
12. M.V. Berry, J.F. Nye: Nature 267, 34-36 (1977)
13. C. Upstill: Proc. Roy. Soc. Lond. In press (1978)
14. J.J. Duistermaat: Comm. pure App. Math. 27, 207-281 (1974)
15. V.I. Arnol'd: Usp. Mat. Nauk. 30, no.5, 3-65 (1975) (English translation: Russ. Math. Surv. 30, no.5, 1-75 (1975))
16. M. del C. Romero Fuster: M.Sc. dissertation, University of Warwick (1978)
17. V.I. Shishov: Izv. Vuz. Radiofiz. 14, 85-92 (1971)
18. R. Buckley: Aust. J. Phys. 24, 351-371 (1971)
19. E. Jakeman, J.G. McWhirter: J. Phys. A. 10, 1599-1643 (1977)
20. K.G. Wilson: Rev. Mod. Phys. 47, 773-840 (1975)
21. K.S. Gochelashvily, V.I. Shishov: J. Opt. Quant. Elect. 7, 524-536 (1975)
22. V.H. Rumsey: Radio Science 10, 1o7-114 (1975)
23. M.V. Berry: The next paper in this volume

Distribution of Modes in Fractal Resonators

M.V. Berry

H.H. Wills Physics Laboratory, University of Bristol, Tyndall Avenue
Bristol BS8 1TL, U.K.

1. Introduction

I shall present a conjecture which if correct would greatly extend our understanding of the distribution of modes in vibrating systems. Consider a region R of D-dimensional space, with a boundary ∂R which is d-dimensional, where $d = D-1$. Let eigenfunctions ψ_n and eigenvalues k_n in R satisfy

$$\left.\begin{aligned}(\nabla^2 + k_n^2)\,\psi_n &= 0 \text{ in } R\\ \psi_n &= 0 \text{ on } \partial R\end{aligned}\right\} . \qquad (1)$$

Let M_D be the measure of R (e.g. the volume if D = 3) and let m_d be the measure of ∂R (e.g. the area if D = 2). Then it is known [1,2,3] that the asymptotic mode number N(k), defined as the number of modes with $k_n < k$ (where $k \to \infty$) is given (after a smoothing whose technical details need not concern us) by

$$N(k) = \frac{M_D k^D}{(D/2)!\,(4\pi)^{D/2}} - \frac{m_d k^d}{4(d/2)!\,(4\pi)^{d/2}} + \ldots \qquad (2)$$

My conjecture is as follows: formula (2) remains valid when the resonator R and/or its boundary ∂R are fractals [4] (that is if D and/or d are not integers) provided D and d are interpreted as the Hausdorff-Besicovitch (fractal) dimensions of R and ∂R, and M_D and m_d are the Hausdorff D- and d-measures [4] of R and ∂R. I shall list some examples indicating the vast scope of this generalisation, and then present a discussion of its meaning and a plausibility argument for its correctness.

2. Examples

These fall into three classes. I: ∂R is a fractal, R is not. II: R is a fractal, ∂R is not. III: R and ∂R are both fractals.

Class I Here D is an integer but d is not. The first example is oscillations of water in a lake, where R (the lake) has D = 2, while ∂R (the lake's coastline) has $1 < d < 2$ (usually $d \simeq 1.3$,[4]). A model for this might be the Koch drum, where R is a planar membrane (D = 2) and ∂R the Koch snowflake curve [4] ($d = \log 4/\log 3 = 1.262$). Another example is vibrations of the whole Earth, where R (the Earth's matter) has D = 3 and ∂R (the Earth's surface) has $2 < d < 3$. Another example is the acoustic modes of a concert hall with fractally irregular walls ($D = 3$, $2 < d < 3$). Finally there are the oscillations of fluid in sponges or fractally porous rock ($D = 3$, $2 < d < 3$).

Class II Here d is an integer but D is not. The first example is the Weierstrass guitar, by which I mean the vibrations of a (long!) wire bent into a segment of a fractal curve such as the Weierstrass function [4,5] or the graph of one-dimensional Brownian motion [4]. In this case d = 0 (because the boundary is simply two points) and $1 < D < 2$. The second example is the Mandelbrot drum of the first kind, by which I mean vibrations of a rigid elastic sheet (e.g. fibre-glass) moulded to fit a fractal surface ($2 < D < 3$) and bounded by a smooth curve ($d = 1$).

Class III Here neither D nor d is an integer. The first example is the Mandelbrot drum of the second kind, which is the same as the drum of the first kind just defined except that its boundary is fractal, as in the case, for instance, of the surface of an island bounded by its coastline ($d = D - 1$). Another example is vibrations of the material of sponges or fractally porous rock (as opposed to the Class I vibrations of their fluid contents), for which $2 < D < 3$. A model for such resonators is the shivering Sierpinski sponge ($D = \log 20/\log 3 = 2.727$). The final set of examples in this class concerns fractal networks: the elastic vibrations of a tree, for instance, have $D < 3$ (because the branches are nearly volume-filling) and $d \lesssim 2$ (because the leaves are nearly area-filling); another problem of this sort is the waves described by Schrödinger's equation in networks [6] such as Cayley trees [7,8], to model the behaviour of electrons in disordered media.

3. Discussion

The fact that the terms in the formula (2) for $N(k)$ retain their meaning when D and/or d are fractional does not, of course, ensure that the formula can be validly employed for fractals. However, I shall now give a scaling argument strongly suggesting that at least the k-dependence of (2) is correct.

This is based on the idea that modes with wave numbers less than k, and hence wavelengths exceeding $\lambda = 2\pi/k$, are unaffected by detail in R and ∂R on scales smaller than λ. Therefore $N(k)$ can be estimated by replacing the fractals R and ∂R by λ-smoothed manifolds R_k and ∂R_k. These smoothed manifolds are not fractals, but have integer dimensionalities equal to the topological dimensions [4] $D_T (\leq D)$ and d_T $(= D_T - 1 \leq d)$ of R and ∂R respectively. Now the conventional (integer-dimension) version of (2) can be employed, provided it is realised that the measures M_{D_T} and m_{d_T} are k-dependent. Thus

$$N(k) = C_1 M_{D_T}(k) k^{D_T} + C_2 m_{d_T}(k) k^{d_T} + \ldots, \tag{3}$$

where C_1 and C_2 are constants.

As k increases, so do M_{D_T} and m_{d_T}, the laws of increase being

$$M_{D_T}(k) \propto M_D k^{D-D_T}, \quad m_{d_T}(k) \propto m_d k^{d-d_T}. \tag{4}$$

(For example, when ∂R is a coastline, $d_T = 1$ and $d \simeq 1.3$, and $m_{d_T}(k)$ is the coast's increasing length as measured on the scale λ.) Substitution of (4) into (3) gives precisely the same k-dependence as (2). By continuity it seems very likely that the constants in (2) are also correct.

This argument shows how the wave equation (1) should be interpreted in the cases where R is a fractal (Classes II and III in Section 2). Obviously ∇^2 cannot then be written in D-dimensional coordinates! But it can be written in D_T-dimensional coordinates, so that (1) can be thought of as a wave equation on the smoothed manifold R_k rather than R. (Dr. F.J. Wright has pointed out to me that this procedure is equivalent to solving (1) using the finite-element method, by discretising ∇^2 on a grid sampling R on the scale λ.) A simple illustrative example is the Weierstrass guitar (Section 2), where R_k is a smooth one-dimensional string with length $l(k)$. Then an explicit approximate 'quantum condition' for the eigenvalues, with an obvious origin, is

$$k_n \simeq \frac{n\pi}{l(k)}, \quad \text{i.e. } N(k) \simeq \frac{l(k)k}{\pi}, \tag{5}$$

which agrees with the first term of (2) for a D-dimensional string.

Suppose now that ψ satisfies not the Dirichlet boundary condition as in (1) but the Neumann condition, namely that the normal derivative of ψ vanishes on ∂R. Then the only effect on (2) is to change the sign of the second term from negative to positive. If ∂R is a fractal, this Neumann condition has no meaning if applied literally, and it seems reasonable to interpret it as applying instead to the smoothed boundary ∂R_k.

Next, I want to suggest that (2) with fractional d and/or D is likely to be a useful approximation in many cases where R and ∂R are not fractals. These are cases where R and/or ∂R possess hierarchical structure which does not however extend to infinitely small scales. Then of course the true asymptotic form of N(k) as $k \to \infty$ is given by (2) with integer D and d. However, for k corresponding to wavelengths $2\pi/k$ in the midst of the hierarchy of R and ∂R it is probable that N(k) will be given by (2) with D and d equal to 'pseudo-fractal dimensions' corresponding to the properties of the hierarchy extrapolated to indefinitely small scales.

The conjectured generalisation of (2) to fractal R and ∂R is radical. This can be appreciated by first considering what is known about smooth (non-fractal) resonators. The simplest case is where ∂R is such that (1) is separable. This corresponds to 'integrable' motion [8,9] of the straight 'rays' in R when reflected specularly from ∂R, the integrability being manifested by rays enveloping caustics in R,[10]. For generic smooth resonators the rays are 'non-integrable,' and typically fill R chaotically without forming caustics. However, in my proposed extension to fractal resonators, the reflection of rays <u>is not defined</u> if ∂R is a fractal, and the rays themselves <u>do not exist</u> if R is a fractal. Therefore no 'geometrical-optics' approximation can be invoked, even in the short-wave limit $k \to \infty$.

The eigenfunctions ψ in fractal resonators are examples of what I have called 'diffractals' [11,12]. It is possible that near a fractal boundary ∂R the form of ψ might in some statistical sense resemble the diffractal that evolves from a random fractal wavefront, whose properties are beginning to be understood [11,12].

References

1. H.P. Baltes, E.R. Hilf: <u>Spectra of finite systems</u> (B-I Wissenschaftsverlag, Mannheim 1978)
2. R. Balian, C. Bloch: Ann. Phys. (NY) <u>60</u>, 401-447 (1970)
3. R. Balian, C. Bloch: Ann. Phys. (NY) <u>64</u>, 271-307 (1971)
4. B.B. Mandelbrot: <u>Fractals</u> (Freeman, San Francisco 1977)
5. M.V. Berry, Z.V. Lewis: to be published
6. M.J. Richardson, N.L. Balazs: Ann. Phys. (NY) <u>73</u>, 308-325 (1972)
7. J. Dancz, S.F. Edwards: J. Phys. <u>C 8</u>, 2532-2548 (1975)
8. V.I. Arnol'd, A. Avez: <u>Ergodic problems of classical mechanics</u> (Benjamin, New York 1968)
9. M.V. Berry: Am. Inst. Phys. Conf. Ser. <u>44</u>, Chap. 2 (Nonlinear Dynamics, ed. by S. Jorna. A.I.P., New York 1978)
10. V.F. Lazutkin: Isv. Akad. Nauk. Mat. Ser. <u>37</u>, 186-216 (1973)
11. M.V. Berry: J. Phys. A.: to be published
12. M.V. Berry: The preceding paper in this volume

Optical Caustics and Diffraction Catastrophes

J.F. Nye

H.H. Wills Physics Laboratory, University of Bristol, Tyndall Avenue
Bristol BS8 1TL, U.K.

Abstract

It is well known that the stable caustics of geometrical optics fall into Thom's classification of the elementary catastrophes of codimension up to 3. The caustics formed by an irregular water drop lens are a good example. By allowing the shape of the drop to change under the control of one or more parameters catastrophes of higher codimension can be formed, in particular, the parabolic umbilic (D_5) and the symbolic umbilic (E_6). They correspond to reactions between umbilic points on the surface of the drop. Each stable caustic singularity of geometrical optics is clothed with a structurally stable diffraction pattern. The elliptic umbilic diffraction catastrophe has been studied in detail and shown to be built on a skeleton of dislocation lines, consisting of rings and hairpins, which is itself structurally stable.

1. Introduction

The understanding of optical caustics in terms of catastrophe theory is one of the earliest, most striking and most direct applications of the theory. A caustic surface, as an envelope of rays, is a concept of geometrical optics. But in real experiments, where we are not at the short wavelength limit, diffraction can play an important part. It softens the sharp singularities of geometrical optics and decorates them with characteristic patterns. This paper summarises some recent work at Bristol on both the ray aspect and the wave aspect of optical caustics.

2. Geometrical Optics and a Model System

In [1] BERRY described a simple model system in which a plane wavefront incident on some object becomes slightly perturbed; for example, a plane wave transmitted through, or reflected from, an undulating pane of glass such as is used for bathroom windows, or transmitted through a thin irregular water drop 'lens' on the surface of a sheet of dirty glass (Fig. 1). The caustics produced by this latter arrangement are easily seen by viewing a distant light through a raindrop on a window pane, placing the eye close to the drop.

These caustics are 'at infinity' -- they are directional caustics. Such a system also produces caustic surfaces in the three-dimensional image space of the droplet lens.

Let W be a wavefront after transmission through the droplet, and let P be a point in the space into which it is propagating. If the length of a straight line from P to an arbitrary point T on the wavefront is ℓ, the condition for PT to be a ray through P is that, with P fixed, ℓ should be stationary as T moves over the wavefront. Thus the rays through P are associated with the critical points of the function ℓ. ℓ plays the role of a potential in catastrophe theory with the two dimensional wavefront surface explored by T being the state space. The three-dimensional space in which P itself can move is the control space; it provides three parameters on which the potential ℓ depends.

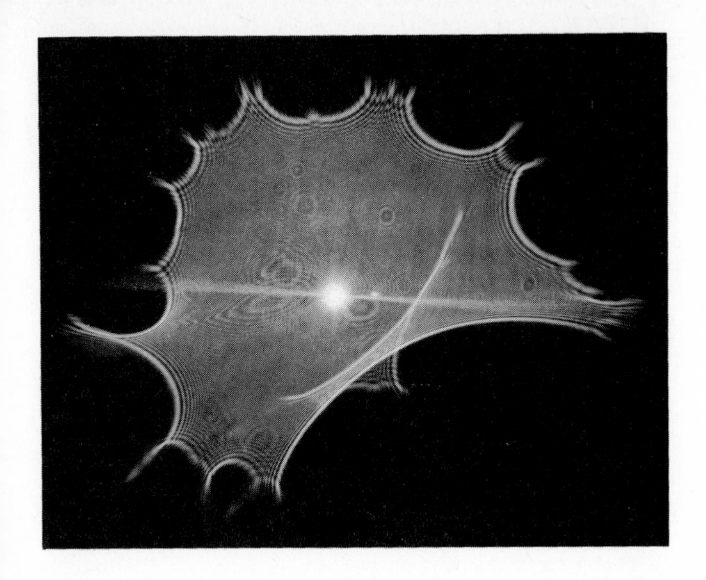

Fig. 1 Caustics formed in the far field by a broadened laser beam incident on a water drop 'lens'.

For fixed P there may be more than one critical point and so more than one ray. For instance in Fig. 2 P receives rays from both N_1 and N_2. But, as P moves to Q, N_1 and N_2 coalesce at N. Q is a centre of curvature and lies on the envelope of the rays, a caustic. This is the most elementary type of focusing and gives rise to a fold catastrophe. It is associated with the coalescene of two critical points. Higher catastrophes give stronger focusing and result from the coalescence of more than two critical points.

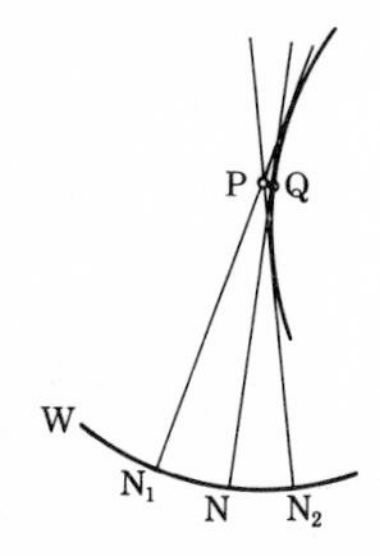

Fig. 2 A caustic in the near field as an envelope of rays.

If we use a fixed screen the dimension of the control space -- the codimension n -- is limited to 2, and the generic singularities are those with n up to 2: the fold and the cusp. If the observation point P can explore the whole of physical space, the generic caustics are the catastrophes with $n \leq 3$; this adds the swallow-tail, elliptic umbilic and hyperbolic umbilic to the list.

If the perturbations of the initially plane wavefronts are produced by passing the light through a wavy piece of glass, there is no restriction on their form, because the glass can have any distribution of thickness. But, if a water drop is used, its surface is governed by surface tension and this restricts the wavefronts to those that satisfy the relevant differential equation. The effect on the caustics is that, if the drop is small enough in its vertical dimension for the pressure inside it to be effectively uniform, all the expected elementary catastrophes (those with n ≦ 3) appear except for the hyperbolic umbilic. However, if the drop is larger, so that the effect of gravity is significant, the hyperbolic umbilic does appear and we have the full complement.

This is an example where the removal of a constraint in the form of a differential equation increases the number of different kinds of singularities. Sometimes removing constraints decreases the number; the flow field examples in [2] show this very well.

In a thin water drop of irregular outline on a horizontal surface there is another type of organisation to be found. Generically the surface of the drop will contain many umbilic points (points where the curvature is locally spherical) and these mark the places which give rise to the elliptic umbilic caustic foci in the space above the drop. But, because the sum of the two principal curvatures of the drop surface is everywhere the same (because the water pressure is uniform), the radius of curvature at each of the umbilic points on the surface is the same. So they all focus at the same height. The result is that, even in a very irregular drop, there is a definite plane of focus (Fig. 3); all the elliptic umbilic caustic singularities lie in this plane. Above and below this organising plane the caustics unfold in a complicated pattern of fold surfaces, ribs (lines of cusps) and swallowtail points which interact in a beautiful way [3].

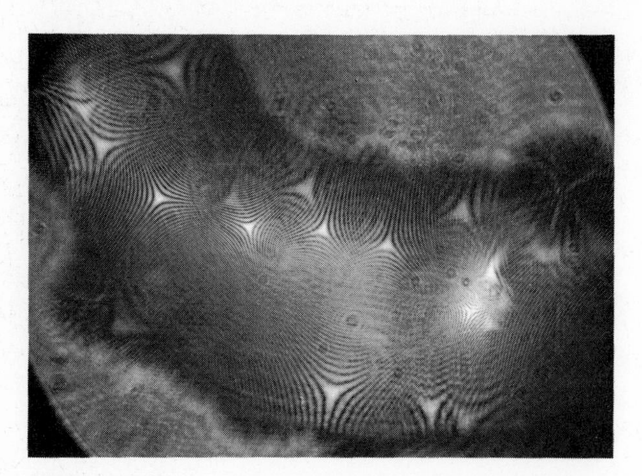

Fig.3 A few millimetres above an irregular drop all the elliptic umbilic foci, seen here as diffraction stars, lie in one plane

If the refracting or reflecting object that is perturbing the wavefront is continuously altered by varying a single parameter, which could be time, the caustics with n ≦ 2 seen on a fixed screen move about, and at certain instants singularities of codimension n = 3 appear. The caustic network formed by the focusing of sunlight on the bottom of a swimming pool is an example of this. The finite size of the sun's disk blurs the detail in the rapidly moving cellular pattern, and even in the laboratory the finest details are lost in diffraction. But, on analysis, the apparent triple junctions seen at low resolution turn out to be illusory. The basic elements of the pattern are simply folds and cusps but organised by higher singularities [4,5].

Another way of observing catastrophes of higher codimension is to pass light through an irregular water drop lens on a glass slide and then gradually tilt the slide so that gravity alters the shape of the drop. Or one can gradually add more water to an existing tilted drop, a droplet at a time, while keeping its outline fixed. Or one can keep its outline fixed and rotate it. Or one can change the outline in a continuous way. All these methods can, in principle, add new control parameters but they may not all be linearly independent. By such means it has proved possible to examine in detail the parabolic umbilic (D_5, n= 4), the symbolic umbilic (E_6, n = 5) and, to some extent, the double cusp (X_9, n = 8) [6].

When one produces a parabolic umbilic focus (Fig. 4) by manipulating a drop, what is actually happening on the drop surface is that an umbilic point that is initially elliptic in character (and therefore is producing an elliptic umbilic caustic figure above the drop) is changed to hyperbolic (producing a hyperbolic umbilic caustic figure) and, at the moment of change, the umbilic point on the surface is parabolic (the contours of constant magnitude of curvature in its vicinity are parabolas). Thus an umbilic reaction has occurred on the surface, organised, in the Thom sense, by the parabolic umbilic.

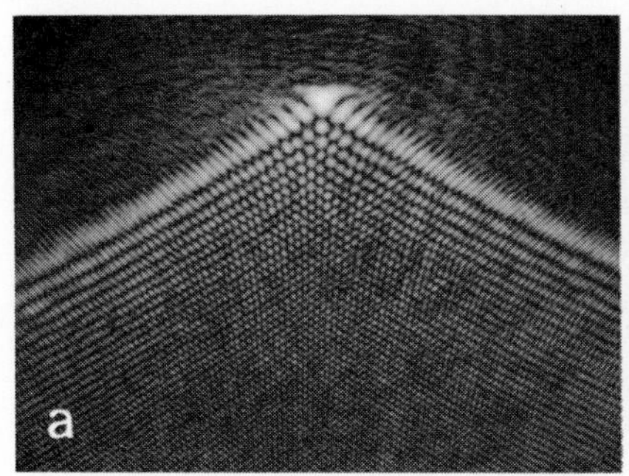

Fig. 4 Parabolic umbilic (D_5) focus. (a) Singular section
(b) Non-singular section: the mushroom

The parabolic umbilic can also organise another type of umbilic reaction, namely the creation or mutual destruction of two hyperbolic umbilic points, but this is an effect of its global rather than of its strictly local structure [6].

In a similar way, when an E_6 caustic is formed (Fig. 5) three umbilic points on the surface of the drop come together and turn into one. In fact they are three

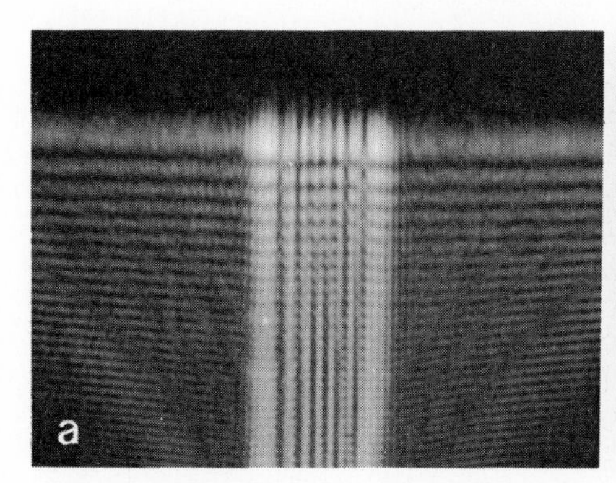

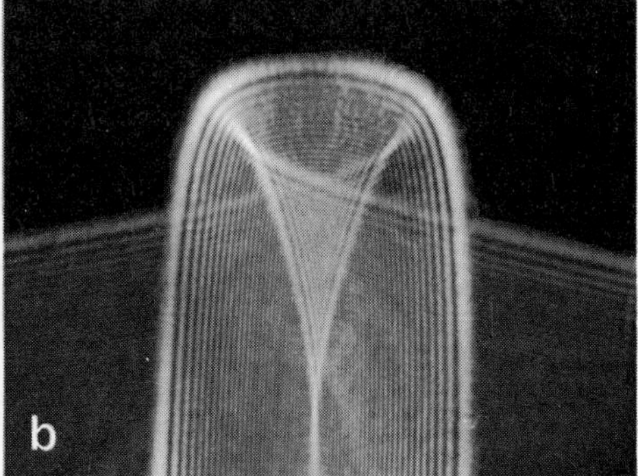

Fig. 5 Symbolic umbilic (E_6) focus. (a) Near the singular section, which would be a T-shaped fold. (b) An unfolding.

hyperbolic points, with singularity indices $-\frac{1}{2}$, $+\frac{1}{2}$, $-\frac{1}{2}$, and they become a single elliptic point with singularity index $-\frac{1}{2}$. E_6 organises this. These examples suggest how one might study umbilic reactions on a surface which is evolving under the control of a certain number of parameters, by regarding the higher catastrophes as organising centres.

3. Diffraction Catastrophes

In Fig. 1 one can see diffraction near the folds and cusps. The fold diffraction pattern, which occurs in the study of the rainbow, was worked out by AIRY in 1838 [7]; the cusp pattern was calculated by PEARCEY in 1946 [8] in a tour de force performed on the mechanical differential analyser machine at Cambridge.

There are analogous patterns, called diffraction catastrophes [9], associated with all the structurally stable caustics, and the essential point is that these patterns are themselves structurally stable [1]. Each catastrophe has its own canonical pattern; that is why it is worth devoting effort to their exploration. When one does this with the catastrophes of codimension 3 one is dealing, of course, with a three-dimensional diffraction pattern; it is therefore a great simplification to realise that such patterns are built on a framework, a scaffolding, of lines whose arrangement determines the main features of the intensity distribution of the diffraction. Thus to a large extent the lines constitute a description of the pattern.

The lines in question are the zeroes of wave intensity. That the zeroes are lines follows from the fact that two conditions must be satisfied (both the real and the imaginary parts of the complex wave amplitude have to be zero). In two dimensions the maxima and minima of intensity are points, and the minima are commonly zeroes (the amphidromic points of tidal theory are an example: at the points the tidal amplitude is zero while its phase is indeterminate). In three dimensions the maxima are points, and the minima are lines (except for isolated points). On these lines the intensity is zero and the phase is indeterminate. They are called dislocation lines [10] by analogy with crystal dislocations, because they interrupt the otherwise smooth wavefronts. The dislocation lines, which largely determine the intensity pattern, can never end; so they either form closed loops or go off to infinity.

At Bristol we have recently completed a full exploration of the elliptic umbilic diffraction catastrophe, both experimentally and theoretically [11]. On the experimental side the technique was to use a broadened laser beam and a water drop lens whose edge was an equilateral triangle. This produces a single elliptic umbilic focus which can then be photographed in successive sections perpendicular to the axis. The central section looks like one of the three-rayed stars in Fig. 3. A non-central section (Fig. 6a) shows a three-cusped caustic triangle as given by geometrical optics, but, interestingly, the corners lie between, rather than along, the rays of the star. Both the Airy pattern (fold) and the Pearcey pattern (cusp) can be seen in Fig. 6a, but broken up by the characteristic elliptic umbilic diffraction. The pattern is repetitive both in the plane of the photograph and in the third dimension. Thus we can speak of the "crystallography" of the diffraction pattern.

The amplitude of the pattern is proportional to a canonical diffraction integral [1,11]. When evaluated numerically the integral gives results (Fig. 6b) which reproduce virtually every detail of the experimental pictures. To gain a deeper understanding of its meaning one can use a stationary phase approximation, which amounts to taking the rays of geometrical optics with appropriate intensities and imagining that they interfere together. For example, the Airy pattern results from two beam interference, the two rays being PN_1 and PN_2 in Fig. 2. Similarly, the Pearcey pattern and the elliptic umbilic pattern arise from three and four beam interference respectively. Outside the caustic in Figs. 6(a) and (b) the parallel fringes are simply the result of two beam interference. At the centre of the pattern, and indeed almost out to the caustic itself, a four beam theory gives all the essential details.

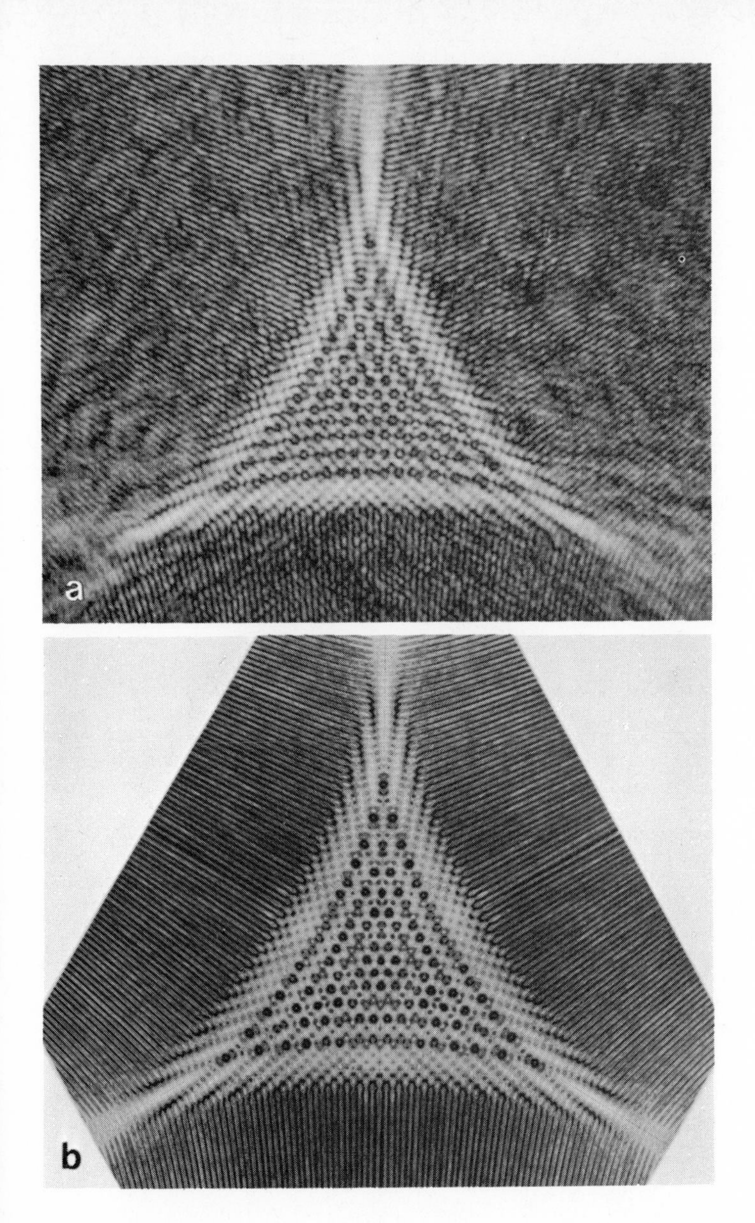

Fig. 6 Section near an elliptic umbilic focus. (a) Observed pattern. (b) Computer simulation (at a slightly greater distance from the focus).

Each S-shape in Fig. 7 is a puckered dislocation ring seen on edge. The puckered rings themselves can be discerned as black hexagons in Figs. 6a and b. Thus, except near the caustic, there is a well-defined crystallographic unit cell, whose contents change only slowly with position; the lattice is curved, it is rhombohedral and the space group is $R\bar{3}m$.

It is interesting to see how the separate dislocation rings near the centre change into long straight lines parallel to a rib, as they must do to achieve the Pearcey

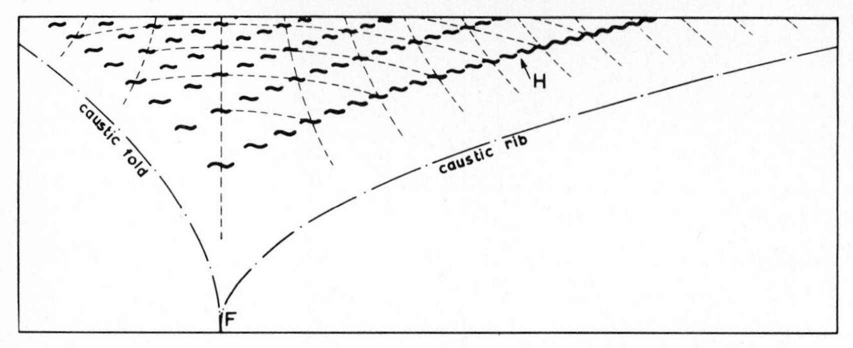

Fig. 7 Elliptic umbilic focus F seen from the side, only the upper half being shown. Broken lines denote unit cells. The two rows of dislocation rings nearest to the caustic are shown at the top of the main diagram, as seen in projection from above.

pattern near to a rib but far from the umbilic centre. They do this, as Fig. 7 suggests, by first approaching one another, en echelon, and then joining up, beyond the point H, so as to produce a long hairpin shape with the two ends sticking out along the rib. There are many such hairpins, each corresponding to a pair of dislocations in the Pearcey pattern.

The notion of structural stability goes deep here. The caustics themselves are structurally stable by Thom's theorem. The diffraction catastrophes are also structurally stable. A dislocation is a structurally stable object -- topology ensures that it survives perturbation -- and, as a direct consequence, the arrays of dislocations which underpin the diffraction catastrophes are themselves structurally stable. By way of illustration, although there are imperfections in our experiment on the elliptic umbilic catastrophe, which make the pattern deviate in minute aspects from the expected symmetry -- nonetheless the dislocation rings and hairpins remain unbroken and in essentially the same arrangement.

Similar considerations and similar interesting dislocation line topology must underly the other diffraction catastrophes. Our laboratory has already made some progress in unravelling their details.

References

1. M.V. Berry: Advances in Physics 25, 1-26 (1976)
2. J.F. Nye: This volume (1979)
3. J.F. Nye: Proc. Roy. Soc. Lond. A 361, 21-41 (1978)
4. M.V. Berry, J.F. Nye: Nature, 267, 34-36 (1977)
5. C. Upstill: Proc. Roy. Soc. Lond. A: In press (1979)
6. J.F. Nye: Phil. Trans. Roy. Soc. Lond.: In press (1979)
7. G.B. Airy: Trans. Camb. Phil. Soc. 6, 379-402 (1838)
8. T. Pearcey: Phil. Mag. 37, 311-317 (1946)
9. H. Trinkaus, F. Drepper: J. Phys. A 10, L11-16 (1977)
10. J.F. Nye, M.V. Berry: Proc. Roy. Soc. Lond. A 336, 165-190 (1974)
11. M.V. Berry, J.F. Nye, F.J. Wright: Phil. Trans. Roy. Soc. Lond.: In press (1979)

Part III

Catastrophes in Infinite Dimensions

Infinite Dimensions and the Fold Catastrophe

Robert Magnus

Science Institute, University of Iceland, Dunhaga 3, Reykjavík, Iceland

Tim Poston

Section de Physique, Université de Genève, 32, bvd. d'Yvoy,
CH-1211 Geneva, Switzerland

Contents

Abstract

A simple example of the manner of breakdown of Thom's theorem in infinite dimensions, but for which catastrophe theoretic methods remain illuminating, is discussed both informally and analytically. The co-dimension 1 character of the set of jets on a Hilbert space which do not admit the usual splitting in Thom's theorem is established and some implications for modelling are discussed.

Supported by Fonds national suisse de la recherche scientifique, grant 2.934-0.77

§1. Introduction

For ($\leq$5)-parameter families of functions $X \to R$, where X is a finite-dimensional real C^∞ manifold, "almost all" bifurcations are locally equivalent, up to sign, to eleven standard types. The usual proof of this theorem (see for instance Trotman and Zeeman [16]) relies heavily on various features peculiar to the finite-dimensional calculus. The extent to which an analogous theorem might be true for X a Hilbert, Banach, or other infinite-dimensional manifold has been somewhat unclear. In the spirit of Thom [15], one would like for instance to be able to draw geometric conclusions about form and its development given only the hypothesis of local minimisation[1] of an 'energy' over X, or of stationarity of a variational quantity on X, without being too detailed about the nature and dimensions of X itself.

For example, suppose X is the space of concentrations of chemicals in a cell, *supposed evenly mixed* -- a common first approximation. Then X has as dimension the number N of chemical species, which is finite (of order 10^4), so that Thom's theorem seems applicable to the "gradient dynamic" bifurcations in X, an important though not universal class. But if the cell occupies a region C, with concentrations at each point given by a function $x:C \to R^N$ (a natural *next* approximation), the space X of such x has infinite dimension. Intuitively, many bifurcations will still behave as for the previous model. Indeed, if much chemical dynamics were not well describable by motions in R^N (for rather low N), its study could hardly have begun until very recently, and most of the mathematics in chemical textbooks would be chemically irrelevant.

Thus, infinite-dimensional behaviour *often* sufficiently resembles finite-dimensional for quantitative indifference between the models, just as Thom's theorem shows that *typically* gradient bifurcations in R^N with ($\leq$5) external parameters resemble (indeed, reduce to) bifurcations in R^2. (For the rigorous "open dense in the Whitney C^∞ topology" meaning of this typicality, alias *genericity*, see Trotman and Zeeman [16]. For detailed exploration of its practical use and limitations in the sciences, see Poston and Stewart [13].) A key step here is that typically a ($\leq$5)-parameter family of functions $R^N \to R$ has only critical points at which the Hessian has corank $\leq$2, upon which various reduction procedures may be applied. The reduction procedures work equally well for smooth functions $X \to R$, X a Banach space, provided no critical points exist where the Hessian fails to satisfy the *Fredholm condition* that makes it split as nicely as all quadratic forms do in finite dimensions. (One can either add parameters to the lemma of Gromoll and Meyer [5] and drag the reduction results up from finite dimensions, or work directly in infinite dimensions as in Magnus [6, 7, 8].) Moreover, subject to this 'Fredholm' restriction, the genericity part of Thom's theorem can also be carried through (Arkeryd [1], Chillingworth [2]) So if this restriction itself were generically satisfied (as for some problems such as scattering caustics it is *necessarily* satisfied), Thom's will would be done in infinite dimensions as it is in finite.

Thus our attention is focused[2] on the quadratic forms on a tangent space to X generically encountered as Hessians: to what extent can we expect them to be Fredholm? Let us concentrate for the moment on X a Hilbert space (the lowest possible infinite-dimensional case).

First, whether we are looking at individual functions $X \to R$ or at r-parameter families of them, the Fredholm condition is an open one (and hence stable) for suitable reasonable choices of topology.

---

[1] This is already a restrictive hypothesis, ruling out applications with limit cycles and other non-point attractors except insofar as their bifurcations may be reduced to those of "generic" Lyapunov functions. But it is often a natural one.

[2] Not perfectly, which would be highly non-generic: it has a stable caustic there.

Secondly, for *individual* quadratic forms on X (or - equivalently and more conveniently - self-adjoint bounded operators on X), it is dense. Indeed, in the space, Q say, of such operators denote by C_m the set of Fredholm ones of corank $m \in \mathbb{N} \cup \{0\}$. Then C_m is a submanifold of Q (of codimension $\frac{1}{2}m(m+1)$) and the set, Z say, of non-Fredholm points (operators) in Q is in the closure of each C_m including

$$C_0 = \{\text{invertibles}\} \simeq \{\text{Morse quadratic forms}\}.$$

This can be proved rather briefly, though it does not seem to be well known. Since the various C_m attach nicely to each other, we seem close to a nice infinite-dimensional 'stratification' of Q (suitably extending the finite-dimensional definitions). With this we might hope to imitate the most elegant density proofs in finite dimensions, such as Mather's proof (most fully expounded by Gibson et al. [3]) of Thom's 'density of topological stability $R^n \to R^p$' conjecture, and prove that finitely-parametrised families generically abide by the Fredholm condition.

But perfectly reasonable families, quite stably, don't.

The easiest way to 'see' the genericity of fold catastrophes of 1-parameter families of functions $R^n \to R$, intuitively, involves consideration of generic curves in the space of quadratic forms on R^n (see Poston and Stewart [13]). Generically, these miss all but forms of corank 0 or 1, with the latter at isolated points only. Such behaviour cannot be generic in Q as above. Consider the curve

$$c:[-1,1] \to Q \qquad t \to tI$$

where I is the identity operator in Q. Any curve c' near c must have an infinite difference in Morse index between c'(-1) and c'(1). This is incompatible with the property of having only isolated points (hence finitely many) at which the index can change, and by only 2 at each point: the property is therefore not dense.

Indeed, the 'bad set' Z of non-Fredholm self-adjoint operators is of codimension one in the following sense. If $A \in Z$ and B is suitably 'small' there is a small real number λ such that $A+B-\lambda I$ is again in Z (see §7 below). A curve b that passes through Z with a non-zero 'I-component of velocity' thus *stably* meets Z: any small enough perturbation of b meets Z too, though perhaps not in a point transformable to A. (Compare the way a 7-parameter family of functions $R^2 \to R$ can stably have a strictly $O(4)$ point, but be unstable with respect to the *type* of $O(4)$ point for cross-ratio 'modularity' reasons.) The geometrical structure of Z, with particular reference to its equivalence classes (orbits) under changes of variable, is a rich field for investigation.

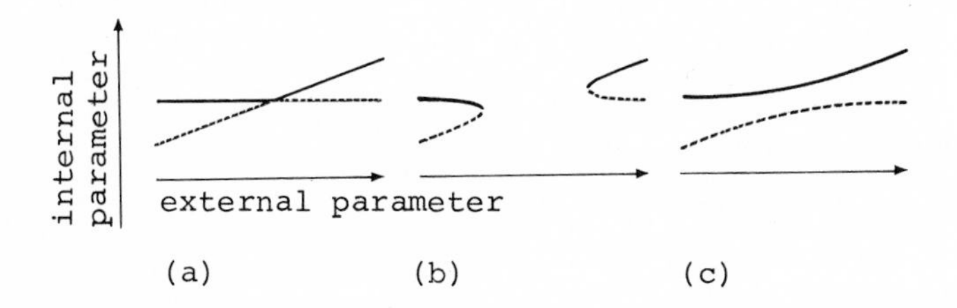

Fig. 1 The unstable 'exchange of stabilities' phenomenon (a) can easily be perturbed to have a disconnected set of critical points (b) which may not even include bifurcation points (c). Thus topological arguments about arcs joining I to -I do not directly suffice for conclusions about critical point sets.

Thus curves in Q do not generically miss Z. (This does not imply that generically they fail to miss it, which is a quite different question.) However, the set of critical points for an unstable family of functions can easily change connectivity under perturbation (even in finite dimensions; Fig. 1), so the possibility remains that any problem giving rise to c as above could be perturbed so as to change c to a *disconnected* union of nicer curves that do not join -I to I. Perhaps any curve meeting Z could be 'broken' there by a small change in the original problem, which then becomes Fredholm after all?

Density cannot be salvaged in this way. The purpose of this paper is to examine a natural class of problems in which

(i) *failure* of the Fredholm condition is generic, but
(ii) generically the bifurcations that occur are structurally stable under perturbations within the class,

and to examine the obstructions to finding a perturbation (even outside the class) which reduces to a structurally stable finite-dimensional problem. The behaviour involved is sufficiently *un*pathological (despite the failure of such reducibility), and accessible enough to the geometric imagination, as to raise hopes of a classification analogous to Thom's. What *does* generically happen in r-parameter infinite-dimensional variational problems, without the Fredholm condition?

The functional analytic aspect of this paper is not novel or deep in classical bifurcation theoretic terms: the intended contribution is the treatment of its relation to genericity and stability questions. These have been almost entirely ignored in the literature on functional analytic bifurcation problems, despite the practical relevance of stability in particular. When the behaviour of a classical one-parameter problem changes radically under arbitrarily small perturbations of seven different kinds [10], all natural to the problem, it is worth recovering stability and robustness of the model by adding the seven parameters. (Though from the Golubitsky-Schaeffer [4] 'imperfect bifurcation' viewpoint, keeping the one and the seven quite distinct, even more are necessary.) The plate buckling problem in question, however, reduces by its Fredholm nature to finite (ten) dimensions. The general non-Fredholm theory of structural stability must develop partly by acquisition of examples, such as the study below.

Our treatment is for the most part fairly informal, as the immediate object is to gain insight from this case rather than to prove general theorems. Sections 2 to 5 explore the problem geometrically, §6 and §7 treat more exactly the functional analytic questions that arise, and §8 discusses the implications for scientific modelling.

§2. The Non-Linear Elastic String

What is the simplest way to bring breaking within the scope of a theory of stretching? (Apart from simply specifying a threshold 'breaking strain', with no explanation within the model for it.) We want to work with a continuum model, for the light shed on such models in general, but it is convenient to motivate it by finite considerations.

Fig. 2
A string of particles

Think of an idealised 1-dimensional string as a row of particles (Fig. 2) with positions on the real line R, held together by conservative forces plus damping terms. (Thus stable static configurations are defined by local minimisation of energy. We rule out plastic behaviour by this hypothesis.)

The binding of real particles like atoms can be adequately described for many purposes by a potential ϕ assigning an energy $\phi(x)$ to each possible separation x, usually a potential of the form shown in Fig. 3. In the absence of external forces, the unique equilibrium separation will be at the minimum x_0 of ϕ. In the presence of a separating force F, there will be equilibria at the two points x_1, x_2 in Fig. 3(b) where $d\phi/dx=F$. Clearly x_1 is stable, x_2 unstable, and as F increases the two approach and vanish at the point of inflexion x_3, generically in a fold catastrophe. The atoms then separate. (We ignore fluctuations, which would usually liberate them earlier.)

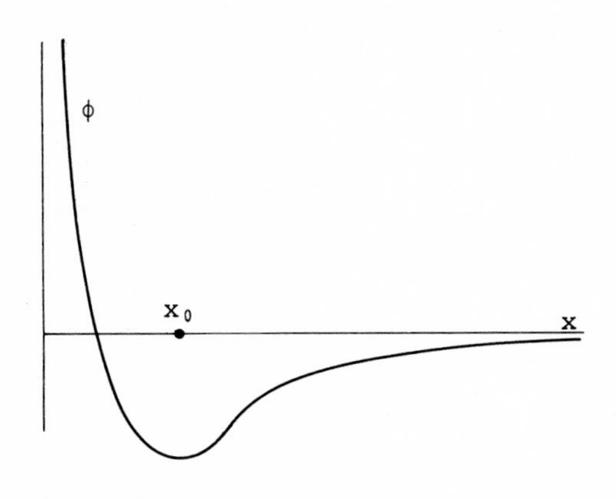

Fig. 3(a)
Interatomic potential with stable equilibrium at x_0

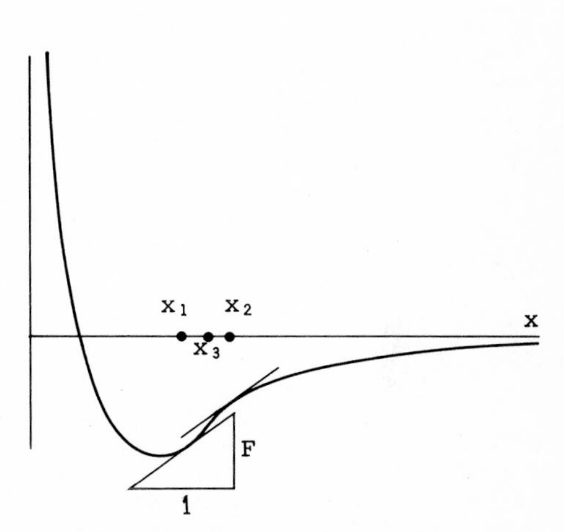

Fig. 3(b)
Stable and unstable equilibria in the presence of a separating force F

Passing, as usual in analytic treatment of elasticity, from considering very small parts of the system to imagining infinitesimal ones, we take the string as a one-dimensional continuum [0,1], defining our unit 1 as the unstressed string length and labelling the points accordingly. Positions f are given as functions from [0,1] to the set of distances below the point of attachment (Fig. 4) with $f(0)=0$ always. The separation of neighbouring particles is replaced by an 'infinitesimal separation' at each x, i.e. by the derivative

$$f'(x) = \frac{df}{dx}(x).$$

By analogy with Fig. 3 we assume at each point x a local relation between extension $f'(x)$ and energy: a 'local energy' function $E^x: R \to R$ such that the integral

$$\int_0^1 E^x(f'(x))dx$$

gives the elastic energy of the whole string.

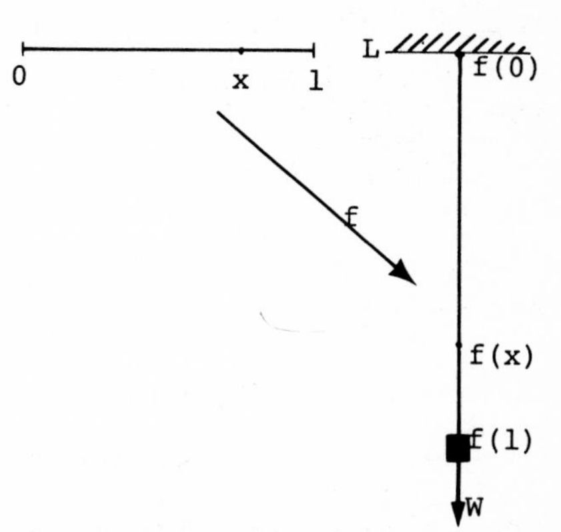

Fig. 4
In string position f, point x of the string is f(x) below L.

A difference from some elastic problem formulations is worth noticing: we do *not* suppose the extension of a small piece of string to be small relative to its unstressed length, which would lead to taking the local stress(strain) function as linear unless its first derivative vanished at the zero-strain point $f'(x)=1$. Motivated by our 'atoms', which may only come unstuck when separated by several times their unstressed distance, we consider substantial stretching and allow our pointwise stress-strain relation, like the interatomic potential, to be nonlinear. Correspondingly, E^x is not quadratic.

Moreover, by stiffness or constraint we take the string as compelled to lie in a given vertical line, and able (like the atoms) to resist compressive forces. A 'totally flexible' elastic string changes its response to tension non-differentiably at 0 (taking compressive forces as negative extensional ones) which would require a smooth variational problem to be defined only on a space of functions with $f'(x)>1$ for all x. (We could in fact allow E^x to have kinks, discontinuities etc. at $f'(x)=1$, since these are irrelevant in all realistic cases to the singularity discussed below. We avoid technical complication by excluding them.) Where $f'(x)<1$, in our model, the string is under compression. If, as in the usual interatomic bond model, *no* compression can bring nearby points into coincidence, each $E^x(y)\to\infty$ as $y\to 0_+$ like the potential in Fig. 3. We suppose this to happen monotonically and without inflexions, as our attention is on tension effects. Otherwise, compressive forces could give 'inward collapse' catastrophes closely analogous to the 'breaking' discussed below.

We suppose that the function

$$E : R^2 \to R$$
$$(x,y) \mapsto E^x(y) = E(x,y)$$

is smooth (arbitrarily differentiable) but we do not make the assumption analogous to treating Fig. 2 as a row of *identical* atoms. We let E^x depend on x, though smoothly. This might reasonably model a string of finite - and hence inevitably not quite constant - thickness, whose behaviour remains 'nonlinearly elastic' rather than 'plastic' up to around breaking or necking point. For simplicity, though it makes little difference to the analysis, we suppose the string weightless, with a concentrated load W at the bottom as in Fig. 4. The total energy in configuration f thus becomes

$$V(f) = \int_0^1 E(x,f'(x))dx - Wf(1). \tag{1}$$

Intuitively, the whole string will be in equilibrium if the elastic resistance to stretching balances the force W creating the tension, at each point. Indeed, without yet worrying about what function space to set the problem in, we differentiate V formally with respect to f. The derivative $V'(f)$ at f is a linear map

$$h \mapsto V'(f).h = \int_0^1 D_2E(x,f'(x))h'(x)dx - Wh(1) \tag{2}$$

where D_2E means the partial derivative of E with respect to its second argument. For equilibrium $V'(f)=0$, that is

$$\int_0^1 D_2E(x,f'(x))h'(x)dx = Wh(1) \quad \text{for } \mathit{all}\ h. \tag{3}$$

This may be rewritten

$$\int_0^1 [D_2E(x,f'(x))-W]h'(x)dx = 0 \quad \text{for all } h, \tag{4}$$

which implies rather obviously

$$D_2E(x,f'(x)) = W. \tag{5}$$

Indeed, any model which did not yield an analogue of (5) would generally be rejected on the grounds, interpreting $D_2E(x,f'(x))$ as 'tension', that somewhere an action or reaction was left unbalanced.

Of course, if we have a solution f' to (5), it may easily be integrated to give an equilibrium position f. Now, (5) has a simple geometric interpretation in the spirit of catastrophe theory. Take the equation

$$\frac{\partial E}{\partial y'}(x,y) = W \qquad (6)$$

which defines a surface S (Fig. 5) in (W,x,y)-space, which a catastrophe theorist would naturally regard as a 'catastrophe manifold' with y as a 'state variable' and x, W as 'controls', in Zeeman's convenient terminology. Here, however, an equilibrium 'state' is not a point in the catastrophe manifold S, but a lifting to it of a straight path

$$[0,1]\to[0,1]\times R, \text{ some fixed } W$$
$$x \mapsto (x,W)$$

in (x,W)-space. Examples are the curves f_1', f_2' in Fig. 5 above the line $W=W_1$. These geometric objects are equivalent (by considering them as graphs $[0,1]\to R$ and integrating the corresponding functions) to equilibria f, but we choose to regard them more in their own right - on geometric grounds here, in §6 on analytic.

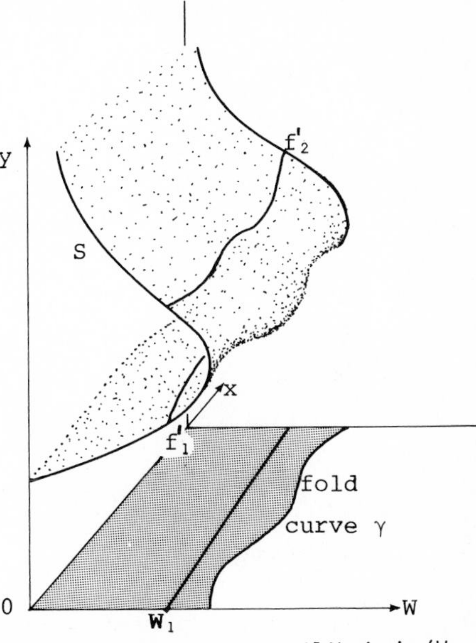

Fig. 5 Surface S of pointwise equilibria in (W,x,y)-space, with folded projection to (W,x)-space

Now we have refused to suppose E(x,y) independent of x, which would have implied (for weightless string) that any fold curve was parallel to the x-axis as in Fig. 6. Such a strong hypothesis of translational symmetry would only be reasonable if it applied uniformly to the whole problem, in particular to f' as well, so that the integrands became constant and the problem reduced to a 1-dimensional fold catastrophe with state variable f(1), x being quite suppressed by the symmetry. The moment variation along the string is allowed for E, Fig. 6 becomes infinitely unstable. (It remains interesting, however, as a realisation of the curve c of §1: the equilibria like f_1' are stable with respect to any possible small perturbation, so that the Hessian is positive definite, while for those like f_2' it is negative definite. The two kinds together give a smooth curve in any space X appropriate to the problem, and their Hessians a curve from plus to minus I through 0 in the space of quadratic forms on X. Nearby, by an arbitrarily small perturbation of E, lie the phenomena discussed below. See §6 for a more precise formulation.)

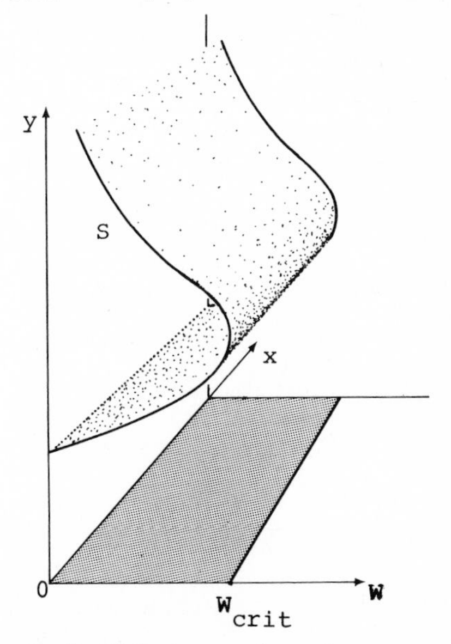

Fig. 6 Equilibrium surface for a weightless, uniform string

§3. Generic 'Breaking'

What, then, generically happens for an arbitrary E? If each $\partial E^x/\partial y$ were monotonic, the string would have a unique, stable equilibrium for each W (Fig. 7) and show no bifurcation. To give a model of breaking, we suppose rather that for each x, $\partial E^x/\partial y$ has a maximum at some y depending on x (and for simplicity, that $\partial^2 E^x/\partial y^2=0$ nowhere else). Moreover $\partial E^x/\partial y \to -\infty$ as $y \to 0_+$, so that no part of the string is compressible to zero length. That is, each E^x looks like Fig. 3(a), though not quantitatively identical to it.

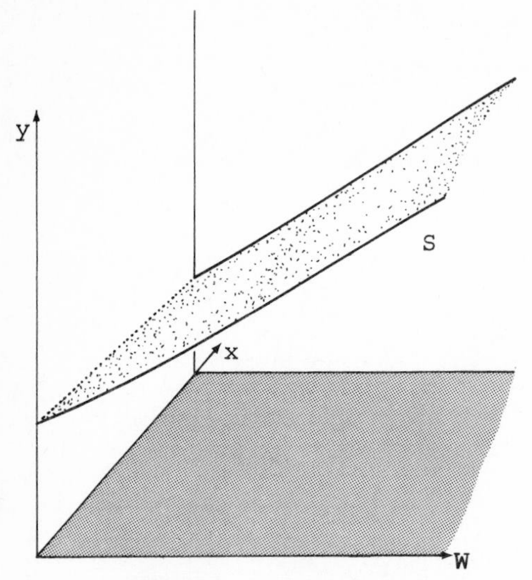

Fig. 7
Equilibrium surface for each dE^x/dy monotonic, giving a unique, stable equilibrium for the string at each load W.

Under these assumptions, we have generically a smooth fold curve γ as in Fig. 5, which has its lowest value W_γ of W either for $0<x<1$ as in Fig. 8(a) or for x=0 or 1 as in Fig. 8(b). The two cases are distinct, but their analyses are so similar that we treat only the interior case.

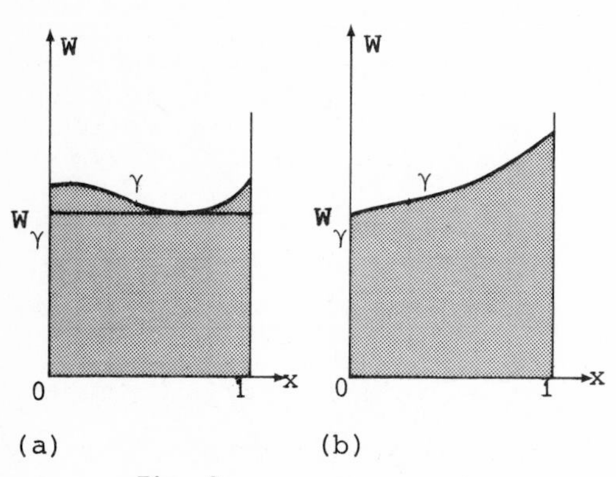

Fig. 8
Fold curves with breaking point (a) at an interior (b) at an end point of the string

It is easy to show that γ generically has parabolic contact with the line $W=W_\gamma$, so that there is a unique (x,y) for which

$$\frac{dE^x}{dy}(y)=\frac{\partial E}{\partial y}(x,y)=0 \ , \quad \frac{d^2E^x}{dy^2}(y)=0$$

but in fact a much stronger statement is possible. Define E and F to be *equivalent* if there exist

(i) a smooth reparametrisation (diffeomorphism) $\phi: R\to R$ of W
(ii) a W-dependent smooth reparametrisation, $\psi^W:[0,1]\to[0,1]$ for each W, of x
(iii) a (W,x)-dependent smooth reparametrisation $\Theta^{(W,x)}: R\to R$ of y
(iv) a (W,x)-dependent 'variable constant' or 'shear function' $\sigma: R\times[0,1]\to R$

transforming one to the other. More concisely, if there are diffeomorphisms ϕ, Ψ, Θ and function σ such that

$$\begin{array}{ccccc} (W,x,y) & \longmapsto & (W,x) & \longmapsto & W \\ R\times[0,1]\times R & \xrightarrow{p^1} & R\times[0,1] & \xrightarrow{p^2} & R \\ \downarrow\Theta & & \downarrow\Psi & & \downarrow\phi \\ R\times[0,1]\times R & \xrightarrow[p^1]{} & R\times[0,1] & \xrightarrow[p^2]{} & R \end{array}$$

commutes, and

$$F(\psi^W(x),\Theta^{(W,x)}(y))-\phi(W)\Theta^{(W,x)}(y)+\sigma(W,x)=E(x,y)-Wy$$

for all W, x, y. It is clear that equivalent E and F give isomorphic bifurcation problems: we defer formal treatment of this to §6. We say E and F are *boundedly equivalent* or B-*equivalent* if ϕ, Ψ, Θ,σ are defined for $|W|$ and $|y|$ less than some B, while the diagram above commutes and the equation holds where the various composites are defined. Since the phenomena of interest concern local equilibria, attention to B-equivalence for large B saves irrelevant arguments to control behaviour at infinity.

Next define E to be *stable* (resp. B-*stable*) if it has a neighbourhood N in an appropriate topology $\mathcal{T}$ such that every $F\in N$ is equivalent (resp. B-equivalent) to E. The conventional choice for $\mathcal{T}$ is the Whitney C^∞ topology, which is convenient for general theorems. Examination of the application of the stability proofs to our particular case shows that we need not control derivatives above order 4. Moreover, if we suppose that no $y>Y$, say, satisfies

$$D_2E(x,y)=W, \quad D_2^2E(x,y)>0$$

for any x, W, then a satisfactory topology for the present discussion is given by a uniform norm on the derivatives of E up to order 4 on the compact domain $[0,1]\times[0,2Y]$, say. (The resulting metric does not make our set of possible integrands into a complete metric space, as we have required E to be C^∞ so as to make V smooth. But in this case the coarser the topology the better the result.) For this problem the two topologies make the same functions 2Y-stable, though the higher singularities that arise with more 'control parameters' than x and W would require controlling more derivatives; hence the convenience of a C^∞ topology in general argument.

Now, the equivalence we have defined corresponds locally to the (1,1)-equivalence of Wasserman [18]. (His use of more general transformations of the codomain R of E than translations by (x,y) does not affect the results in these dimensions.) Fix B larger than the highest W for which any solution to (6) exists, and than any y for which some x has

$$\frac{d^2E^2}{dy^2}(y) = 0.$$

(Such a B exists by our hypotheses above.) This gives a compact region in which proofs of local stability may easily be extended to global stability, combining by a partition of unity the local fields whose flows give local equivalences. Multi-jets and some extra transversality conditions (for example fold curves in (x,W) space must meet transversely) are needed in the proof. We are here barely out of the setting of the original 'only folds and cusps' stability and density theorems of Whitney - apart from distinguishing x and W directions and considering energy values - and anyone equipped to read a rigorous treatment at this point can readily write one down.

Now Wasserman's classification shows that the only stable E for which this problem bifurcates by a development interior to the string involve the parabolically encountered fold, as above, or have the W=constant line meet a cusp point when $W=W_C$ (Fig. 9).

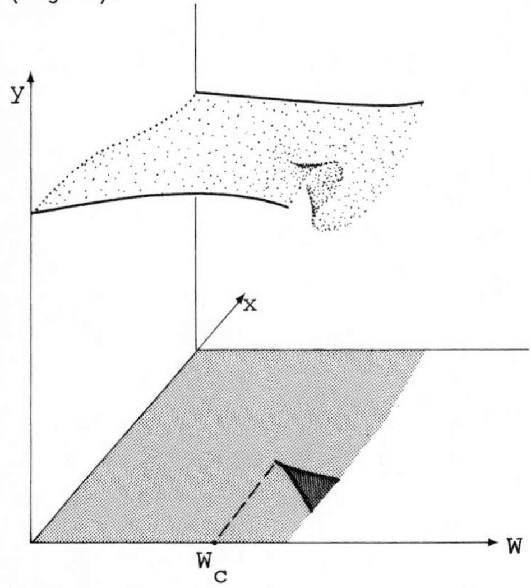

Fig.9 A stable geometry (not permitted by our present hypotheses) which forces a discontinuity in f' for $W > W_c$.

To these we should add the boundary phenomenon of Fig. 10, by which the string breaks at the end (as we would expect for a more-or-less-uniform *heavy* string) but we omit the obvious additions to our discussion to cover end effects. (We could eliminate them entirely by making the string a loop in the plane, with W the result of internal pressure.)

Routine extension of the usual transversality arguments (for a formal treatment of these see [16, 17, 18], for an intuitive account [13]) shows the set of stable E to be *dense* in the topologies described above. Thus generically (E-Wy) is equivalent (in the above fibred sense), in a neighbourhood of any interior point (W,x,y), to one of the functions

$$\tilde{y},\ \pm\tilde{y}^2,\ \tilde{y}^3+(\tilde{x}+\tilde{W})\tilde{y},\ \tilde{y}^3+\tilde{x}\tilde{y}^2+\tilde{W}\tilde{y},\ \tilde{y}^4+\tilde{W}\tilde{y}^2+(\tilde{x}+\tilde{W})\tilde{y},$$

where the ~ variables are appropriate reparametrisations of the original ones.

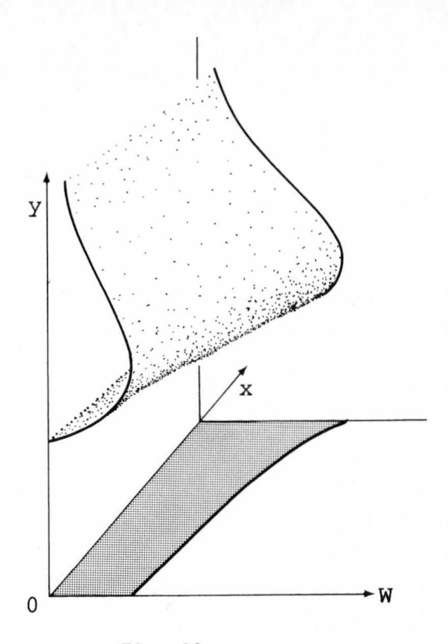

Fig. 10
An equilibrium surface corresponding to Fig. 8(b)

Given our earlier assumptions on the string (which rule out the last case, exemplified in Fig. 9) we have not only the genericity of parabolic contact as in Fig. 8(a), but a canonical local form for E around the point of contact. (Clearly, also, it is generic to have just *one* such point of first contact.)

§4. Failure of "Failure Mode" Equations

Now, for $W=W_\gamma$ in Fig. 11, it is plain that there are still two continuous liftings f_1', f_2' giving solutions to (5). For $x \neq x_0$, we take it that $y=f_1'(x)$ gives a local minimum value of $E^x(y)-Wy$, and $y=f_2'(x)$ a maximum. (Let us denote $E^x(y)-Wy$ by $F^x(y)$ for short.) At $y=f_1'(x_0)=f_2'(x_0)$, F^{x_0} has a point of inflexion.

Now, any continuous perturbation of f_1' will include (though not consist of) a movement, all up or all down in y, of the image of some closed interval $\mathcal{J}$ not containing x_0. Intuitively, such a piece $\mathcal{J}$ of the string is in stable equilibrium (like the whole string for $W<W_c$) in the sense of being in a strict local minimum for energy - though 'local' may not be very large. Thus for any family $f_1'+tg'$, $t\varepsilon R$, given by a reasonable 'perturbation' function g', we expect the total energy $t \mapsto V(f_1+tg)$ to have a quadratic minimum at t=0; no g or g' gives a 'failure mode' or 'buckling mode'. However, if g' vanishes outside a set U containing x_0, the smaller we require U to be the smaller the quadratic term at t=0 will be, because the quadratic minima of F^x at $f_1'(x)$ for $x \neq x_0$ are weaker and weaker as $x \to x_0$. (At a fold point, the quadratic term reaches 0.) But 'in the limit' as $U \to \{x_0\}$ we would perturb only the single value $f_1'(x_0)$, without effect on the integral. The 'failure mode' has escaped from the picture, but the associated bifurcation of the problem has not. For $W_s>W_\gamma$ there are *no* lifting f_1'. For any f' we have infinitely many independent g' (non-zero in distinct sub-intervals of $[x_1, x_2]$ in Fig. 12) such that V(f+tg) has non-zero slope at t=0. We do not locate a one-dimensional 'failure mode' from this side either.

This irreducible infinity of directions is inevitably encountered in a meeting with the *essential spectrum* of the second derivative of V. We discuss the essential spectrum, and more subtle analogues to failure modes, further in §§ 6 and 7.

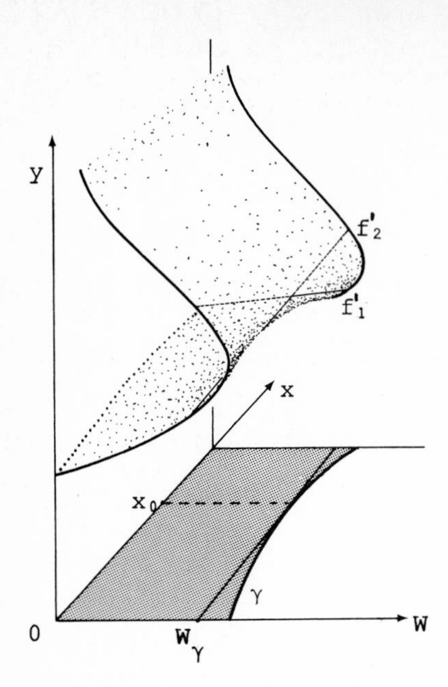

Fig. 11. The two continuous equilibria at the breaking point $W=W_\gamma$

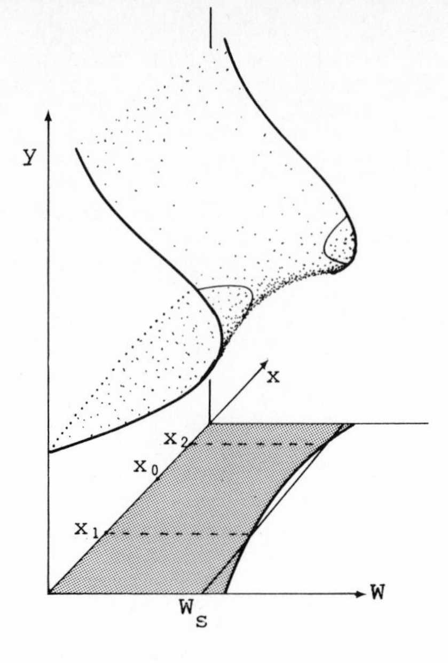

Fig. 12. With a load of $W_s > W_\gamma$, the string can decrease energy by stretching anywhere between x_1 and x_2

We have here, then, something very like the fold catastrophe, in that a stable and an unstable solution disappear together. Note that they approach contact rather than coincidence, and infinitely many discontinuous 'saddle' solutions--agreeing with f_1' on a measurable set and with f_2' on its complement--disappear too. The bifurcation is not the fold catastrophe, but has a close relation to it. However, no separation is possible of the space of state variables f or f' into 'essential' and 'inessential' or 'active' and 'passive'. This is clear equally when one tries to apply an elementary formal failure mode analysis, as follows.

The equivalent of considering the second derivative or quadratic part of V is to linearise (5) with respect to f' (or f), since linearisation is just replacement by the derivative, and (2) was already found by differentiating once. (Cf. §6 for more exact treatment.) Rather trivially, the result is

$$D_2^2 E(x, f_1'(x))\, f'(x) = 0, \tag{7}$$

where f_1' is the 'known solution' about which we linearise, f' the unknown to be solved for. Now, by the above topological discussion, when f_1' reaches its form for $W=W_\gamma$

$$D_2^2 E(x, f_1'(x)) = 0 \text{ if and only if } x = x_0. \tag{8}$$

Hence (7) has only the solution f' = 0 if we assume continuity (even measurability and discard sets of zero measure). The linearised 'Euler-Lagrange equation' (7), whose analogues detect failure modes admirably in cases where they exist, does not detect a singularity in the same way here. The singularity of the second derivative at f_1' of V, considered as a symmetric operator on a suitable space, is that though it takes no directions to 0 it is not surjective--a possible behaviour only for an operator in infinite dimensions. To state this kind of singularity requires unavoidably that one mention the space that the operator's image does not fill (§6). The simplicity of this example thus makes it worth promulgating in 'pragmatic' quarters where formal calculations remain a habit, and specifying spaces, domains and ranges still seems needless sophistication.

§5. Obstructions to Finite-Dimension-Like Stability

The behaviour discussed above occurs for 'generic' E (more strictly, it is a generic property) and is *stable* under small perturbations of E, since (1,1)-stability is generic. Thus we have a class of problems with its own 'Thomist' theorem, in which the finite-dimensional 'Thom's theorem' does not apply. To what extent is this an artefact of the restriction to the class? By way of analogy, restriction to even energy functions $R \longrightarrow R$ makes fold (cubic) points at 0 impossible. But an arbitrarily small perturbation, going outside the restricted class, can bring any one-parameter family of such functions into line with Thom's theorem: is some such perturbation possible here? Many ways of moving out of the present class exist. For instance, we could add a small interaction energy between different points in the string, or a term in f", to the pointwise integral (1). What effects might a general, arbitrary perturbation have?

Lacking analogues for the 'unfolding' machinery applicable in the finite-dimensional and Fredholm cases, we cannot answer this in general. It is clear that much will depend on the definition of 'small' perturbation. (Even within our special class, an integrated-over-x instead of a uniform norm would put pictures like Fig. 13, with infinitely many 'failure modes', arbitrarily close to Fig. 8a.) However, it is certain that the problem can *not* be perturbed into a structurally stable one with only ordinary fold catastrophe bifurcations.

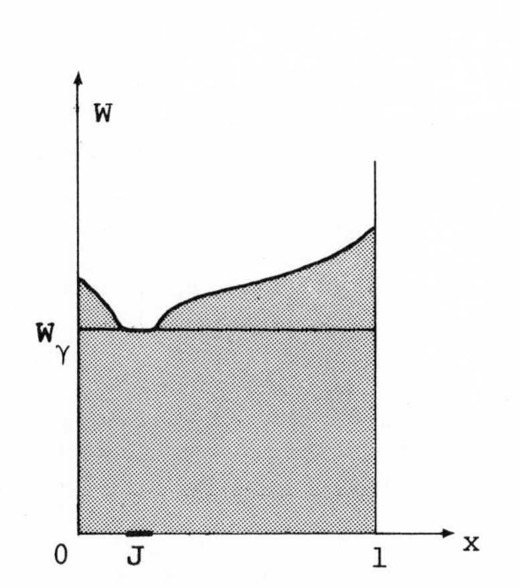

Fig. 13. A problem for which any f' with support in J is a 'failure mode'

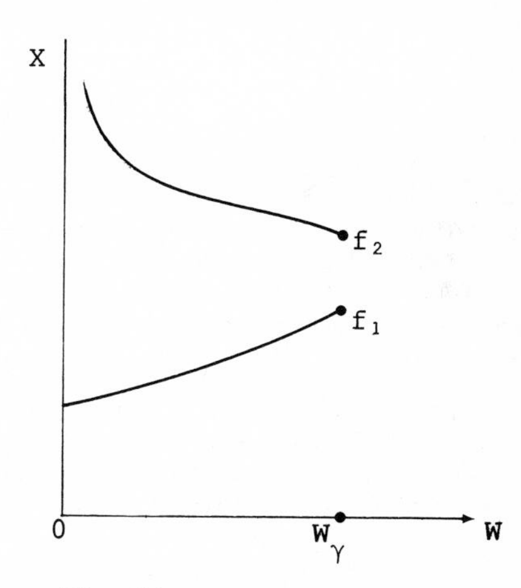

Fig. 14a. Original problem

If we schematically represent the space X of possible f' by a single axis, we can represent the case analysed above by Fig. 14a. (Recall that $f_1' \neq f_2'$ at their moment of disappearance.) A small perturbation should not perturb the curves of Morse equilibria except near a bifurcation point (the smaller, the nearer) so topological change should be confined to the product XxK of X with a small closed interval K around W_γ. In the generic situation given in finite dimensions by Thom's theorem, any path in the control space (here the W-axis) can be lifted uniquely to the set of equilibria until it meets a fold point, and this is the only type of obstruction.

Now, how can the set of equilibria continue from p in Fig. 14b? It cannot be an arc from p to q with finitely many fold points, by the discussion in §1. The equilibria at p and q have positive and negative definite infinite-dimensional Hessians respectively. An arc with finitely many folds that did not join p to q would have a point $w \in K$ above which it 'ended' (even if it tended to no limit point(s) in X above w), incompatibly with Thom's theorem. Thus either way, for compatibility there would be an arc of equilibria with infinitely many fold points $\{F_j\}$ in XxK projecting to points in K. (Any drawing would suggest converging wiggles like sin(1/x), which in infinite dimensions there is room to avoid.) But the images of these F_j in compact K must have a cluster point. Around it we can alter their relative positions (in a manner irreparable by reparametrisation of W) by arbitrarily small perturbations of the total energy function in the neighbourhood of a single F_j. Hence, *global* structural stability is impossible for any small perturbation of $\tilde{V}$ which *locally* satisfies the conclusions of the finite-dimensional Thom's theorem. So even if such perturbations are possible we lose stability, rather than gaining it, by making them.

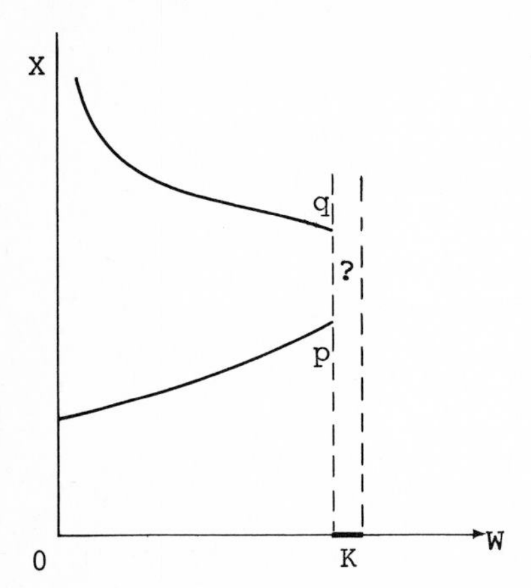

Fig. 14b
Perturbed problem

§6. Functional Analytic Setting

We have avoided, above, being too specific about the function space in which to treat the problem, first for accessibility to those readers more familiar with finite dimensions, second to indicate the generality of the argument. The behaviour found does not depend on a particular and possibly perverse choice of setting for the problem: the 'formal differentiations' etc. above have their rigorous analogue in any space of sufficiently tame functions. We give here a particular choice of setting which emphasises the naturality of f' rather than f as a 'state' description, and illustrate in it the way in which the topological equivalence of §3 can be carried over to the analysis.

Since the energy is given by the formula

$$V(f) = \int_0^1 [E(x, f'(x)) - W f'(x)]\, dx \tag{9}$$

in which only f' appears on the right, it seems natural to consider the function f', rather than f, as defining a state of the string. Accordingly, we take a space X of functions u representing possible f', with energy V defined by

$$V(u) = \int_0^1 [E(x, u(x)) - W u(x)]\, dx. \tag{10}$$

We do not want to insist that u be continuous. (We see in §8 the relevance of discontinuous f' to the relation between the finite and infinitesimal element problems, and $W>W_\gamma$ in the related problem represented in Fig. 9 would *compel* f' to be discontinuous.) We do want V to be a C^∞ function if possible, to resemble finite-dimensional catastrophe theory as much as practical.

There are two rather natural Banach spaces on which V is C^∞. These are C[0,1], the continuous functions on [0,1] with the supremum norm, and $L^\infty[0,1]$, the almost-everywhere-bounded measurable functions with the 'almost everywhere supremum' norm. The latter allows for the discontinuous functions mentioned above. The discussion turns out very similar, however, for the two spaces: we exemplify it with X = C[0,1].

The derivative of V is then given by

$$\langle V'(u), h\rangle = \int_0^1 [D_2E(x, u(x)) - W]\, h(x)\, dx. \tag{11}$$

Since X is not reflexive we cannot make use of its dual X*. Instead we exploit a more general theory (Magnus [9]), setting X into duality with itself by means of the bilinear form

$$\{u, v\} = \int_0^1 u(x)\, v(x)\, dx. \tag{12}$$

This leads to the identification of V' with the mapping

$$t:X \longrightarrow X,\ t(u)(x) = D_2E(x,u(x)) - W. \tag{13}$$

The 'Hessian' of V at u is the linear operator $t'(u) = T_u$,

$$T_u:X \longrightarrow X,\ (T_uh)x = D_2^2E(x,u(x))\, h(x).$$

That is, the 'Hessian' is just multiplication by the function

$$x \longmapsto \frac{\partial^2E}{\partial y^2}(x, u(x))$$

in the space C[0,1]. A similar result holds for $L^\infty[0,1]$.

Thus T_u is an isomorphism provided $D_2^2E(x,u(x)) \neq 0$ for all $x \in [0,1]$. (For $L^\infty[0,1]$, $D_2^2E(x,u(x))$ has to be bounded away from 0 on the complement of a set of measure zero). If $V'(u) = 0$ the Morse Lemma holds [9], with its implications of stability under a class of perturbations in the integrand. We have a quadratic minimum if $D_2^2E(x,u(x)) > 0$, a quadratic maximum if $D_2^2E(x,u(x)) < 0$. For $L^\infty[0,1]$ saddles are possible, since $D_2^2E(x,u(x))$ may change sign at a discontinuity of u.

For the critical W_Y, the solution u_0 of $V'(u) = 0$ which gives a minimum has $D_2^2E(x_0, u_0(x_0)) = 0$. The 'Hessian' T_{u_0} remains injective, but surjectivity fails. Let $a(x) = D_2^2E(x,u_0(x))$, and set $Y = \{u \in X : u(x_0) = 0\}$. Clearly $R(T_{u_0}) \subset Y$ and codim Y = 1. In fact $R(T_{u_0})$, though not closed, is dense in Y. For the function $\sqrt{|a|}$ belongs to Y but division by a(x), attempting to invert T_{u_0}, gives a function with an infinite discontinuity going as $(x-x_0)^{-\frac{1}{2}}$ at x_0 in the generic case, thus not in C[0,1] nor even in $L^\infty[0,1]$. On the other hand, if $g \in Y$, we may approximate g by continuous functions which vanish in a neighbourhood of x_0, and these are in $R(T_u)$. If anything might be called a failure mode, it would be the quotient X/Y, which may be thought of as the one-dimensional space $C(\{x_0\})$ of continuous functions on the set $\{x_0\}$.

In the case of $L^\infty[0,1]$ the range of T_{u_0} remains unclosed, by the same counter-example, and its closure is the space M of functions $v \in L^\infty[0,1]$ such that lim v(x) = 0 $(x \to x_0)$ after adjustment on a set of measure zero. The quotient L^∞/M can be described in a way that brings out the idea that it is somehow localised at x_0. For each $f \in L^\infty[0,1]$ let

$$|f|_{x_0} = \lim_{\varepsilon\to 0} \text{ess. sup.}\ \{|f(x)| : x \in (x_0 - \varepsilon, x_0 + \varepsilon)\}. \tag{15}$$

This defines a semi-norm on $L^\infty[0,1]$ which depends only on the germ f_{x_0} of f at x_0. Now consider two germs f_{x_0} and g_{x_0} to be equivalent if $|f - g|_{x_0} = 0$. For an equivalence class $[f_{x_0}]$ of germs we can define unambiguously a norm

$$\|[f_{x_0}]\| = |f|_{x_0} \tag{16}$$

which makes the set of classes $[f_{x_o}]$ into an infinite-dimensional Banach space (actually a Banach algebra). In fact all we have done is to give a description of L^∞/M which emphasises germs.

On a speculative level we could think of the configuration space not as $C[0,1]$ but as the sheaf of germs of continuous functions over $[0,1]$ (or of locally bounded measurable functions over $[0,1]$). Then the 'failure mode' is the cokernel of the morphism [multiplication by $D_2^2E(x,u_o(x))$]. This cokernel is a sheaf having support at the single point x_o, and this is obviously related to the fact that the bifurcation 'happens' at x_o rather than being spread over the whole interval $[0,1]$.

We have seen, then, that a bifurcation can occur without the 'Hessian' acquiring a null-direction ('failure mode'), something which cannot happen in finite dimensions. The resulting bifurcation does not seem to be structurally stable under arbitrary small smooth perturbations of V. (We did not establish this instability in §5, since we did not show that the topological perturbations discussed there were realisable: only that no stable essentially finite-dimensional behaviour was nearby.) However, it *is* generically stable under small smooth perturbations of the integrand function E.

To make this kind of stability precise, we define an appropriate local equivalence for functions

$$V(W,u) = \int_0^1 [F(x,u(x)) - Wu(x)]\, dx \tag{17}$$

with $F:[0,1]\times R \longrightarrow R$ smooth. Let us call such V *energy functions*. Suppose V_1, V_2 are energy functions and (u_1,W_1), (u_2,W_2) points in $X\times R$. We shall say that (V_1,u_1,W_1) is *locally equivalent* to (V_2,u_2,W_2) if for i = 1,2 there are neighbourhoods N_i of u_i in X and M_i of W_i in R, with smooth mappings

$$\Phi: M_1\times N_1 \longrightarrow N_2, \qquad \Psi: M_1 \longrightarrow M_2$$

such that

(a) $\Phi(W_1,u_1) = u_2$
(b) $\Psi(W_1) = W_2$
(c) $\Phi(W,\ )$ is a diffeomorphism for each $W \in M_1$
(d) $\Phi(W,u)$ is a critical point of $V_2(\Psi(W),\ )$ if and only if u is a critical point of $V_1(W,\ )$.

If every energy function obtained from an integrand near that of V is locally equivalent to V, we say that V is *stable* around (W,u). (We shall not attempt to globalise stability here relative to x and W--note that we are already global relative to position on the string--as our present concern is with local bifurcation structure.) It is clear that this stability notion is strong enough to examine bifurcation geometry with.

If E(x,y) and F(x,y) are B-equivalent integrands, then with $|W|$ and $|y|$ less than B we have

$$F(\Psi(W,x),\ \Theta(W,x,y)) - \phi(W)\ \Theta(W,x,y) + \sigma(W,x) = E(x,y) - Wy. \tag{18}$$

Then

$$\begin{aligned} V_1(W,u) &= \int_0^1 [E(x,u(x)) - Wu(x)]\, dx \\ &= \int_0^1 [F(\Psi(W,x),\ \Theta(W,x,u(x)) - \phi(W)\ \Theta(W,x,u(x)) + \sigma(W,x)]\, dx. \end{aligned}$$

Hence

$$\begin{aligned} D_uV_1(W,u) = 0 &\Leftrightarrow D_yF(\Psi(W,x),\ \Theta(W,x,u(x))) - \phi(W) = 0, \quad 0 \leqslant x \leqslant 1 \\ &\Leftrightarrow D_x\Psi(W,x)\ [D_yF(\Psi(W,x),\ \Theta(W,x,u(x))) - \phi(W)] = 0 \end{aligned}$$

(since $D_x\Psi(W,x) \neq 0$ for $0 \leqslant x \leqslant 1$, $\Psi(W,\ )$ being a diffeomorphism)

$$\Longleftrightarrow \int_0^1 D_x\Psi(W,x)\,[D_yF(\Psi(W,x),\ \Theta(W,x,u(x))) - \phi(W)]\ v(x)\ dx = 0 \quad \text{for all } v \in X$$

$$\Longleftrightarrow \int_0^1 [D_yF(z,\ \Theta(W,\zeta(W,z),\ u(\zeta(W,z)))) - \phi(W)]\ v(\zeta(W,z))\ dz = 0 \quad \text{for all } v \in X.$$

where $z \mapsto \zeta(W,z)$ is inverse to $x \mapsto \Psi(W,z)$, for each W. The last equation above holds if and only if

$$z \longmapsto \Theta(W,\zeta(W,z),\ u(\zeta(W,z)))$$

is a critical point of $V_2(\phi(W),\ )$, where

$$V_2(W,u) = \int_0^1 [F(x,u(x)) - Wu(x)]\ dx. \tag{19}$$

It remains to be shown that the transformation of X, whereby a function u(x) becomes the function

$$x \longmapsto (W,\zeta(W,x),\ u(\zeta(W,x))),$$

is a diffeomorphism. This transformation is the composition of two others. The first replaces u(x) by $x \longmapsto \Theta(W,x,u(x))$. It is easily seen that this maps X to itself and is smooth, surjective and infuctive, these last three properties following at once from the corresponding properties of $y \mapsto \Theta(W,x,y)$. The second is one where u(x) is replaced by $x \longmapsto u(\zeta(W,x))$. This is obviously an isomorphism of x with itself.

§7. The "Bad Set" of Hilbert Space Operators

In this section we prove the assertion made in §1 about the set Z of non-Fredholm self-adjoint bounded linear operators in a Hilbert space H. (Generalisations to more general spaces would be of considerable interest.) Let Q stand for the space of bounded self-adjoint operators on H.

Proposition 1

Let $A \in Z$ and $r > 0$. For any $B \in Q$ such that $\|B\| < r$, $\exists\ \lambda \in (-r,r)$ such that $A + B - \lambda I \in Z$. Intuitively, Z has 'codimension one' in Q.

Proof

Let Y be the C*-algebra of bounded linear operators on H, and C be the closed ideal of compact operators. Then the quotient Y/C is again a C*-algebra. The *essential spectrum* of any $A \in Y$ is by definition the set $\lambda \in \mathbb{C}$ such that $A - \lambda I$ fails to be a Fredholm operator index 0. The *real* elements of Y/C are just the canonical images of the real (=self-adjoint) elements of Y. We denote the image of A by [A].

In any C*-algebra the spectral radius of a real element equals its norm. If x is a real unit of a C*-algebra and y is another element, then x + y is a unit if

$$\|y\| < \|x^{-1}\|^{-1}.$$

But $\|x^{-1}\|$ is the spectral radius of x^{-1}, which is $(d(0,\sigma(x)))^{-1}$ where $d(0,\sigma(x))$ is the distance from 0 of the spectrum $\sigma(x)$ of x. Hence x + y is a unit if

$$\|y\| < d(0,\sigma(x)).$$

We now introduce a lemma whose proof we defer.

Lemma

If $A \in Q$, $[A]$ is a unit of Y/C if and only if A is a Fredholm operator. The spectrum of $[A]$ for any $A \in Q$ is just the essential spectrum of A.

(The second sentence is not logically needed, but follows easily from the first.) Assuming this, we prove the proposition as follows.

Let $A \in Z$. Then $[A]$ is not a unit of Y/C. That is, $0 \in \sigma([A])$. Let $B \in Q$. If $A + B - \lambda I$ is a Fredholm operator for $-r < \lambda < r$, then $[A + B - \lambda I] = [A] + [B] - \lambda[I]$ is a unit of Y/C for $-r < \lambda < r$, so that $d(0,\sigma([A] + [B])) \geqslant r$. Thus we cannot have $\|[B]\| < r$, since this would imply that $[A]$ is a unit of Y/C. Hence $\|[B]\| > r$. Now suppose that we choose $B \in Q$ such that $\|B\| < r$. Then $\|[B]\| < \|B\| < r$, hence for some $\lambda \in (-r,r)$, $A + B - \lambda I$ fails to be a Fredholm operator. ■

Proof of Lemma

If $A \in Q$ is a Fredholm operator, then

$$A = S + T$$

where S is invertible and T compact. Hence

$$[A][S^{-1}] = [AS^{-1}] = [I + TS^{-1}] = [I],$$
$$[S^{-1}][A] = [S^{-1}A] = [I + S^{-1}T] = [I].$$

Therefore, $[A]$ is a unit of Y/C.

Conversely, let $[A]$ be a unit of Y/C, with inverse $[B]$. Then both AB and BA are Fredholm operators of index 0.

Now, recall that a Fredholm operator is one with finite-dimensional kernel, and finite-codimensional range. Usually the range is required to be closed, but this follows from the other two conditions. In fact, if $T \in Y$ and $R(T)$ has finite codimension, there is a bounded linear extension of T to $H \times R^n$ for some n. (Just map R^n to an algebraic complement of $R(T)$.) By the open mapping theorem this extension maps closed sets to closed sets, hence $R(T)$ is closed.

Now AB and BA are both Fredholm operators of index 0. Since $\ker A \subset \ker BA$, $\dim(\ker A) < \infty$. Since $R(AB) \subset R(A)$, codim $R(A) < \infty$. Hence A is a Fredholm operator. (Similarly, B is Fredholm and both have index 0, A because it is self-adjoint, B because $\mathrm{ind}(AB) = \mathrm{ind}(A) + \mathrm{ind}(B) = \mathrm{ind}(B)$.) ■

Note. This proof of the proposition, though long, is quite simple. It amounts to proving that in a C*-algebra, the spectrum is a continuous set-valued function (using the Hausdorff metric on the set of closed subsets of $\mathbb{C}$) on the *real* elements. The other part of the proof is summarised neatly in the second sentence of the lemma.

A slightly shorter proof is possible using the concept of 'measure of non-compactness', but it gives less insight.

We conclude this section with an obvious corollary of Proposition 1.

Proposition 2

Let $T \in Q$ be invertible and $A \in Z$. Given $\varepsilon > 0, \exists \delta > 0$ such that for any $B \in Q$ with $\|B\| < \delta$, $\exists \lambda \in (-\varepsilon,\varepsilon)$ such that $A + B - \lambda T \in Z$. ■

Intuitively, the transverse directions to Z include all invertibles in Q, and these form an open subset of Q.

It is clear that the geometry of Z (and of the corresponding non-Fredholm operators encountered in the non-Hilbert case, such as those important in our string model) must be better understood before a clear picture of genericity among functions on such spaces can emerge. Just so, Mather's [11] stratifying jet spaces $R^n \longrightarrow R^p$ was the key to establishing Thom's conjecture on topological stability.

§8. Finite Elements and the Nature of Solutions

Of course, our motivating 'particle' model at the beginning of §2 would generically break (fluctuations aside) by an ordinary fold catastrophe at the weakest bond. This might be felt more 'realistic' in some sense than the behaviour of our continuum model and it certainly raises subtle questions about the nature of such modelling. *However small* the finite elements (in this case, the inter-atomic bonds), the bifurcation is reducible to one dimension. But in the obvious limiting problem it is essentially infinite-dimensional. We have a great 'qualitative' gulf between the two kinds of model, in that the topology is utterly different, though the numerical predictions from a particular E would be well matched by those of a many-finite-element version approximating it. ("Quantitative is just poor qualitative, unless quantities are the qualities you are after." Matthew Pordage.)

Moreover the nature of the fold bifurcation in the finite version points up the relevance of the 'saddle' solutions, which mix f_1' and f_2', to the problem. In the finite case, the equilibrium which approaches and annihilates the stable one as W increases is not a local maximum (corresponding to f_2') but a 1-saddle, with only the weakest bond in its local maximum, the rest in minima. To this in the continuous string model would correspond a discontinuous f', agreeing with f_1' except on a set P of positive measure within a small interval J around the weakest point x_0-- but how small an interval? 'In the limit' we again find only x_0 itself, and again the value at one point makes no difference. But for any P of positive measure, $f'|P = f_2'|P$ means that from f' there are infinitely many decreasing directions for V.

There are many ways to approach the limit. For example, f' could agree with f_2' on

$$J \cap \{x : \frac{1}{2k} < |x - x_0| < \frac{1}{2k+1}, \quad k \in \mathbb{N}\}$$

and with f_1' elsewhere, as J shrinks to $\{x_0\}$. This range of possibilities is equally apparent as finite elements tend to infinitesimal. (For example, the pair of neighbours of the weakest link themselves tend to the break point, and provide another conceptual 'limiting failure mode' as reasonable as single-infinitesimal-element breaking.) It is reflected in the infinite dimension (§6) of L^∞/M.

The naturalness of discontinuity in the finite element picture reinforces the claims of $L^\infty[0,1]$ as our 'state space' (despite its greater subtlety of definition) for this deceptively innocent-looking problem, as does Fig. 9 which clearly forces a discontinuous f'. It is interesting to note that minimising global energy over possible points for such a discontinuity leads to a 'Maxwell convention' of choosing the deeper minimum at each point. This minimisation selects among uncountably many Morse minima, able to be comfortably apart (as Morse points must be) only because $L^\infty[0,1]$ is not second countable. The sheets of the catastrophe manifold (or, since our control is one-dimensional, we may be classical and call them 'branches'), all entangled in the bifurcation point for either Fig. 9 or Fig. 11, are a multitude that no man can number.

The problem of infinite versus finite dimensions in relation to genericity arguments substantially antedates catastrophe theory. Transversality arguments by counting equations against unknowns are traditional in physics, but require finite dimensions to reduce meaningfully to this form. Hence in order to discuss the genericity of the phenomenon life, in 1961 Wigner [19] supposed that the relevant Hilbert spaces for his quantum theoretical discussion were of finite dimensions N and R. He justified this as follows: "Since no assumptions will be made concerning the magnitude of N and R, both of which must surely be assumed to be very large, this assumption appears harmless enough. It is made to make the mathematical analysis easy and can be justified since, as the total energy available is finite, both parts of the system are restricted to a finite number of states."

But in the physics of bulk matter the states available to a macroscopic sample can be so many, and so thickly distributed in some parts of the energy range, that the most practical approach is to allow them to be infinite in number by approximating the finite sample with an infinite one. (The radiative spectrum--representing the set of gaps between the energies of states--of an atom in vacuo is obervably discrete That of the sun is for all practical purposes continuous.) As soon as infinite dimension becomes in some respect qualitatively/computationally the most practical description of a system, phenomena generic in such a model must be considered. The 'real' system is liable to approximate infinite-dimensional behaviour in that respect, too. A 'generic' operator on 4-dimensional space has 4 distinct eigenvalues, which one may expect to be *comfortably* distinct. But with 10^4-dimensional space, 10^4 eigenvalues can easily look for most purposes like an interval, and ∞ like a better approximation to 10^4 than 4 is. Conversely, most numerical routines for computing with continuous models use numbers less than 10^4 as approximations to ∞.

Thus neither in Wigner's tentative argument that life is non-generic, nor in Zeeman's discussion [20] of what life is generically like:

> "For example we can implicitly assume X is large enough to describe the states of a cell in the embryo, with at least 10,000 dimensions for representing the concentrations of the various chemicals involved . . . in each case [Thom's] theorem explicitly hands us back the same simple surface",

should the question of what is generic in infinite dimensions have been evaded[3]. The strictly infinite, like the strictly infinitesimal, can be an essential tool in understanding the behaviour of the finite.

Finally, we remark that this paper illustrates that a P.D.E. equivalent to extremisation of an appropriate integral may sometimes be geometrised as a lifting problem within catastrophe theory. This clearly has interest for far more subtle problems in elasticity, biochemistry, etc. (For instance, catastrophe theoretic methods may reduce a multi-dimensional argument of the integrand to a one-dimensional y like that used here, making the infinite-dimensional phenomena encountered more tractable. But such uses of the catastrophe theory viewpoint, if they develop successfully, will be quite different from simple 'appeals to Thom's theorem'--as indeed are most successful applications of catastrophe theory to date.

The authors thank David Chillingworth and Leif Arkeryd for very illuminating discussions.

References

1 Arkeryd, L., Thom's theorem for Banach spaces, Tech. Report, Math. Dept., University of Gothenburg 1978

2 Chillingworth, D.R.J., A global genericity theorem for bifurcations in variational problems, preprint, Math. Dept., University of Southampton 1978

3 Gibson, C.G., Wirthmüller, K., du Plessis, A.A. and Looijenga, E., Topological Stability of Smooth Mappings, Lecture Notes in Math. 552, Springer 1977

4 Golubitsky, M. and Schaeffer, D., A Theory for Imperfect Bifurcation via Singularity theory, Comm. Pure and Appl. Math., to appear

5 Gromoll, D. and Meyer, W., On differentiable functions with isolated critical points, Topology 8, 361-370, 1969

---

[3] One might add that since concentrations are intrinsically non-negative, Zeeman should anyway have appealed not to the standard Thom's theorem but to its extension by the list of 'constraint catastrophes' ([12,13,14]).

6 Magnus, R.J., On universal unfoldings of certain real functions on a Banach space, Math. Proc. Camb. Phil. Soc. 81, 91-95, 1977

7 Magnus, R.J., Determinacy in a class of germs on a reflexive Banach space, Math. Proc. Camb. Phil. Soc., to appear

8 Magnus, R.J., Universal unfoldings in Banach spaces: reduction and stability, Math. Report 107, Battelle Geneva 1977

9 Magnus, R.J., A splitting lemma for non-reflexive Banach spaces, preprint, Science Institute, University of Iceland (in preparation)

10 Magnus, R.J. and Poston, T., On the full unfolding of the von Kārmān equations at a double eigenvalue, Math. Report 109, Battelle Geneva 1977

11 Mather, J.,"How to stratify mappings and jet spaces," in Singularités d'Applications Différentielles, Plans-sur-Bex 1975 (O. Burlet, F. Ronga, eds.), Lecture Notes in Math. 535, 128-176, Springer 1976

12 Pitt, D.H. and Poston, T., Determinacy and unfoldings in the presence of a boundary, to appear.

13 Poston, T. and Stewart, I.N., Catastrophe Theory and its Applications, Pitman, London and San Francisco 1978

14 Siersma, D., Singularities of functions on boundaries, corners, etc., Dept. of Math., University of Utrecht, preprint 92, 1978

15 Thom, R., Stabilité Structurelle et Morphogénèse, Benjamin, N.Y. 1972, trans. D. Fowler as Structural Stability and Morphogenesis, Benjamin - Addison Wesley, N.Y. 1975

16 Trotman, D.J.A. and Zeeman, E.C., "Classification of elementary catastrophes of codimension ≤ 5," in Structural Stability, the Theory of Catastrophes, and Applications in the Sciences (P.J.Hilton, ed.) Lecture Notes in Math. 525, 263-327, Springer 1976, and in [21] 497-561

17 Wassermann, G., Stability of Unfoldings, Lecture Notes in Math. 393, Springer 1974

18 Wassermann, G., Stability of unfoldings in space and time, Acta Math. 135, 57-128, 1975

19 Wigner, E.P., "The Probability of the Existence of a Self-Reproducing Unit," in The logic of Personal Knowledge: Essays in Honor of Michael Polanyi, ch. 19, Routledge and Kegan Paul, London 1961, and in E.P.Wigner, Symmetries and Reflections, 200-208, Indiana Univ. Press 1967

20 Zeeman, E.C., "Catastrophe Theory," in [21] 1-64: published with editorial changes in Scientific American vol. 234 part 4, 65-83, 1976

21 Zeeman, E.C., Catastrophe Theory: Selected Papers 1972-77, Addison - Wesley 1977

Catastrophes and Bifurcations in Variational Problems

G. Dangelmayr

Institute for Information Sciences, University of Tübingen
D-7400 Tübingen, Fed. Rep. of Germany

Contents

Summary

Using an extension of catastrophe theory to infinite dimensional spaces, the bifurcation properties of the solutions of nonlinear operator equations, deriving from variational problems, are shown to be determined by those of the stationary points of unfoldings of singularities and so are classifiable in terms of catastrophe polynomials. A stability analysis of the bifurcating solutions is carried out. The results are applied to the buckling of columns and lead here to a nonversal unfolding of a double-cusp. Implications of the techniques for nonequilibrium thermodynamics are indicated.

1. Introduction

Catastrophes and bifurcations are terms which generally refer to qualitative changes in the behaviour of systems due to a change in the parameters on which they depend. Catastrophe theory, discovered by Thom [1] and generalized into singularity theory by Arnold [2], describes the generic bifurcations of parameter-dependent families of smooth functions $f:X\to\mathbb{R}$, where X is a finite-dimensional real C^∞ manifold. If the number of parameters λ is not too large (≤ 5[1], resp. ≤ 16[2]), almost all singularities of f are locally equivalent to a set of "standard forms". This theory has found a great number of applications in modeling, on the same geometrical basis, a variety of "critical" phenomena encountered in physics (Zeeman and Poston & Stewart [3]). However, most, if not all, phenomena of continuum physics require for their description an infinite-dimensional manifold X. Examples range from hydrodynamics to elasticity theory, nonequilibrium thermodynamics to general relativity. In the general spirit of Thom [1], one might expect that an extension of singularity theory to systems with infinitely many degrees of freedom should exhibit the standard topological forms at a physically relevant level.

Concurrent with the development of catastrophe theory and singularity theory has been the extensive theory of bifurcations which, in a general setting, deals with the multiplicity and branching of solutions of nonlinear operator equations of the type $F(u,\lambda)=0$, where $F:B_1\times\mathbb{R}^k\to B_2$, B_1 and B_2 being Banach spaces and $\mathbb{R}^k$ is the space of parameter λ. In this field, motivated by studies of bifurcations in continuum physics, topological methods (e.g., the Leray-Schauder degree theory [4]), methods of nonlinear functional analysis (e.g., iteration schemes [5] and the Banach space version of the implicit function theorem [6]) intermingled and led Hale and his collaborators [7] to introduce the notion of "generic bifurcation". Although the latter notion appears not yet completely precisized, it corresponds in a sense to that of "finite determinacy" of singularity theory and possibly might provide a link between singularity and bifurcation theory. For, in practice, Hale's notion is designed to ensure a truncation of the series expansion of the Lyapunov-Schmidt branching equations, thus establishing some sort of analogy with finite determined k-jets in singularity theory.

Generalizing singularity theory to reflexive Banach spaces B_1 has been pursued by Magnus [8] with the result that fundamental statements of the former theory, in particular classification theorems, can be carried over to infinite-dimensional spaces, if a "Fredholm restriction" is satisfied. However, as we shall see, this condition is too strong for many variational problems.

Therefore, in Section 2, we relax the Fredholm restriction by introducing a generalized gradient operator, the range of which is an appropriately chosen Banach space B_2 rather than the dual B_1^*. In this way we show that the bifurcation properties of solutions of nonlinear branching equations given by this operator are determined by those of the stationary points of unfoldings of singularities. Furthermore, a stability analysis of the bifurcating solutions is carried through.

In Section 3 we apply these results to the buckling of columns. This problem has been treated by Keener [9] by means of an iteration scheme within the frame of bifurcation theory, however, under physically unrealistic assumptions. We show here, using the methods of Section 2, that the bifurcation of the properly formulated physical buckling problem is determined by a double-cusp.

Finally in Section 4 we point out how the techniques of Section 2 can be applied to problems of nonequilibrium thermodynamics.

2. Catastrophe Theory and Nonlinear Branching Equations

Let B_i (i=1,2) be separable, reflexive Banach spaces over the reals with duals B_i^* and B_1 continuously and densely embedded in B_2 and assume that B_2 is continuously and densely embedded in a Hilbertspace H. Assume further that the inner product $<\cdot,\cdot>$ in H is well defined for $u \epsilon B_2$, $u^* \epsilon B_2^*$ and that $u^*(u)=<u,u^*>$, i.e., B_2 and B_2^* are dual with respect to the inner product of H. Then every inclusion in $B_1 \subset B_2 \subset H \subset B_2^* \subset B_1^*$ is continuous. In applications H is usually L_2, B_2 is L_p, $p \geq 2$, and B_2^* is L_q with $1/p+1/q=1$. For B_1 one usually chooses some Sobolev-space, depending on the specific problem at hand.

Let U resp. V be a neighbourhood of 0 in B_1 resp. $\mathbb{R}^k$ and suppose that the nonlinear operator $F: U \times V \mapsto B_2$ depends smoothly on $(u,\lambda) \epsilon U \times V$ with $F(0,0) = 0$. Let us suppose that F is the gradient of a smooth function $R: U \times V \to \mathbb{R}$ with $R(0,\lambda) = 0$ in the sense that

$$D_1 R(u,\lambda)\tilde{u} = <F(u,\lambda),\tilde{u}> \tag{2.1}$$

holds for any triple $(u,\lambda,\tilde{u}) \epsilon U \times V \times B_1$, D_1 denoting differentiation with respect to B_1. Eq. (2.1) means that the derivative map $D_1 R: U \times V \mapsto B_1^*$ can be expressed in the form $D_1 R = I \circ F$, where I is the inclusion map $B_2 \hookrightarrow B_1^*$, so that the range of $D_1 R$ is contained in the dense subset B_2 of B_1^*. Let us denote by L the linear and continuous operator $L: B_1 \mapsto B_2$ defined by $L:=D_1 F(0,0)$. We suppose that L is a Fredholm operator with index 0, i.e., having a finite-dimensional null space spanned by $\{v_1,\ldots, v_m\} \subset B_1$, $<v_i, v_j>= \delta_{ij}$, and a closed range $E_2:=LB_1 \subset B_2$. We further assume that the annihilator of E_2 is spanned by $\{v_1,\ldots,v_m\}$, too. In applications to differential equations L is usually an elliptic differential operator, self-adjoint with respect to the inner product in an L_2-space, so that the latter condition is in general satisfied. Let P_i be the projection of B_i into the finite dimensional space $N:= sp\{v_1,\ldots,v_m\}$,

$$P_i u: = \sum_{j=1}^{m} <u,v_j>v_j \quad (u \epsilon B_i),$$

and set $Q_i:=1-P_i$, $E_i:=Q_i B_i$ (i=1,2). Then the inclusions in $E_1 \subset E_2 \subset E \subset E_1^* \subset E_2^*$ are continuous , E being the orthogonal complement of N in H. Let $L_o: E_1 \to E_2$ be the linear operator $L_o:=Q_2 L|_{E_1}$. The above assumptions imply that L_o is an isomorphism.

For $\lambda=0$, u=0 is a solution of the equation

$$F(u,\lambda) = 0. \tag{2.2}$$

Considering the bifurcation problem corresponding to Eq. (2.2), we look for all small solutions of Eq. (2.2) when λ variies in a neighbourhood of 0. (For given λ we call a solution u_λ of Eq. (2.2) a small solution, if $u_\lambda \to 0$ for $\lambda \to 0$.) Since F has the gradient property (2.1) one expects that solving Eq. (2.2) can be reduced to the problem of finding the stationary points of some function g, defined on a finite dimensional space, so that catastrophe theory can be applied. This is, indeed, asserted by the following theorem:

Theorem 1: Under the above conditions there exists a smooth function h, mapping some neighbourhood $\tilde{U} \subset \mathbb{R}^m \times \mathbb{R}^k$ of (0,0) to E_1, with $h(0,0)= 0$, $D_y h(0,0)=0$ $(y=(y_1,\ldots,y_m) \epsilon \mathbb{R}^m)$, and, furthermore, a smooth function $g: \tilde{U} \mapsto \mathbb{R}$ satisfying

$$g(0,0) = \partial g(0,0)/\partial y_i = \partial^2 g(0,0)/\partial y_i \partial y_j = 0 \quad (1 \leq i,j \leq m), \tag{2.3}$$

such that, locally, in a neighbourhood of (0,0) in $B_1 \times \mathbb{R}^k$, every small solution $u=u_\lambda$ of Eq. (2.2) is uniquely determined by a stationary point of g, i.e., by a solution $y_\lambda = (y_{1\lambda},\ldots,y_{m\lambda})$ of the equation

$$\nabla_y g(y,\lambda) = 0 \tag{2.4}$$

according to

$$u_\lambda = \sum_{i=1}^{m} y_{i\lambda} v_i + h(y_\lambda,\lambda). \tag{2.5}$$

Proof: For $u \in B_1$ set $u=v+w$, where $v = P_1 u \in N$, $w = Q_1 u \in E_1$. Since the finite-dimensional space N is isomorphic to $\mathbb{R}^m$ by means of the map $v \mapsto y = (y_1,\ldots,y_m)$, where $y_i = \langle v,v_i \rangle$, we can identify N with $\mathbb{R}^m$. Eq. (2.2) may be split into the pair of equations

$$P_2 F(\sum_{i=1}^{m} y_i v_i + w,\lambda) = 0 \tag{2.6a}$$

$$Q_2 F(\sum_{i=1}^{m} y_i v_i + w,\lambda) = 0. \tag{2.6b}$$

Since

$$D_{E_1} Q_2 F(0,0) = L_o$$

and since L_o is an isomorphism, the implicit function theorem [10] tells that Eq. (2.6b) can be uniquely solved locally for $w \in E_1$. This means that there exists a smooth function $h: \tilde{U} \mapsto E_1$ with $h(0,0) = 0$ and

$$Q_2 F(\sum_{i=1}^{m} y_i v_i + h(y,\lambda),\lambda) = 0. \tag{2.7}$$

From Eq. (2.7) we obtain

$$D_y Q_2 F(\sum_{i=1}^{m} y_i v_i + h(y,\lambda),\lambda)\Big|_{(y,\lambda)=(0,0)} = L_o D_y h(0,0) = 0.$$

Hence, it follows that $D_y h(0,0) = 0$. By inserting $w=h$ into Eq. (2.6a), the small solutions of Eq. (2.2) are uniquely determined by the real system of equations

$$g_j(y,\lambda) = \langle F(\sum_{i=1}^{m} y_i v_i + h(y,\lambda),\lambda), v_j \rangle = 0, \quad 1 \le j \le m. \tag{2.8}$$

Consequently the bifurcation problem (2.8) has been reduced to a finite-dimensional system of equations. Eq. (2.8) is the standard Lyapunov-Schmid branching equation system. It remains to show that the g_j are the partial derivatives of a function g. This follows immediately from Eq. (2.1): Defining

$$g(y,\lambda) := R(\sum_{i=1}^{m} y_i v_i + h(y,\lambda),\lambda) \tag{2.9}$$

and differentiating Eq. (2.9) with respect to y_j, one obtains

$$\begin{aligned} \partial g(y,\lambda)/\partial y_j &= D_1 R(\sum_{i=1}^{m} y_i v_i + h(y,\lambda),\lambda) v_j \\ &= \langle F(\sum_{i=1}^{m} y_i v_i + h(y,\lambda),\lambda), v_j \rangle = g_j(y,\lambda). \end{aligned} \tag{2.10}$$

Hence, from Eqs. (2.8)-(2.10) every small solution u_λ of Eq. (2.2) is uniquely determined by a solution of the "gradient equation" (2.4) via (2.5). Eq. (2.3) follows at once from $R(0,\lambda) = 0$, $F(0,0) = 0$, and $\partial^2 g(0,0)/\partial y_i \partial y_j = \langle Lv_i, v_j \rangle$. ∎

The problem of solving Eq. (2.2) has thus been reduced to the problem of finding the stationary points of an unfolding $g(y,\lambda)$ of the singularity $g_0(y):=g(y,0)$. Under the condition, that g_0 is finitely determined and that g is a universal unfolding of g_0, the theory of catastrophes can be applied to g, so that the classification theorems of singularity theory [1,2] can be used to classify the branching properties of Eq. (2.2). The problem is now to derive a splitting-lemma for R such that R is equivalent (in the sense of singularity theory) to some function $\tilde{R}$, which is the sum of a regular quadratic form and g. If this would turn out to be true, one could expect that most of the notations of catastrophe theory, like "versal, resp. universal unfoldings", "(right-)equivalence", "codimension" etc. can be generalized to infinite dimensions so that the classification theorems still remain valid. However, the answer is, in general, negative. Under the hypothesis of theorem 1 no splitting-lemma for R can be derived because the operator $T:B_1 \to B_1^*$, defined by $(Tu)(\tilde{u}) = D_1^2R(0,0)u\tilde{u}$ $(u,\tilde{u}\in B_1)$ doesn't have a closed range: Since T=IL, T is a Fredholmoperator only if $B_2 = B_1^*$. Therefore, when attempting to derive a splitting-lemma stronger conditions than the above ones must be imposed.

Theorem 2: Let $B_1=B_2=H$ and denote by γ the isomorphism $\gamma:H\times\mathbb{R}^k \mapsto \mathbb{R}^m\times E\times\mathbb{R}^k$, defined by $\gamma(u,\lambda)=(y,w,\lambda)$ with $y_i:=\langle u,v_i\rangle$ and $w:=(1-P)u$, P being the orthogonal projection $P:H\mapsto N$. Then there exists a local diffeomorphism Φ in $\mathbb{R}^m\times E\times\mathbb{R}^k$, $\Phi(y,w,\lambda)= (y,\Phi_1(y,w,\lambda),\lambda)$, such that the following transformation formula holds:

$$R\circ\gamma^{-1}\circ\Phi(y,w,\lambda) = ||\tilde{P}w||^2 - ||(1-\tilde{P})w||^2 + g(y,\lambda), \tag{2.11}$$

with an orthogonal projection $\tilde{P}$ in E.

Theorem 2 is essentially an extension of the splitting lemma of Gromoll and Meyer [11] with additional parameters added [12]. Magnus [8] has proved a weaker form of theorem 2 for the case that H is a reflexive Banach space, with the quadratic terms in Eq. (2.11) replaced by $\langle Tu,u\rangle$. He generalized the basic definitions of singularity theory to reflexive Banach spaces and showed, that most of the theorems of singularity theory remain valid, provided that T is a Fredholm operator. (See also Magnus and Poston [13].) The main result of his papers is, that $R(u,\lambda)$ is a versal unfolding of $R(u,0)$ iff $g(y,\lambda)$ is a versal unfolding of $g_0(y):=g(y,0)$. However, his theorems do not apply to R under the hypothesis of theorem 1, because the Fredholm property is not satisfied if $B_2 \neq B_1^*$. In this case R cannot be classified by catastrophe theory, despite the fact that the bifurcation properties of Eq. (2.2) are in one to one correspondence to those of the stationary points of g. There are in fact variational problems, where theorem 2 does not apply, whereas theorem 1 still works.

Applications to Variational Problems

In order to apply the above theorems to variational problems, we have to look for appropriate function spaces, such that the conditions for the theorems are satisfied: Let Ω be a bounded domain in $\mathbb{R}^n$ with boundary $\partial\Omega$ and closure $\bar{\Omega}$. Let further $\partial_i:=\partial/\partial x_i$ $(1\le i\le n)$, and for any multiindex $\alpha=(\alpha_1,\dots,\alpha_n)$ with nonnegative integers α_i the corresponding differential operator of order $|\alpha|=\alpha_1+\dots+\alpha_n$ will be denoted by $\partial^\alpha:=\partial_1^{\alpha_1}\cdots\partial_n^{\alpha_n}$. If necessary, differentiation has to be taken in the generalized sense. $C^1(\bar{\Omega})$ is the Banach space of all functions $u=u(x)$ $(x\in\bar{\Omega})$, such that $\partial^\alpha u$ is continuous for $|\alpha|\le 1$ ($\partial\Omega$ is assumed to be sufficiently regular). $L_p(\Omega)$ $(p\ge 1)$ is the Banach space of all functions $u(x)$ on Ω (up to a set of measure zero) such that $\int_\Omega |u(x)|^p dx<\infty$, with the standard norm $||\cdot||_p$. The Sobolev space $W_p^1(\Omega)$ is the space of all functions $u(x)$, whose generalised derivatives are in $L_p(\Omega)$ for $|\alpha|\le 1$. Define $C_0^1(\bar{\Omega})$ to be the set of all $u\in C^1(\bar{\Omega})$ with $\partial^\alpha u(|\alpha|\le 1)$ vanishing on $\partial\Omega$ and let $\overset{\circ}{W}{}_p^1(\Omega)$ be the completion of $C_0^1(\Omega)$ in $W_p^1(\Omega)$. The norm in $W_p^1(\Omega)$ is

$$||u||_{1,p} := [\sum_{|\alpha|\le 1} ||\partial^\alpha u||_p^p]^{1/p} \quad .$$

For $p>1, W_p^1(\Omega)$ is reflexive and $W_2^1(\Omega)$ is a Hilbertspace. In $\mathring{W}_2^1(\Omega)$ there exists an inner product $\langle\cdot,\cdot\rangle_1$, which is equivalent to that of $W_2^1(\Omega)$:

$$\langle u,v\rangle_1 := \sum_{|\alpha|=1} (\partial^\alpha u, \partial^\alpha v),$$

where $(\cdot,\cdot)$ denotes the inner product in $L_2(\Omega)$.

Consider now a variational problem $\delta R = 0$ extremizing the functional

$$R(u,\lambda) = \int_\Omega (\tfrac{1}{2}(\nabla u)^2 + f(u,\lambda,x))dx, \quad \lambda \in \mathbb{R}^k \tag{2.12}$$

with respect to u, where $u=u(x)$ is defined on Ω and vanishing on $\partial\Omega$. Suppose f is smooth and $f(0,\lambda,x) = f'(0,0,x) = 0$, the prime denoting $\partial/\partial u$. The Euler Lagrange equations associated with (2.12) are given by

$$F(u,\lambda) := -\Delta u + f'(u,\lambda,x) = 0,$$

with Dirichlet's boundary conditions. For simplicity we suppose that $f''(0,0,x)=\mu$, independent of x, where μ is an eigenvalue of Δ (subject to Dirichlet's boundary conditions) with multiplicity 1 and eigenfunction $v=v(x)$, $(v,v)=1$. Hence, for $\lambda=0$, $u=0$ is a solution of Eq. (2.13) and the linearization of (2.13) around $u=0$ has a nontrivial solution v. If R is differentiable with respect to u in some function space, the second derivative of R at $(u,\lambda)=(0,0)$ is given by

$$Bu\tilde{u} := D_1^2 R(0,0)u\tilde{u} = \int_\Omega (\nabla u\cdot\nabla\tilde{u} + \mu u\tilde{u})dx.$$

To apply theorem 2, we have to find a Hilbert space H of functions vanishing on $\partial\Omega$, such that the operator $T:H\mapsto H$, defined by $\langle Tu,\tilde{u}\rangle = Bu\tilde{u}$ is a Fredholmoperator. Obviously the only choice we have is $H=\mathring{W}_2^1(\Omega)$. T is then given by $T=-1+\Delta^{-1}$, which is, of course, a Fredholmoperator in H, since Δ^{-1} maps $\mathring{W}_2^1(\Omega)$ to $\tilde{W}_2^3(\Omega)$ and the embedding of $\tilde{W}_2^3(\Omega)$ in $\mathring{W}_2^1(\Omega)$ is compact. (We use the notation $\tilde{W}_p^l(\Omega) := W_p^l(\Omega)\cap\mathring{W}_p^1(\Omega)$). However, for R to be welldefined on $\mathring{W}_2^1(\Omega)\times\mathbb{R}^k$, $f_\lambda(u,x) := f(u,\lambda,x)$ must be in $L_1(\Omega)$ for $u\in\mathring{W}_2^1(\Omega)$. This imposes a growth restriction on f, which is satisfied, if, for example,

$$|f_\lambda(u,x)| \le K(\lambda)\ (1+|u|^{2n/(n-2)}) \quad \text{for } n \ge 3$$

$$|f_\lambda(u,x)| \le K(\lambda)e^{u^2} \quad \text{for } n = 2$$

$$\text{unrestricted} \quad \text{for } n = 1$$

The condition that the derivatives of R of any order exist imposes additional conditions on f. If, for example, we suppose that f is a polynomial in u of degree r, R can be regarded as a smooth function on $\mathring{W}_2^1(\Omega)\times\mathbb{R}^k$ only if $r\le 2n/(n-2)$ for $n\ge 3$. As another example consider the function $f = f_1(\lambda)$ sinu. R is not smooth in this case, because its derivatives of order $\ge 2n/(n-2)$ cease to exist. We see, therefore, that theorem 2 is applicable only to a restricted class of variational problems (cf. [12]), whereas theorem 1 works for arbitrarily smooth f: It has been shown [12] in a more general context by means of Sobolev's embedding theorem, that one can always find some $p \geq 2$ such that R is smooth on $\tilde{W}_p^2(\Omega) \times \mathbb{R}^k$ and that F is a smooth operator from $\tilde{W}_p^2(\Omega) \times \mathbb{R}^k$ to $L_p(\Omega)$. Since the differential operator $L := -\Delta + \mu$, regarded as an operator from $\tilde{W}_p^2(\Omega)$ to $L_p(\Omega)$, satisfies the Fredholm condition, theorem 1 can be applied with $B_1 = \tilde{W}_p^2(\Omega)$, $B_2 = L_p(\Omega)$ and $H = L_2(\Omega)$. Assuming, for example, that $\lambda \in \mathbb{R}$ and

$$2a := \int_\Omega v(x) \partial f'(0,0,x)/\partial\lambda dx \neq 0 \quad , \quad 3b := \int_\Omega v(x) f'''(0,0,x) dx \neq 0,$$

it follows that $g(y,\lambda) = a\lambda y + by^3 +$(terms of higher order), i.e., g is equivalent to the fold catastrophe.

Perturbation Analysis and Catastrophe Theory

In general it is not possible to solve Eq. (2.7) exactly. In most cases it is only possible to calculate the coefficients of the power series of h and g by a perturbation analysis.The problem is then where to truncate the power series without changing the qualitative bifurcation properties. Here catastrophe theory is very helpful. Let us assume that $g_o(y) := g(y,0)$ is l-determined, $l \geq 3$. Then, as a first approximation, it is sufficient to determine the l-th Taylor polynomial $j^l g_o(y)$ of g_o (the l-jet), which can be calculated if the (l-1)-th Taylor polynomial of h is known. In practice one could first determine the 3-jet and test the sufficiency of it. If $j^3 g_o$ is not finitely determined, one calculates the 4-jet, and so on. After l-3 steps one arrives at the finitely determined l-jet. In a next step one has to test whether the unfolding g of g_o is versal (or transversal). This can be achieved if $j^l g_o$ and all terms of the form

$$\tilde{g}_i(y) := \frac{\partial}{\partial\lambda_i} j^{l+2} g(y,\lambda)\Big|_{\lambda=0} , \quad 1 \leq i \leq k,$$

i.e., only those terms in $j^{l+2} g(y,\lambda)$, which are linear in the λ_i are known. How to test finite determinacy and transversality is discussed in Poston and Stewart [14]. If one has convinced oneself that g_o is l-determinate and that g is versal (i.e., $c := \text{codim } g_o \leq k$), the polynomial

$$\hat{g}(y,\lambda) := j^l g_o(y) + \sum_{i=1}^{k} \lambda_i \tilde{g}_i(y)$$

has the same qualitative bifurcation properties as g. Hence, catastrophe theory tells us that we have to calculate only a finite number of coefficients in the Taylor expansion of g for investigating the qualitative bifurcation behaviour of Eq. (2.2) and to find the normal form $f_o(x)$ ($x \in \mathbb{R}^m$) in Thom's [1] and Arnold's [2] lists, which is equivalent to $g_o(y)$ resp $j^l g_o(y)$. Then g and $\hat{g}$ are, as unfoldings, equivalent to the universal unfolding $f(x,d)$ of f_o,

$$f(x,d) = f_o(x) + \sum_{i=1}^{c} d_i f_i(x) \ , \quad d = (d_1, \ldots, d_c) \quad ,$$

with polynomials f_i. The morphism from g to f can be determined to first order in (y,λ) from $\hat{g}$ and the bifurcation problem (2.2) is qualitatively solved if the bifurcation properties of f are known. If g is not versal, a similar procedure yields the result (cf. [12]) that the bifurcation set of g is mapped to a section of the bifurcation set of f.

Stability Considerations

We regard a small solution u_λ of Eq. (2.2), determined from a stationary point y_λ of g via theorem 1, as a stationary solution of the differential equation

$$\frac{du}{dt} = -F(u,\lambda). \tag{2.13}$$

A map $t \to u(t)$ ($t \in \mathbb{R}_+$) is called a solution of Eq. (2.13), if (1) u(t) is continuous in the norm of B_1 and (2) in the norm of B_2 u(t) is differentiable and satisfies Eq. (2.13). The following stability analysis is a formal one in that we do not utilize nonlinear semigroups in Banach spaces (cf. Segal, Chernoff and Marsden [15]), but rather confine ourselves to the linearized stability problem for u_λ. To this end we suppose that the domain of the linear operator $K(y,\lambda): B_1 \to B_2$, defined by

$$K(y,\lambda) := D_1 F(\sum_{j=1}^{m} y_j v_j + h(y,\lambda),\lambda) \ , \quad K(0,0) = L,$$

can be extended from B_1 to some Hilbertspace $\tilde{H}$, where $B_1 \subset \tilde{H} \subset H$ with dense and topological inclusions (Friedrichs' extension). We further suppose that $K(y,\lambda)$, considered as a self-adjoint operator in H with dense domain, has a discrete spectrum $\mu_j(y,\lambda)$ ($j \in \mathbb{N}$) with $\mu_j(0,0) > 0$ for $j > m$ and $\mu_j(0,0) = 0$ for $1 \leq j \leq m$. Then, by continuity, $\mu_j(y,\lambda)$ is positive for $j > m$ and sufficiently small (y,λ). By linear stability analysis, u_λ is (asymptotically) stable, if $\mu_j(y_\lambda,\lambda) > 0$ for $1 \leq j \leq m$. Denoting the Jacobimatrix of g with respect to y by

$$g^{(2)}(y,\lambda) := \{g_{ij}^{(2)}(y,\lambda)\}_{1 \leq i,\ j \leq m}$$

we have the following

Lemma: The number of negative (positive, vanishing) eigenvalues of $K(y,\lambda)$ is equal to the number of negative (positive, vanishing) eigenvalues of $g^{(2)}(y,\lambda)$.

Proof: Let $\tilde{E}$ be the completion of E_1 in $\tilde{H}$ and set

$$\tilde{K}(y,\lambda) := Q\,K(y,\lambda)\big|_{\tilde{E}}$$

where Q is the orthogonal projection $H \to E$. $\tilde{K}(y,\lambda)$ is positive for sufficiently small (y,λ). Differentiating the equation

$$F(\sum_{i=1}^{m} y_i v_i + h(y,\lambda),\lambda) = \sum_{i=1}^{m} v_i \frac{\partial g(y,\lambda)}{\partial y_i}$$

with respect to y_j it follows that

$$K(y,\lambda)\cdot(v_j + h_j(y,\lambda)) = \sum_{i=1}^{m} g_{ji}^{(2)}(y,\lambda) v_i \ , \tag{2.14}$$

where $h_j(y,\lambda) := \partial h(y,\lambda)/\partial y_j$. From Eq. (2.14) it follows for $w \in \tilde{E}$ that

$$\langle K(y,\lambda)w, v_j \rangle = -\langle \tilde{K}(y,\lambda)w,\ h_j(y,\lambda) \rangle. \tag{2.15}$$

Let us define a linear operator

$$J(y,\lambda)u := Qu + \sum_{i=1}^{m} \langle u,v_i\rangle (v_i + h_i(y,\lambda)). \tag{2.16a}$$

$J(y,\lambda)$ is invertible in H and $\tilde{H}$ for sufficiently small (y,λ). The adjoint of $J(y,\lambda)$ is given by

$$J^+(y,\lambda)u = Qu + \sum_{i=1}^{m} (\langle u,v_i\rangle + \langle h_i(y,\lambda), Qu\rangle)v_i \quad . \tag{2.16b}$$

From Eqs. (2.14) to (2.16) $u\epsilon\tilde{H}$ is seen to satisfy

$$J^+(y,\lambda)K(y,\lambda)J(y,\lambda)u = \tilde{K}(y,\lambda)Qu + \sum_{i,j=1}^{m} g_{ij}^{(2)} (y,\lambda)\langle u,v_i\rangle v_j$$

and, therefore,

$$\langle K(y,\lambda)\ J(y,\lambda)u,\ J(y,\lambda)u\rangle = \langle\tilde{K}(y,\lambda)Qu,Qu\rangle + \sum_{i,j=1}^{m} g_{ij}^{(2)} (y,\lambda)\langle u,v_i\rangle\langle u,v_j\rangle \ . \tag{2.17}$$

This proves the Lemma. As a consequence only the minima of g correspond to stable stationary solitons of Eq. (2.13).

3. Buckling of Columns

We apply theorem 2 to the buckling of columns: Consider a pinned uniform column of length l, bending stiffness EI, subjected to an axial load P and restrained against lateral displacement $W(S)$ $(0\leq S\leq l)$ by a foundation that provides a restoring force $k_1W+k_3W^3$ $(k_1,k_3>0)$ per unit length. In the presence of an initial displacement $W_0(S)$ the free energy F is given by [16]

$$F = \frac{1}{2} \int_0^l \{EI(\frac{d^2W}{dS^2})^2 - P\ \frac{dW}{dS}\ [\frac{dW}{dS} +2\ \frac{dW_0}{dS}] + k_1W^2 + \frac{1}{2}k_3W^4\}dS. \tag{3.1}$$

With the dimensionless quantities

$$S = \frac{ls}{\pi},\quad W = (k_1/k_3)^{1/2}u,\quad W_0 = \tau(k_1/k_3)^{1/2}u_0,\quad \lambda = (l/\pi)^2 P/2EI,$$

$$\mu = (l/\pi)^4 k_1/EI,\quad R = (l/\pi)^3 Fk_3/k_1EI,$$

it follows that

$$R = R(u,\lambda,\tau,\mu) = \int_0^\pi \{\frac{1}{2}(u'')^2 - \lambda u'(u'+2\tau u_0') + \frac{1}{2}\ \mu u^2(1+\frac{1}{2}u^2)\}ds \tag{3.2}$$

where $' := d/ds$. The imperfection parameter τ [17] varies in a neighbourhood of $\tau=0$ (the ideal column) and for u_0 we choose a harmonic expansion

$$u_o(s) = (2/\pi)^{1/2} \sum_{j=1}^{\infty} U_j \sin js$$

with constants U_j. The equilibrium condition ($\delta R=0$) yields the following Euler-Lagrange equation for u:

$$u'''' + 2u'' + \mu(u+u^3) = -2\lambda\tau u_o'' \tag{3.3a}$$

with boundary conditions

$$u(0) = u(\pi) = u''(0) = u''(\pi) = 0. \tag{3.3b}$$

Let H be the completion, in the Hilbert space $W_2^2((0,\pi))$, of all functions u(s), $s\varepsilon[0,\pi]$, which are continuously differentiable and satisfy the boundary conditions (3.3b) and suppose that $u_o \varepsilon H$. The inner product in H is given by $\langle u,v\rangle = (u'',v'')$, where $(\cdot,\cdot)$ is the standard inner product in $H_0 := L_2((0,\pi))$. According to Sobolev's embedding theorem every $u\varepsilon H$ is continuously differentiable and the derivative u' has a generalized square integrable derivative. Hence R, as a polynomial in u, is analytic in $H \times \mathbb{R}^3$. The second derivative of R for $\tau=0$, u=0, is given by

$$D_H^2 R(0,\lambda,\mu,0)vw = (L(\lambda,\mu)v,w),$$

where $L(\lambda,\mu)$ is the differential operator

$$L(\lambda,\mu) := d^4/ds^4 + 2\lambda\, d^2/ds^2 + \mu ,$$

resulting from linearizing Eq. (3.3a) for $\tau=0$ around u=0. Thus, the linear operator $T(\lambda,\mu):H\to H$, defined by

$$D^2 R(0,\lambda,\mu,0)vw = \langle T(\lambda,\mu)v,w\rangle,$$

is given by

$$T(\lambda,\mu) = (d^4/ds^4)^{-1} L(\lambda,\mu)$$

and is, therefore, by Rellich's theorem, a Fredholmoperator with index 0. The eigenfunctions of $L(\lambda,\mu)$ satisfying the boundary conditions (3.3b) are $v_j(s) = (2/\pi)^{1/2} \times \sin js$ and represent an orthonormal basis in H_o, being complete in H, too. By definition [16, 17], the buckling load λ_c is the smallest value of λ for which one eigenvalue

$$\nu_j(\lambda,\mu) = j^4 - 2\lambda j^2 + \mu \quad (j\varepsilon \mathbb{N})$$

of L vanishes and all remaining eigenvalues are nonnegative. If $m^2(m+1)^2 > \mu > m^2(m-1)^2$ for some $m\varepsilon\mathbb{N}$, the buckling load is given by $\lambda_c = \lambda_c(\mu) := (m^2+\mu/m^2)/2$ with $\nu_m(\lambda_c,\mu) = 0$ and $\nu_j(\lambda_c,\mu) > 0$ for $j \neq m$, i.e., we have bifurcations from a simple eigenvalue. However, if $\mu=\mu_m:=m^2(m+1)^2$ we obtain $\lambda_c=\lambda_m:=(m^2+(m+1)^2)/2$ with $\nu_m(\lambda_m,\mu_m)=\nu_{m+1}(\lambda_m,\mu_m) = 0$ and $\nu_j^m:= \nu_j(\lambda_m,\mu_m) > 0$ for $j \neq m,m+1$, so that in this case there occurs bifurcation from a double eigenvalue. The bifurcation properties of Eq. (3.3) have been studied in detail in [12] by means of a perturbation analysis. Here we summarize the results.

A model different from the one presented here only in that k_3 is replaced by $-k_3$ has been considered by Keener [9], but with a fixed length $l=(EI/k_1)^{1/4}\pi$, so that the term $\mu(u+u^3)$ in Eq. (3.3a) is replaced by $u-u^3$. Hence, no double eigenvalue occurs in that model. Furthermore the restoring force is physically unrealistic because it reinforces the lateral displacement if $|W|>\sqrt{(k_1/k_3)}$. Keener calculated (by using an iteration procedure) the bifurcation set in the λ-τ-plane and obtained a cusp. He did not analyse the stability of the bifurcating solutions, which turn out to be physically not realistic (cf. [12]). Especially for $\tau= 0$ Keener obtained for $\lambda<1$ two nontrivial solution branches, which turn into the trivial solution for $\lambda\to 1$ and vanish for $\lambda>1$. However, both nontrivial solutions are unstable, whereas the trivial solution is stable for $\lambda<1$ and unstable for $\lambda>1$, too. Hence in that model there would not exist any stable solution for $\lambda>1$, what contradicts experiment. In contradistinction, the results obtained for the model presented here are physically plausible.

Bifurcation from a Simple Eigenvalue

We investigate the case of a simple eigenvalue, i.e., we suppose that $m^2(m+1)^2>\mu>m^2(m-1)^2$ for some m. Theorem 2 ensures that, locally, the small solutions of the boundary value problem (3.3) are uniquely determined by the stationary points of an (in that case) analytic function $g(y,\rho,\tau)$, where $y\in \mathbb{R}$ and $\lambda=\lambda_c(\mu)+\rho$. First we analyze the bifurcations for $\tau=0$. We find that

$$g(y,\rho,0) = -m^2\rho y^2 + dy^4 + \sum_{j=3}^{\infty} \hat{g}_j(\rho)y^{2j} \tag{3.4}$$

with analytic functions $\hat{g}_j(\rho)$ and

$$d = 3\mu/8\pi \ , \quad e := -\hat{g}_3(0) = \mu/8\pi^2\nu_{3m}(\lambda_c,\mu).$$

The stationary points of $g|_{\tau=0}$ are given by y=0 and

$$y_\pm(\rho) = \pm m(\rho/2d)^{1/2}[1+ \frac{3e}{4d}\rho+O(\rho)] \ , \quad \rho\geq 0. \tag{3.5}$$

The solutions of (3.3), which correspond to y=0 and $y=y_\pm(\rho)$, are the trivial solution u=0 and

$$u_\pm(\rho,s) = \pm \{m(\rho/2d)^{1/2}[1 + \frac{3e}{4d}\rho]v_m(s) + \frac{m\mu}{2\pi\nu_{3m}(\lambda_c,\mu)} (\rho/2d)^{3/2}v_{3m}(s) + O(\rho^{3/2})\} \ , \quad \rho\geq 0. \tag{3.6}$$

A stability analysis shows that the trivial solution is stable for $\rho<0$ and unstable for $\rho>0$, whereas the nontrivial solution branches are stable. This gives the familiar bifurcation diagram of Fig. 1, [7].

We look now for bifurcations including variations of the imperfection parameter τ.

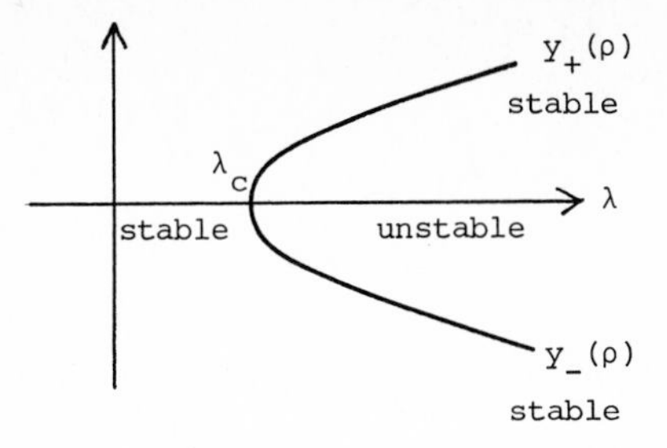

Fig. 1 Bifurcation diagram of Eq. (3.3) in the y-λ-plane for $\tau=0$. The trivial solution u=0 (associated to y=0) is stable for $\lambda<\lambda_c$ and unstable for $\lambda>\lambda_c$. The solutions $u_\pm(\rho,s)$ (associated to $y_\pm(\rho)$) are stable.

For g one obtains

$$g(y,\rho,\tau) = -c_1\tau y - m^2\rho y^2 - c_3\tau y^3 + dy^4 - ey^6 + \text{terms of higher order} \quad (3.7)$$

where

$$c_1 = 2m^2U_m\lambda_c, \quad c_3 = 9m^2\lambda_c U_{3m}/\pi\nu_{3m}(\lambda_c,\mu).$$

Assuming that $U_m \neq 0$, g is equivalent to the cusp catastrophe: There exists a diffeomorphism $(y,\rho,\tau) \to (x(y,\rho,\tau), a(\rho,\tau), b(\rho,\tau))$, such that

$$g(y,\rho,\tau) = f(x(y,\rho,\tau), a(\rho,\tau), b(\rho,\tau)) + \alpha(\rho,\tau) \quad , \quad (3.8)$$

where α is a translation and f is the cusp catastrophe polynomial,

$$f(x,a,b) = \frac{1}{4}x^4 - ax - \frac{1}{2}bx^2 .$$

x,a,b can be calculated by a perturbation expansion:

$$x = -x_\tau\tau + x_1y + x_3y^3 + \ldots$$

$$a = 2(2\pi/3\mu)^{1/4}m^2U_m\lambda_c\tau + \ldots, \quad b = (2\pi/3\mu)^{1/2}m^2\rho + \ldots ,$$

where

$$x_\tau = (2\mu/3\pi)^{3/4}m^2\lambda_c(U_m + 27U_{3m})/3\pi\nu_{3m}(\lambda_c,\mu)$$

$$x_1 = (2\mu/3\pi)^{1/4}, \quad x_3 = (2\mu/3\pi)^{3/4}\mu/4\pi^2\nu_{3m}(\lambda_c,\mu).$$

The bifurcation properties of Eq. (3.3) are now in a one-to-one correspondence to those of f, the bifurcation set of which is given by the familiar equation

$$27a^2 - 4b^3 = 0 \quad . \quad (3.9)$$

Inserting $a = a(\rho,\tau)$, $b = b(\rho,\tau)$ into Eq. (3.9) and solving for ρ we obtain

$$\rho(\tau) = [\mu/m^2]^{1/3}(9U_m\lambda_c\tau/4\sqrt{\pi})^{2/3} + O(\tau^{2/3}). \quad (3.10)$$

Eq. (3.10) is the equation for the bifurcation set of g in the ρ-τ-plane, shown in Fig. 2.

Inside the cusp there exist three solutions of Eq. (3.3), two of which are stable, whereas the third one is unstable. Outside the cusp there exists only one stable solution, which reduces to the trivial solution if $\tau = 0$.

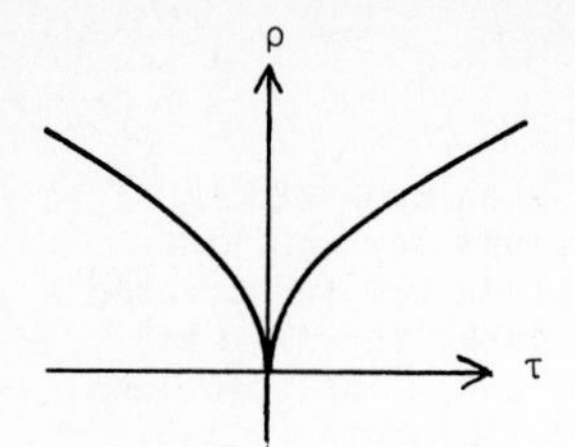

Fig. 2: The bifurcation set of $g(y,\rho,\tau)$ (Eq. (3.7))

Bifurcation from a Double Eigenvalue: The Double Cusp

Next we consider the case of a double eigenvalue by investigating the bifurcations of Eq. (3.3) if (ρ,λ,τ) variies in a neighbourhood of $(\mu_m,\lambda_m,0)$ for some $m\in\mathbb{N}$. Set $\kappa := \mu-\mu_m$, $\rho := \lambda-\lambda_m$. The function g, defined in theorem 2, depends now on two "order parameters" (y_1,y_2). First we look for bifurcations for $\tau=0$. It can be shown that

$$\tilde{g} := g\big|_{\tau=0} = \frac{1}{2}(\kappa-2m^2\rho)y_1^2 + \frac{1}{2}(\kappa-2(m+1)^2\rho)y_2^2 \tag{3.11}$$

$$+ (d+3\kappa/8\pi)(y_1^4+y_2^4+4y_1^2y_2^2) + \sum_{j+k\geq 3} \hat{g}_{jk}(\rho,\kappa)y_1^{2j}y_2^{2k} \quad ,$$

with $d = 3\mu_m/8\pi$ and analytic functions $\hat{g}_{jk}(\rho,\kappa)$. Elementary though lengthy calculations (which make use of the fact, that g depends only on even powers of y_1,y_2) lead to the following bifurcation properties of $\tilde{g}$:

There are four analytic curves $\kappa_j(\rho)$ $(1\leq j\leq 4)$, which divide the κ-ρ-plane into various regions, each region giving rise to different stationary points of $\tilde{g}$, resp. to different solution branches of Eq. (3.3). The $\kappa_j(\rho)$ are given by

$$\begin{aligned}
\kappa_1(\rho) &= 2m^2\rho \\
\kappa_2(\rho) &= 2(m+1)^2\rho \\
\kappa_3(\rho) &= [4(m+1)^2-2m^2]\rho + O(\rho), \ \rho\leq 0 \\
\kappa_4(\rho) &= [4m^2-2(m+1)^2]\rho + O(\rho), \ \rho\geq 0 \quad .
\end{aligned} \tag{3.12}$$

In a first approximation the $\kappa_3(\rho)$, $\kappa_4(\rho)$ are straight lines and $\kappa_4(\rho)\leq 0$ for m=1,2, $\kappa_4(\rho)\geq 0$ for $m\geq 3$. Furthermore, for small $\rho>0$, one finds that $\kappa_4(\rho)<\kappa_1(\rho)<\kappa_2(\rho)<-\kappa_3(-\rho)$. There exist nine different stationary points of $\tilde{g}$, i.e., nine different solutions of the equations $\partial\tilde{g}/\partial y_1 = \partial\tilde{g}/\partial y_2 = 0$, which are given by

(1) $y_1 = y_2 = 0$
to which corresponds u=0 as a solution of Eq. (3.3). This solution exists for all values of (ρ,κ) and is stable for $\kappa\geq\kappa_1(\rho)$ and $\kappa\geq\kappa_2(\rho)$.

(2) $y_1 = 0,\ y_2 = \pm\{(\kappa_2(\rho)-\kappa)/d\}^{1/2}\,(1+\alpha_1(\rho,\kappa))$, (3.13)
with an analytic function $\alpha_1(\rho,\kappa)$, $\alpha_1(0,0)=0$. The two solutions (3.13) exist for $\kappa\leq\kappa_2(\rho)$. They are stable for $\rho\geq 0$ and (ρ,κ) with $\rho\leq 0$ and $\kappa\leq\kappa_3(\rho)$.

(3) $y_2 = 0,\ \ y_1 = \pm\{(\kappa_1(\rho)-\kappa)/d\}^{1/2}\,(1+\alpha_2(\rho,\kappa)),\ \ \alpha_2(0,0) = 0$ (3.14)

The two solutions (3.14) exist for $\kappa\leq\kappa_1(\rho)$. They are stable for $\rho\leq 0$ and (ρ,κ) with $\rho\geq 0$ and $\kappa\leq\kappa_4(\rho)$.

(4) $y_1 = \pm\,\{(\kappa_3(\rho)-\kappa)/3d\}^{1/2}\,(1+\alpha_3(\rho,\kappa)),\ \alpha_3(0,0) = 0$ (3.15a)

$$y_2 = \pm \{(\kappa_4(\rho)-\kappa)/3d\}^{1/2} \quad (1+\alpha_4(\rho,\kappa)),\ \alpha_4(0,0) = 0 \tag{3.15b}$$

The four solutions (3.15) exist for $\kappa \leq \kappa_3(\rho)$ and $\kappa \leq \kappa_4(\rho)$, i.e., in that domain, where both solution pairs (3.13) and (3.14) exist and are stable. For $\kappa = \kappa_3(\rho)$ Eq. (3.15) reduces to (3.13) and for $\kappa = \kappa_4(\rho)$ to (3.14). This means, that secondary bifurcation occurs [18]. A stability analysis shows that the associated solutions of Eq. (3.3) are unstable.

In Fig. 3 the bifurcation diagram of g is shown in linear approximation.

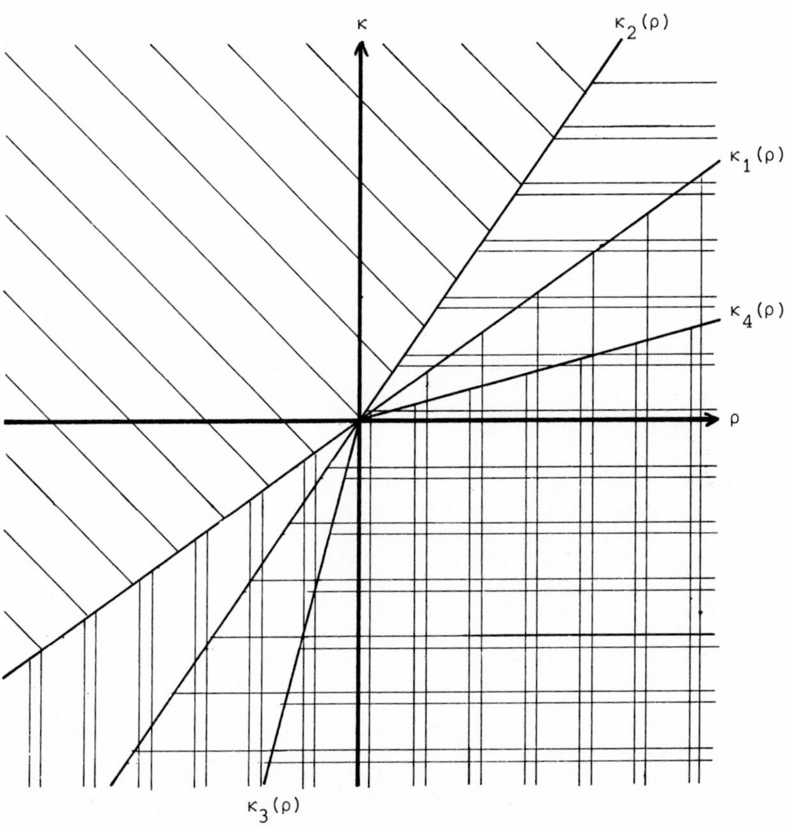

Fig. 3: The bifurcation diagram of $\tilde{g}$ in κ-ρ-plane in linear approximation for $m \geq 3$. The bifurcation set consists of the lines $\kappa_1(\rho)$, $\kappa_2(\rho)$ and of the curves $\kappa_3(\rho)$ $(\rho \leq 0)$ and $\kappa_4(\rho)(\rho \geq 0)$.

\ : Stability domain of $y_1 = y_2 = 0$ $(u(s) = 0)$, in the remaining domain the trivial solution is unstable.

— : Unstable existence domain of (3.13) = : stable existence domain of (3.13)

| : Unstable existence domain of (3.14) || : stable existence domain of (3.14)

: Existence domain of the unstable solutions (3.15)

From the preceding analysis it is clear that the solutions (3.6) of Eq. (3.3) correspond to the stationary points (3.13) resp. (3.14), if $\mu_{m-1} < \mu < \mu_m$, resp. $\mu_m < \mu < \mu_{m+1}$. In addition we obtained stability boundaries for (3.13) and (3.14) ($\kappa_3(\rho)$ for (3.13) and $\kappa_4(\rho)$ for (3.14)), which are not obtainable from the bifurcation analysis of a simple eigenvalue.

For $\tau \neq 0$ g is given by

$$g(y_1,y_2,\rho,\kappa,\tau) = -2\tau[m^2U_m y_1+(m+1)^2U_{m+1}y_2] + \frac{1}{2}(\kappa-2m^2\rho)y_1^2$$
$$+ \frac{1}{2}(\kappa-2(m+1)^2\rho)y_2^2 -\tau(g_{30}y_1^3+g_{21}y_1^2y_2+g_{12}y_1y_2^2+g_{03}y_2^3) \tag{3.16}$$
$$+ d(y_1^4+y_2^4+4y_1^2y_2^2) + \text{terms of higher order,}$$

with constants g_{ik} (i+k=3). The singularity $g_o(y_1,y_2) := g(y_1,y_2,0,0,0)$ is equivalent to

$$f_o(x_1,x_2) = x_1^4+x_2^4+4x_1^2x_2^2 \quad ,$$

which belongs to the unimodal family $T_{2,4,4}$ in Arnold's notation [2], sometimes also called X_9 or double cusp.

f_o has codimension eight with the universal unfolding

$$f(x_1,x_2;\alpha_j) = f_o(x_1,x_2) + \alpha_1x_1+ \alpha_2x_2+ \alpha_3x_1^2 +\alpha_4x_1x_2 + \alpha_5x_2^2$$
$$+ \alpha_6x_1^3+ \alpha_7x_2^3+ \alpha_8x_1^2x_2^2 \quad .$$

If U_m or U_{m+1} is different from zero, the bifurcation set of g turns out to represent a three-dimensional submanifold of the bifurcation set of f, which is a seven-dimensional manifold in the space $\{\alpha_j : 1 \le j \le 8\}$ of unfolding parameters, and therefore very difficult to analyze, because odd powers of x_1,x_2 have to be taken into account. We refer the reader to the papers by Zeeman [19] and Callahan [20], where the double cusp is analyzed to some extent. In ρ-κ-τ-space the bifurcation set Σ_g of g consists of several surfaces intersecting the ρ-κ-plane in the curves $\kappa_j(\rho)$ $(1 \le j \le 4)$. In Fig. 4 those surfaces are shown, which intersect ρ-κ-plane in the straight lines $\kappa_1(\rho)$, $\kappa_2(\rho)$ (cf., also Callahan [20]).

Poston and Magnus [21] have analyzed the bifurcation of a rectangular buckling plate from a double eigenvalue. They also obtained a double cusp and considered its universal unfolding, identifying various unfolding parameters as imperfection parameters. This can, of course, also be done in our column model. It is obvious then, that the double cusp plays an important role in buckling problems and will occur in all those cases where the dominant buckling mode changes.

4. Catastrophe Theory in Nonequilibrium Thermodynamics

In the nonlinear regime of nonequilibrium thermodynamics, systems develop order or dissipative structures [22,23] with possibility of nonequilibrium phase transitions. For certain critical values of the exteriour parameters the state of the system undergoes rapid changes, which can be understood as generalized phase transitions [24, 25]. On the other hand catastrophe theory describes discontinuous changes in geometric terms alone. Hence, there should be an intimate connection between catastrophe theory and generalized phase transitions in nonequilibrium thermodynamics. In this section we sketch how a generalization of Onsager's theory may, by means of the methods described in section 2, lead to a classification of nonequilibrium phase transitions via catastrophe and bifurcation theory.

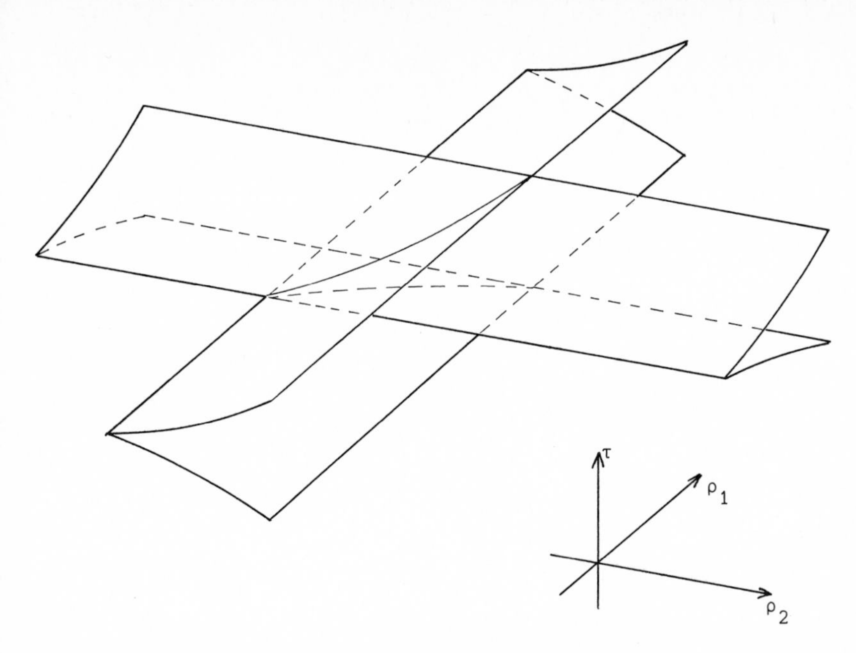

Fig. 4: The section of Σ_g, which intersects the ρ-κ-plane in the lines $\kappa_1(\rho)$, $\kappa_2(\rho)$, after a linear coordinate transformation $\rho_1 = \kappa-\kappa_1(\rho)$, $\rho_2 = \kappa-\kappa_2(\rho)$.

Consider a thermodynamic system characterized by n extensive variables $\alpha_i = \alpha_i(x,t)$ $(1 \leq i \leq n)$, $x \in V \subset \mathbb{R}^3$, where t is the time. Assuming local equilibrium [22], there exists an entropy density function $s = s(\alpha_1,\ldots,\alpha_n,x)$. Generally, a continuity equation for the α_i is valid,

$$\frac{\partial \alpha_i}{\partial t} = -\nabla \cdot \vec{A}_i + \sigma_i \quad , \tag{4.1}$$

where $\vec{A}_i$ is the flow and σ_i the source of the variable α_i. Differentiating s with respect to time one obtains from Eq. (4.1) the continuity equation for s:

$$\frac{\partial s}{\partial t} = -\nabla \cdot \sum_{i=1}^{n} p_i \vec{A}_i + \sum_{i=1}^{n} (\nabla p_i \cdot \vec{A}_i + p_i \sigma_i) \tag{4.2}$$

where $p_i := \partial s/\partial \alpha_i$ is the intensive variable conjugate to α_i. Hence, the entropy flow is given by $\vec{A}[s] = \Sigma p_i \vec{A}_i$ and for the entropy source $\sigma[s]$ one obtains

$$\sigma[s] = \sum_{i=1}^{n} (\nabla p_i \cdot \vec{A}_i + p_i \sigma_i) =: \sum_{l} X_l J_l \quad , \tag{4.3}$$

where $\sigma[s]$ has been splitted into generalized fluxes $\{J_l\} := \{\vec{A}_i, \sigma_i\}$ and generalized forces $\{X_l\} := \{p_i, \nabla p_i\}$. (The splitting of $\sigma[s]$ in fluxes and forces is, of course, not unique, one can also use linear combinations which have to be chosen such that no mixture of tensorial quantities of different rank will occur [26] in the definition of J_l, X_l. The splitting used here has been chosen for simplicity.) In the linear approximation to nonequilibrium thermodynamics (close to equilibrium) one assumes a linear dependence of the fluxes J_l on the forces X_l,

$$J_l = \sum_r C_{lr} X_r \quad , \tag{4.4}$$

with only local dependences taken into account. The Onsager coefficients C_{lr} satisfy the reciprocity relations $C_{rl}=C_{lr}$ [27], so that the generalized thermodynamic fluxes become

$$J_l = \frac{1}{2} \frac{\partial}{\partial X_l} \{ \sum_{i,j} C_{ij} X_i X_j \} \quad . \tag{4.5}$$

A generalization of Eq. (4.5) has been developed by Edelen [28]. He postulated the existence of a function $f=f(X_l,x,\lambda)$, such that $J_l=\partial f/\partial X_l$, where $\lambda=(\lambda_1,..,\lambda_k)\varepsilon\ \mathbb{R}^k$ is a set of exterior parameters (e.g., fixed concentrations in chemical reactions). For the fluxes it follows that $\partial J_r/\partial X_l = \partial J_l/\partial X_r$, i.e., a generalization of Onsager's reciprocity relations. Defining the functional $R = R[p,\lambda]$ by $R = \int_V f dx$, Edelen proved the inequality

$$0 \leq P \leq R[2p,\lambda] - R[p,\lambda], \qquad (p:=(p_1,\ldots,p_n))$$

under the assumption that R is a convex functional, with $P:=\int_V \sigma[s]dx$ being the total entropy production. Let us now deduce some further implications which follow from the existence of f:

(1) The general evolution criterion set up by Glansdorff and Prigogine [22] reads

$$d_X P/dt := \int_V \sum_l J_l \frac{\partial X_l}{\partial t} dx \leq 0. \tag{4.6}$$

Differentiating R with respect to time one obtains

$$dR/dt = \int_V \frac{\partial f}{\partial t} dx = d_X P/dt \quad , \tag{4.7}$$

hence, the evolution criterion takes the simple form $dR/dt \leq 0$. R has, therefore, a specific physical meaning: The temporal evolution of nonequilibrium systems is related to a decreasing R.

(2) Suppose that the matrix $s_{ij} := \partial^2 s/\partial\alpha_i\partial\alpha_j$ is negative definite. From Eq. (4.1) one obtains continuity equations for the intensive variables p_i:

$$\frac{\partial p_i}{\partial t} = \sum_{j=1}^{n} s_{ij} \left(\frac{\partial f}{\partial p_j} - \nabla \cdot \frac{\partial f}{\partial \nabla p_j}\right) \quad . \tag{4.8}$$

From Eq. (4.8) it follows that the stationary states of the system ($\partial p_i/\partial t=0$) are determined by the Euler-Lagrange equations associated to R (resp. f). We assume that the intensive state variables $p = (p_1,\ldots,p_n)$ are subjected to some nonhomogeneous boundary conditions $Bp=u$ on ∂V, where B is some operator (in case of Dirichlet's boundary conditions $B\equiv 1$, in the von Neumann case B is the normal derivative) and u is a given function on ∂V. Let, for given (p,λ), the linear second order differential operator $L[p,\lambda]$ be defined by

$$\begin{aligned}(L[p,\lambda]v)_i := \sum_{j=1}^{n} \Big\{ & \frac{\partial^2 f}{\partial p_i \partial p_j} v_j + \frac{\partial^2 f}{\partial p_i \partial \nabla p_j} \cdot \nabla v_j \\ & - \nabla \cdot \Big[\frac{\partial^2 f}{\partial \nabla p_i \partial p_j} v_j + \frac{\partial^2 f}{\partial \nabla p_i \partial \nabla p_j} \cdot \nabla p_j\Big]\Big\} \quad ,\end{aligned} \tag{4.9}$$

where the derivatives of f are taken at $(p(x),\nabla p(x),\lambda)$ and v satisfies the homogeneous boundary conditions Bv=0 on ∂V.

From Eq. (4.8) it is seen that the stationary states are obtainable from a variational principle, so that the methods described in section 2 can be applied. Let, for example, for some $\lambda=\lambda_0$, $p_0=p_0(x)=(p_{10}(x),\ldots,p_{n0}(x))$ be a solution of the Euler-Lagrange equations

$$\frac{\partial f}{\partial p_j} - \nabla \cdot \frac{\partial f}{\partial \nabla p_j} = 0 \quad , \quad 1\le j\le n \quad , \tag{4.10}$$

and suppose that the differential operator $L[p_0,\lambda_0]$ has an m-dimensional null space, spanned by $\{v_1(x),..,v_m(x)\}$. Then, by means of theorem 1, for given λ close to λ_0, all solutions p_λ of Eq. (4.10) are in a one-to-one correspondence to the stationary points y_μ of a function $g(y,\mu)(\mu:=\lambda-\lambda_0)$, so that, locally, the bifurcation properties of the solution of Eq. (4.10) are obtained precisely from those of g. Hence, applying catastrophe theory to g, the classification theorems can be used to classify nonequilibrium steady states phase transitions, if f exists.

(3) Let y_λ be a stationary point of g and let $p_\lambda(\lambda=\lambda_0+\mu)$ be the associated stationary solution of Eq. (4.10). Glansdorff and Prigogine [22] deduced from a macroscopic point of view a stability criterion, which has been extended by Schlögl [24] and Pfaffelhuber [29] using the concept of information gain: p_λ is an asymptotically stable stationary state if for small deviations ΔX_l, ΔJ_l, the relation

$$\int_V \sum_l \Delta X_l \Delta J_l > 0. \tag{4.11}$$

holds. If we consider small deviations $p_\lambda \to p_\lambda + v$ (v sufficiently small) it is, under appropriate boundary conditions, easily seen that the stability criterion (4.11) takes the form

$$\int_V \sum_{i=1}^{n} v_i(L[p_\lambda,\lambda]v)_i dx > 0. \tag{4.12}$$

Hence, p_λ is asymptotically stable, if all eigenvalues of $L[p_\lambda,\lambda]$ are positive, and unstable, if one eigenvalue is negative. Assuming that all "noncritical" eigenvalues of $L[p_\lambda,\lambda]$ are positive, it follows from the lemma of section 2 that only those stationary states p_λ are stable, for which the corresponding stationary point y_μ yields a local minimum. If y_μ yields a saddle point, the corresponding p_λ turns out to be unstable.

Fruitfull discussions with Prof. W. Güttinger, Dr. H. Eikemeier and Dr. E. Dilger are gratefully acknowledged.

References:

1. R. Thom, "Structural Stability and Morphogenesis", Benjamin, New York 1975.

2. V.I. Arnold, "Critical Points of Smooth Functions", in"Proceedings of the International Congress of Mathematicians", Vol. I, Vancouver 1972.

3. E.C. Zeeman, "Catastrophe Theory, Selected Papers 1972-1977", Advanced Book Program, Reading, Addison-Wesley 1977.
T. Poston and I. Stewart, "Catastrophe Theory and its Applications", Pitman 1978.

4. M.A. Krasnoselskii, "Topological Methods in the Theory of Nonlinear Integral Equations", Pergamon Press, Oxford 1964.

R.J. Magnus, "Applications of Topological Degree to the Theory of Branching", Math. Report 90, Battelle, Geneva 1974.

5. J.P. Keener and H.B. Keller, "Perturbed Bifurcation Theory", Arch. Rat. Mech. Anal. 50 (1973), 159.

J.P. Keener, "Perturbed Bifurcation Theory at Multiple Eigenvalues", Arch. Rat. Mech. Anal. 56 (1974), 348.

6. M. Vainberg and V. Trenogin, "Theory of Branching of Solutions of Nonlinear Equations", Noordhoff 1974.

7. S. Chow, J. Hale and J. Mallet-Paret, "Application of Generic Bifurcation", Arch. Rat. Mech. Anal. 59(1975), 159 and Arch. Rat. Mech. Anal. 62 (1976), 209.

J. Hale, "Generic Bifurcations with Applications", in "Nonlinear analysis and mechanics: Heriot-Watt Symposium Vol. I" (R.J. Knops, Ed.) Research Notes in Math. 17, Pitman 1977.

8. R.J. Magnus, "On Universal Unfoldings of Certain Real Functions on a Banach Space", Math. Rep. 100, Battelle, Geneva 1976.

- "Determinacy in a Certain Class of Germs on a Reflexive Banach Space", Math. Rep. 103, Battelle, Geneva 1976.

- "Universal Unfoldings in Banach Spaces: Reduction and Stability", Math. Rep. 103, Battelle, Geneva 1977.

9. J.P. Keener, "Buckling Imperfection Sensitivity of Columns and Spherical Caps", Quarterly of Appl. Math., July 1974.

10. S. Lang, "Differential Manifolds", Addison-Wesley 1972.

11. D. Gromoll and W. Meyer, "On Differentiable Functions with Isolated Critical Points", Topology 8 (1969), 361.

12. G. Dangelmayr, Thesis, University of Tübingen 1978.

13. R.J. Magnus and T. Poston, "Infinite Dimensions and the Fold Catastrophe", this volume.

14. T. Poston and I.N. Stewart, "Taylor Expansions and Catastrophes", Research Notes in Math. 7, Pitman 1976.

15. I. Segal, "Nonlinear Semigroups", Ann. Math. 78 (1963), 339.

P.R. Chernoff and J.E. Marsden, "Properties of Infinite Dimensional Hamiltonian Systems", Lecture Notes in Math. 425, Springer 1974.

16. L. Landau und E. Lifschitz, "Theoretische Physik" Bd. VII (Elastizitätstheorie), Akademie Verlag Berlin 1970.

17. J. Thompson and G. Hunt, "A General Theory of Elastic Stability", Wiley-Interscience 1973.

18. L. Bauer, H. Keller and E.L. Reiss, "Multiple Eigenvalues lead to Secondary Bifurcation", SIAM Rev. 17 (1975), 101.

19. E.C. Zeeman, "The Umbilic Bracelett and the Double-Cusp Catastrophe", in "Structural Stability, the Theory of Catastrophes and Applications in the Sciences", (P. Hilton, Ed.) Lecture Notes in Math. 525, Springer 1976.

20. J. Callahan, "Special bifurcations of the double cusp", University of Warwick, Preprint, 1978.

21. R. Magnus and T. Poston, "On the Full Unfolding of the Von Karman Equations at a Double Eigenvalue", Math. Rep. 109, Battelle, Geneva 1977.

22. P. Glansdorff and I. Prigogine, "Thermodynamic Theory of Structures, Stability and Fluctuations", Wiley-Interscience 1971.

23. G. Nicolis, "Dissipative Structures with Applications to Chemical Reactions", in "Cooperative Effects, Progress in Synergetics", (H. Haken, Ed.), North-Holland 1974.

24. F. Schlögl, "Fluctuations in Thermodynamic Non Equilibrium States", Z. Physik 249 (1971), 1.

25. H.K. Janssen, "Stability of Transport", Z. Physik 253 (1972), 176.

26. I. Prigogine, "Introduction to Thermodynamics of Irreversible Processes", Wiley Interscience 1967.

27. S.R. DeGroot and P. Mazur, "Non-Equilibrium Thermodynamics", North-Holland 1962.

28. D. Edelen, "A Nonlinear Onsager Theory of Irreversibility", Int. J. Engng. Sci. 10 (1972), 481.

29. E. Pfaffelhuber, "Information-Theoretic Stability and Evolution Criteria in Irreversible Thermodynamics", J. Stat. Phys. 16 (1977), 69.

Semiclassical Path Integrals in Terms of Catastrophes

G. Dangelmayr, W. Güttinger, and W. Veit

Institute for Information Sciences, University of Tübingen
D-7400 Tübingen, Fed. Rep. of Germany

Summary

Using a splitting procedure for action functionals, a semiclassical approximation of path integrals, remaining finite on caustics, is constructed. It is governed by oscillatory generalized Airy integrals involving catastrophe polynomials, whose bifurcation properties reflect those of real and complex classical paths.

1. Introduction

During the past years semiclassical approximations have been extensively developed and applied to various quantum phenomena [1], [2], [3], [4], starting from Feynman's path integral [5] for the quantum mechanical propagator K connecting P_1 with P_2,

$$K(P_2,P_1) = \int D_{12}(q)\exp\{iS_{12}(q)/\hbar\}. \tag{1.1}$$

In Eq. (1.1), $P_1=(q_1,t_1)$ and $P_2=(q_2,t_2)$ represent two "space-time points" with times $t_1<t_2$ and generalized coordinates $q_i=(q_{i1},\dots,q_{in})^T$ $(i=1,2)$ of a system of particles with n degrees of freedom. $D_{12}q$ denotes the path differential and $S_{12}(q)$ the action functional, defined on paths $q=q(t)(t_1\le t\le t_2)$ connecting P_1 with P_2 (indicated by the subscripts 1,2 in S_{12}). The path integral approach to quantum mechanics has the advantage that it gives great insight into the role played by classical paths if Planck's constant $\hbar$ is small compared to the classical action, i.e., in the semiclassical limit $\hbar\to 0$ [6]. In this limit, the main contribution to the path integral comes from a neighborhood of the classical paths $q_c=q_c(t;P_1,P_2)$ connecting P_1 with P_2, i.e., from solutions of the Euler-Lagrange equations derived from the variational principle $\delta S_{12}=0$. In conventional semiclassical approximations to K the action functional is approximated by a quadratic form,

$$S_{12}(q) \simeq S(P_1,P_2) + \frac{1}{2}\delta^2 S_{12}^c(\tilde{q}) \tag{1.2}$$

where $\tilde{q} := q-q_c$. Here, $S(P_1,P_2) := S_{12}(q_c)$ is the action along q_c, i.e., Hamilton's principal function, and $\delta^2 S_{12}^c$ the second variation of S_{12} along q_c, a quadratic form defined on all paths $\tilde{q}=\tilde{q}(t)$ with $\tilde{q}(t_1)=\tilde{q}(t_2)=0$. Upon inserting the approximation (1.2) into the expression (1.1), the following approximation formula for K is obtained [1], [7], [8]:

$$K(P_2,P_1) \simeq \Sigma(2\pi i\hbar)^{-n/2}|\det \partial^2 S/\partial q_1\partial q_2|^{1/2} \exp\{iS/\hbar + \text{phases}\} \tag{1.3}$$

summing over all distinct classical paths connecting P_1 with P_2. The procedure leading to formula (1.3) can be interpreted as a generalization of the method of stationary phase to infinite dimensions, because (1.3) is based on the quadratic approximation (1.2) of the action functional around the classical paths, which take over the role of the stationary points in the stationary phase method in finite dimensions. De Witt-Morette [7] has shown that the r.h.s. of (1.3) is just the leading term of an expansion of K in powers of $\hbar$, what confirms the validity of (1.3) in the semiclassical limit.

However, the approximation (1.3) suffers from the following defects: (1) The existence of at least one classical path is required: In classically forbidden regions the path integral cannot be approximated by the expression (1.3). To overcome this difficulty, complex classical paths have been introduced by Miller [2] and Koeling and Malfliet [3], which give rise to a complex classical action whose imaginary part appears as an attenuation factor for the wave amplitude in the classically forbidden region. This procedure can be understood as a generalization of the saddle point method to infinite dimensions. (2) The approximation (1.3) breaks down on the caustics, since the Hessian $\det\partial^2 S/\partial q_1 \partial q_2$ in Eq. (1.3) diverges if $(P_1, P_2) = (P_1^*, P_2^*)$ are conjucate points with respect to the classical path (a definition of conjugate points is given in Section 2). This is in striking contrast to experiment which yields finite intensities on caustics. The infinity arises from the fact that, in general, there are several different classical paths, which coincide on the caustics. This implies that it is no longer possible to treat the classical paths separately, because their neighborhoods, which essentially contribute to the path integral, are overlapping. It is, indeed, a complicated interference of the classical paths which causes the observed oscillatory intensity diffraction patterns near the caustics (cf. Berry [9]).

The objective of this paper is to present a general semiclassical approximation formula for K which, in accordance with experiment, remains finite on and near conjugate points, i.e., on caustics. This approximative path integral turns out to be proportional to an oscillatory "generalized Airy integral" which is governed by catastrophe polynomials in the sense of Thom [10] and whose bifurcation properties correspond to those of the classical paths.

The basis for our approximation formula is developed in Section 2. We start from fixed space-time points P_1^*, P_2^*, which are assumed to be conjugate points with respect to a classical path $q_0(t)$. The original action functional S_{12} is transformed into a modified functional S^* depending on functions $u=u(s)$ $(0 \le s \le 1)$ of a fixed function space with the space-time points playing the role of parameters. We present a splitting Lemma for S^* based on Fourier expansions of the functions u. This enables us to determine an approximation to S^*, which contains terms of order higher than two in a finite number of "essential" Fourier coefficients in an appropriate catastrophe polynomial. The dependance on the remaining infinite number of "non-essential" Fourier coefficients is approximated quadratically. This approximate S^* is used in Section 3 to derive a semiclassical approximation of the path integral, valid and finite in a whole neighborhood of the conjugate points P_1^*, P_2^*. In order to achieve this, an integration technique is used which replaces the "time-lattice-integration" [11] by an integration over Fourier coefficients. We shall further show that our approximation formula reduces to the conventional one given by Eq. (1.3) for space-time points away from the caustic. Additional terms arise from complex paths. Both, the real and the complex classical paths are uniquely determined by the stationary points of the catastrophe polynomial. Finally, in Section 4, we discuss some examples.

2. Splitting of the Action Functional

We start from a Lagrangean of the form

$$\mathcal{L}(q,\dot{q},t) = \frac{1}{2}\dot{q}^T M(t)\dot{q} + \dot{q}^T A(q,t) - V(q,t) \quad (\dot{\ } = d/dt), \tag{2.1}$$

where M(t) is a symmetric and positive "mass-matrix", V(q,t) is the potential and A(q,t) is a vector, taking, for example, magnetic effects into account. The classical paths are determined by the Euler-Lagrange equations

$$\frac{d}{dt}(\partial\mathcal{L}/\partial\dot{q}) - \partial\mathcal{L}/\partial q = 0, \tag{2.2}$$

forming a system of second-order differential equations. If a triple $(t_1, q_1, \dot{q}_1)$ of initial conditions is prescribed, there is a unique classical path $q(t;t_1,q_1,\dot{q}_1)$ which starts from $P_1=(q_1,t_1)$ with initial velocity $\dot{q}_1$.

In the semiclassical approximation of the path integral, however, we need classical paths $q_c(t;P_1,P_2)$ connecting two given space time points P_1,P_2. If $\det\partial q(t_2;t_1,q_1,\dot{q}_1)/\partial\dot{q}_1 \neq 0$, the equation

$$q(t_2;t_1,q_1,\dot{q}_1) = q_2 \tag{2.3}$$

is uniquely solvable for $\dot{q}_1=\dot{q}_1(P_1,P_2)$ and we obtain a unique classical path $q_c(t;P_1,P_2) := q(t;t_1,q_1, \dot{q}_1(P_1,P_2))$ connecting P_1 with P_2. Let now $P_i^* = (q_i^*,t_i^*)$ ($i=1,2$, $t_1^*<t_2^*$) be two fixed space-time points, for which $q_0(t_2^*)=q_2^*$, where $q_0(t)=q(t;t_1^*,q_1^*,\dot{q}_1^*)$ with some initial velocity $\dot{q}_1^*$, and suppose that the matrix $Q(t_2^*)$,

$$Q(t) := \partial q(t;t_1,q_1^*,\dot{q}_1^*)/\partial\dot{q}_1$$

has an m-dimensional null-space ($1\leq m\leq n$). Then equation (2.3) is no longer uniquely solvable for $\dot{q}_1^*$ with $(P_1,P_2,\dot{q}_1)$ in a neighborhood of $(P_1^*,P_2^*,\dot{q}_1^*)$. To a given pair (P_1,P_2) close to (P_1^*,P_2^*) there exist, in general, several branches of classical paths connecting P_1 with P_2, which become all identical with $q_0(t)$ for $(P_1,P_2) \to(P_1^*,P_2^*)$. P_1^*, P_2^* are called conjugate (focal) points of multiplicity m with respect to $q_0=q_0(t)$. This statement is clearly equivalent to the fact that the linearized Euler-Lagrange equation around $q_0(t)$ (the Jacobi equation) has precisely m linearly independent solutions $\tilde{q}_j(t)$ ($1\leq j\leq m$) satisfying $\tilde{q}_j(t_1^*)=\tilde{q}_j(t_2^*)=0$, which are given by $\tilde{q}_j(t)=Q(t)b_j$ where $\{b_1,\dots,b_m\}$ is a basis of the null space of $Q(t_2^*)$. The caustic $C(P_1)$ relative to some initial point P_1 is defined as the envelope of all classical paths starting from P_1, i.e., by

$$C(P_1) := \{P_2=(q_2,t_2)\colon q(t_2;t_1,q_1,\dot{q}_1)= q_2,\ \det\partial q(t_2;t_1,q_1,\dot{q}_1)/\partial\dot{q}_1 = 0\}.$$

Hence, it follows that P_2^* is just the point on the caustic relative to P_1^*, where $q_0(t)$ contacts the caustic. (We consider here only a special type of caustic. In general caustics and conjugate points are defined in terms of the evolution of some initial Lagrangean manifold in phase space under the classical flow [12]. But it is precisely on the caustic defined above where the conventional semiclassical propagator (1.3) diverges.) In Fig. 1 three caustics are shown, each of them representing a section of the bifurcation set of the swallowtail catastrophe.

Since there are several classical paths connecting given space-time points P_1, P_2 in a neighborhood of (P_1^*,P_2^*), Hamilton's principal function S is a multivalued function in (P_1,P_2)-space, each branch being the action along a classical path. On the other hand, the classical paths are just those in the space of all paths which make the first variation of S_{12} vanishing. The multivalued principal function can, therefore, be regarded as the action functional, calculated at its multiple "stationary points". Hence, by applying the methods of catastrophe theory in infinite-dimensional spaces [13, 14] to the action functional, one might expect that only a finite number of "essential variables" in infinite-dimensional function space suf-

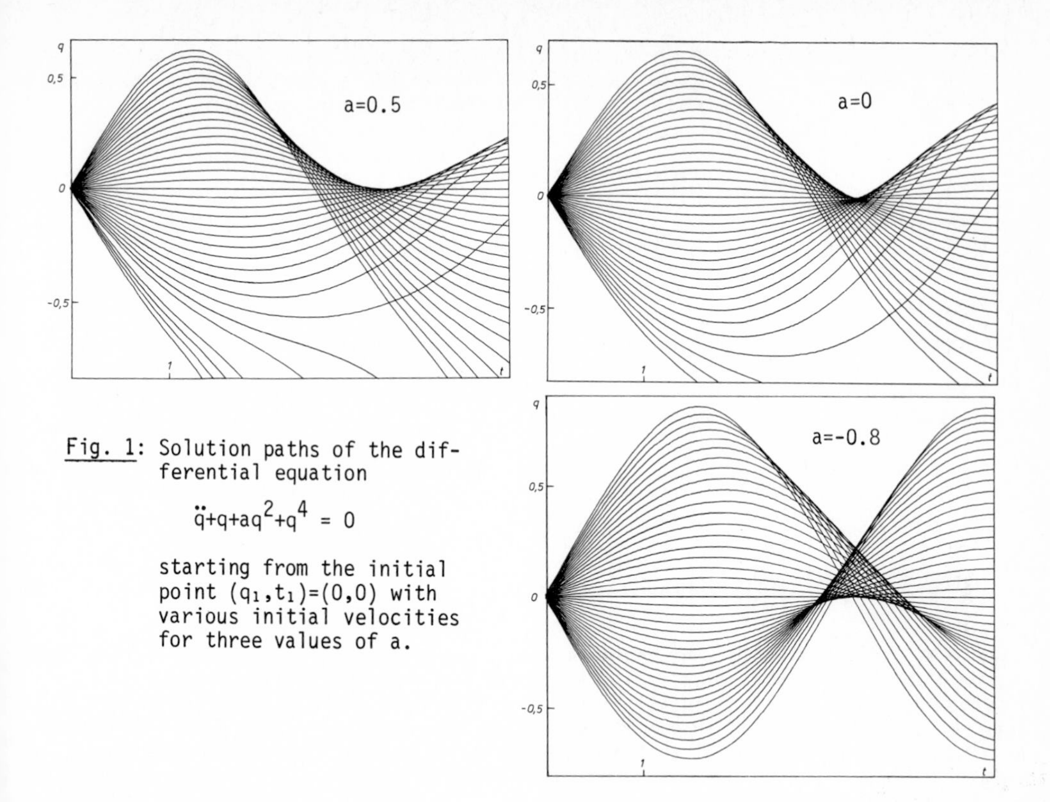

Fig. 1: Solution paths of the differential equation

$$\ddot{q}+q+aq^2+q^4 = 0$$

starting from the initial point $(q_1,t_1)=(0,0)$ with various initial velocities for three values of a.

fices to characterize the multiple "stationary points". However, to apply these methods, we have to transform the original action functional into a modified one, which is defined on a fixed function space with the space-time points playing the role of parameters. This can be achieved in the following way:

Let $\hat{q}(t) \equiv \hat{q}(t;P_1,P_2)$ be an arbitrary path satisfying $\hat{q}(t_i) = q_i$ $(i=1,2)$ and $\hat{q}(t;P_1^*,P_2^*)=q_0(t)$. For example,

$$\hat{q}(t;P_1,P_2) = q_0(t) + \Delta q_1 + \tilde{M}(t;t_1,t_2)(\Delta q_2-\Delta q_1), \tag{2.4}$$

where $\Delta q_i := q_i-q_0(t_i)$, and

$$\tilde{M}(t;t_1,t_2) := \left(\int_{t_1}^{t} M^{-1}(\tau)d\tau\right)\left(\int_{t_1}^{t_2} M^{-1}(\tau)d\tau\right)^{-1}, \tag{2.5}$$

Let $W_2^{1,n}((t_1,t_2))$ be the Sobolev space of all continuous functions $q=q(t) =(q_1(t),\dots,q_n(t))^T$, such that the generalized derivatives $\dot{q}(t)$ are square integrable and denote by $Z(P_1,P_2)$ $(P_i:=(q_i,t_i))$ the manifold of all $q\in W_2^{1,n}((t_1,t_2))$ with $q(t_i)=q_i$ $(i=1,2)$. Define Z by

$$Z := \bigcup_{\{(P_1,P_2):t_1<t_2\}} Z(P_1,P_2).$$

Let furthermore H be the Sobolev space of all continuous functions $u=u(s)= (u_1(s),\dots,u_n(s))^T$ $(0\leq s\leq 1)$ with square integrable derivatives u' $(':=d/ds)$ and

$u(0)=u(1)=0$. We denote by Y the set of all $\lambda=(\tilde{P}_1,\tilde{P}_2)\varepsilon R^{2n+2}$ with $\tilde{P}_i=(\tilde{q}_i,\tilde{t}_i)\varepsilon R^{n+1}$ and $\tilde{t}_1<\Delta t^*+\tilde{t}_2$, where $\Delta t^*:=t_2^*-t_1^*$. An invertible map $\gamma:Z\mapsto Y\times H$ is constructed in the following way: For $q\varepsilon Z(P_1,P_2)$ we set $\lambda:=(P_1-P_1^*,P_2-P_2^*)\varepsilon Y$ and define $u\varepsilon H$ by

$$u(s) := q(t_1+s\Delta t) - \hat{q}(t_1+s\Delta t;P_1,P_2) \ , \quad \Delta t:= t_2-t_1 . \qquad (2.6)$$

If we not set

$$S^*(u,\lambda) := S_{12}(\gamma^{-1}(\lambda,u))\Delta t \ , \qquad (2.7)$$

we arrive at a new action functional S^*, which is defined on the fixed Hilbertspace H with the space-time points P_1,P_2 (collected in λ) playing the role of parameters. It is convenient to split S^* into $S^*(u,\lambda)=S_0^*(\lambda)+W(u,\lambda)$, where $S_0^*(\lambda):=S^*(0,\lambda)$. W can be expressed in terms of a new Lagrangean $\mathcal{L}_1$,

$$W(u,\lambda) = \int_0^1 \mathcal{L}_1(u,u',s,\lambda)ds \ , \qquad (2.8)$$

with

$$\mathcal{L}_1 = \frac{1}{2}u'^T M_1(\sigma,s)u'+u'^T A_1(u,\lambda,s)-V_1(u,\lambda,s), \ (\sigma:=(t_1-t_1^*,t_2-t_2^*)) \qquad (2.9)$$

where $M_1(\sigma,s):=M(t_1+s\Delta t)$. The vector A_1 and the new potential V_1 are easily obtained from $\hat{q}$ and M,A,V. Every classical path connecting P_1 with P_2 is, clearly, uniquely determined by a solution $u=u_\lambda(s)$ of the Euler-Lagrange equations corresponding to $\mathcal{L}_1$ for $\lambda=(P_1-P_1^*, P_2-P_2^*)$. In particular u=0 is, for $\lambda=0$, the solution determining $q_0(t)$. It can be shown [15] that W can be regarded as a smooth function on some neighborhood of (0,0) in $H\times R^{2n+2}$, provided that M,A,V are smooth in a domain of R^{n+1} containing the set $\{(q_0(t),t): t_1^*\leq t\leq t_2^*\}$. The second derivative of W with respect to H at $(u,\lambda)=(0,0)$ can be written in the form

$$D_H^2 W(0,0)u\tilde{u} = (Lu,\tilde{u}), \ (u,\tilde{u}\varepsilon H) \qquad (2.10)$$

where $(\cdot,\cdot)$ is the standard inner product in $H_0:=L_2((0,1))^n$ and L is the differential operator

$$L :=-\frac{d}{ds}(M_0(s)\frac{d}{ds}+A_0(s))+A_0^T(s)\frac{d}{ds} - V_0(s) \ , \qquad (2.11)$$

with the matrices in Eq. (2.11) being defined by

$$M_0(s):=M_1(0,s), \quad A_0(s):=\partial A_1(0,0,s)/\partial u, \quad V_0(s):=\partial^2 V_1(0,0,s)/\partial u\partial u.$$

Let μ_j ($j\varepsilon N$) be the eigenvalues of the Sturm-Liouville eigenvalue problem $Lv=\mu v$ with boundary conditions $v(0)=v(1)=0$, each eigenvalue counted according to its multiplicity, and let $\{v_j=v_j(s)\}$ be the corresponding eigenfunctions which can be supposed to form an orthonormal basis in H_0. Since P_1^*, P_2^* have been assumed to be conjugate points of multiplicity m with respect to $q_0(t)$, it follows that precisely m eigenvalues vanish. We suppose, therefore, that $\mu_1=\ldots\mu_m=0$ and $\mu_j\neq 0$ for j>m. Let us expand $u\varepsilon H$ in terms of the v_j,

$$u = \sum_{i=1}^{m} y_i v_i + \sum_{j=m+1}^{\infty} a_j v_j \ , \qquad (2.12)$$

where $y_i=(u,v_i)$ $(1\leq i\leq m)$ and $a_j=(u,v_j)$ $(j>m)$. Then, with the notations $y=(y_1,..,y_m)$, $a=\{a_j : j>m\}$, S^* and W become functions of (y,a,λ). By means of Theorem 2 in [13]it is easy to prove [15] the following splitting Lemma for W:

Lemma: There exist functions $h(y,\lambda)=\{h_j(y,\lambda): j>m\}$ with $h_j(0,0)=\partial h_j(0,0)/\partial y_i=0$ and a symmetric matrix $\{B_{ij}(y,a,\lambda): i,j>m\}$ satisfying $B_{ij}(0,0,0)=\mu_i\delta_{ij}$, and, furthermore, a function $g(y,\lambda)$ such that by substituting a+h for a in W the following formula holds:

$$W(y,a+h(y,\lambda),\lambda) = \frac{1}{2} \sum_{i,j>m} B_{ij}(y,a,\lambda)a_i a_j+g(y,\lambda) \tag{2.13}$$

The h_j,B_{ij} and g, defined in a neighborhood of $(y,a,\lambda)=(0,0,0)$, are smooth and g is completely degenerate, i.e.,

$$g(0,0) = \partial g(0,0)/\partial y_i = \partial^2 g(0,0)/\partial y_i\partial y_j=0 \quad (1\leq i,j\leq m).$$

In the Appendix a qualitative deduction of Eq. (2.13) is given. A rigorous proof in terms of properties of the Hilbertspace H can be found in [15] and a construction prescription for g and h is contained in the proof of Theorem 1 in [13]. From Eqs. (2.7) and (2.12) it follows that, to a given $\lambda=(P_1-P_1^*, P_2-P_2^*)$, every solution $q_\lambda(t)$ of the Euler-Lagrange equation with $q_\lambda(t_i)=q_i$ $(i=1,2)$ is uniquely determined by a solution $y_\lambda = (y_{1\lambda},\ldots,y_{m\lambda})$ of the gradient system

$$\nabla_y g(y,\lambda) = 0, \tag{2.14}$$

thus yielding the classical path $q_\lambda(t)$ in terms of y_λ:

$$q_\lambda(t) = \hat{q}(t;P_1,P_2) + \sum_{i=1}^{m} y_{i\lambda}v_i((t-t_1)/\Delta t) + \sum_{j>m} h_j(y_\lambda,\lambda)v_j((t-t_1)/\Delta t). \tag{2.15}$$

Fixing the initial point $P_1=\overline{P}_1$ and varying the endpoint P_2, the caustic relative to $\overline{P}_1$ is locally given by the manifold

$$C(\overline{P}_1) = \{P_2: \nabla_y g(y,\lambda) = 0,\ \det g^{(2)}(y,\lambda) = 0, \lambda =(\overline{P}_1-P_1^*,P_2-P_2^*)\} ,$$

where $g^{(2)}=g^{(2)}(y,\lambda)$ denotes the Jacobi-matrix of g with respect to y. The caustic can, therefore, be regarded as a catastrophe or bifurcation manifold in the sense of Thom [10]. A similar result has been obtained by DeWitt-Morette [17], who showed by means of the Hamilton Jacobi equation that caustics coincide with catastrophes without, however, proving the existence of the unfolding $g(y,\lambda)$ of the singularity $g_0(y) := g(y,0)$. Let us now assume that $g_0(y)$ is finitely determined, i.e., that $k:=\operatorname{codim} g_0<\infty$. This implies that the Euler-Lagrange equations are nonlinear, because in the linear case it follows that $g_0(y)=0$ [15]. It is a familiar result of catastrophe theory [10,15] that g can be expressed in terms of some appropriate catastrophe polynomial (a normal form) $f=f(x,d)$

$$f(x,d) = f_0(x) + \sum_{i=1}^{k} d_i f_i(x),\ (d=(d_1,\ldots,d_k)\varepsilon R^k) \quad , \tag{2.16}$$

where f_0 and the f_j are polynomials in x with $\partial^2 f_0(0)/\partial x_i\partial x_j =0$. For g one obtains

$$g(y,\lambda) = e(\lambda) + f(x,d(\lambda)). \tag{2.17}$$

Here, $x=x(y,\lambda)$ with $\det\partial x/\partial y\neq 0$, i.e., the map $(y,\lambda)\mapsto(x(y,\lambda),\lambda)$ is a diffeomorphism. The unfolding parameters d are functions of λ satisfying $d(0)=0$ and e is a translation,

$$e(\lambda) = g(0,\lambda) - f(x(0,\lambda),d(\lambda)).$$

For small (y,λ) the matrix $C_{ij}(y,\lambda) := B_{ij}(y,0,\lambda)$ is regular. Therefore, in virtue of Eq. (2.13), the approximation $B_{ij}(y,a,\lambda)\approx C_{ij}(y,\lambda)$ implies the following approximation for $S^*(y,a,\lambda)$:

$$S^*(y,a,\lambda) \approx S^*_0(\lambda)+e(\lambda)+\frac{1}{2}\sum_{i,j>m} C_{ij}(y,\lambda)(a_i-h_i(y,\lambda))(a_j-h_j(y,\lambda))+g(y,\lambda). \quad (2.18)$$

The essential difference between Eq. (2.18) and the conventional approximation formula (1.2) lies in the fact that in (2.18) we have accounted for terms of order higher than two with respect to the "singular" Fourier coefficients $y_1,\ldots,y_m$, whereas the remaining "regular" Fourier coefficients $a_{m+1},a_{m+2},\ldots$ have been taken into account only up to second order. This is natural because

$$D^2_H W(0,0,0)(y,a)^2 = \sum_{j>m}\mu_j a_j{}^2$$

is a regular quadratic form in the space $\{y=0,a\}$ and vanishes for $\{y,a=0\}$. In the next section we shall see that Eq. (2.18) yields a finite expression for the propagator in contrast with the infinity produced by the approximation (1.3) based on (1.2).

3. Semiclassical Approximation of Path Integrals in a Neighborhood of Conjugate Points

In what follows we derive a finite semiclassical approximation formula for the path integral in a neighborhood of the conjugate points P^*_1, P^*_2. In [15] an integration technique for the path integral is developed in which the time lattice integration [11] is replaced by integration over Fourier coefficients of functions $u=u(s)$ $(0\leq s\leq 1)$ satisfying $u(0)=u(1)=0$, with respect to a complete orthonormal system. Using the modified action functional S^* and the orthonormal system $\{v_j\}$ introduced in Section 2, this technique leads to the following expression for K:

$$K(P_2,P_1) = \lim_{N\to\infty} K_N(P_2,P_1) \quad (3.1)$$

where

$$K_N := J^*_N(\sigma)\int dy_1\ldots dy_m\int da_{m+1}\ldots da_{nN}\exp\{iS^*(y,a_{(nN)},\lambda)/\hbar\Delta t\} \quad (3.2)$$

with $\sigma:=(t_1-t^*_1,\ t_2-t^*_2)$, and

$$J^*_N(\sigma) := \{\det(2\pi i\hbar\int_{t_1}^{t_2} M^{-1}(t)dt)\}^{-1/2}\prod_{j=1}^{nN}(\mu_{j0}(\sigma)/2\pi i\hbar\Delta t)^{1/2} \quad . \quad (3.3)$$

In Eq. (3.2) $a_{(nN)}$ stands for $a=\{a_j:j>m\}$ with $a_j=0$ for $j>nN$. The $\mu_{j0}(\sigma)$ in Eq. (3.3) are the eigenvalues of the eigenvalue problem $L_1(\sigma)v=\mu v$ with $v(0)=v(1)=0$, where L_1 is the differential operator

$$L_1(\sigma) := -\frac{d}{ds} M_1(\sigma,s) \frac{d}{ds}$$

(cf. Eq. (2.8)).

In the semiclassical limit the main contribution to the path integral comes from a neighborhood of the classical paths. Hence we can replace the infinite integration domain in Eq. (3.2) over the y_i by one extending only over a neighborhood of $y=0$. Then the approximation (2.16) can be used which, when inserted into Eq. (3.2) yields after a transformation $y \leftrightarrow x$ the following expression for K:

$$K(P_2,P_1) \simeq K_1(\lambda)\Psi(\lambda) \ , \tag{3.4}$$

where

$$K_1(\lambda) = (2\pi i\hbar\Delta t)^{-(n+m)/2}\exp\{i(S_0^*(\lambda)+e(\lambda))/\hbar\Delta t - m'\pi i/2\} . \tag{3.5}$$

Here, Ψ is an oscillatory integral,

$$\Psi(\lambda) = \int_\Omega dx_1 \ldots dx_m H(x,\lambda)\exp\{if(x,d(\lambda))/\hbar\Delta t\}, \tag{3.6}$$

where the integration goes over some neighborhood Ω of $x=0$, which contains all stationary points of f. In Eq. (3.5) m' is the number of negative eigenvalues μ_j, i.e., the number of conjugate points, crossed by $q_0(t)$ for $t_1^* \le t < t_2^*$. The function $H(x,\lambda)$ in Eq. (3.6) is defined in the following way: Let $\kappa_j(y,\lambda)$ $(j>m)$ be the eigenvalues of the matrix $\{C_{ij}(y,\lambda)\}$ and define the function $F(y,\lambda)$ by

$$F(y,\lambda) = \left|\left(\prod_{j=1}^{m} \mu_{jo}(\sigma)\right)\left(\int_0^1 M_1(\sigma,s)^{-1}ds\right)^{-1}\left(\lim_{N\to\infty} \prod_{j=m+1}^{nN} \mu_{jo}(\sigma)/\kappa_j(y,\lambda)\right)\right|^{1/2}$$

Then $H(x,\lambda)$ is given by

$$H(x,\lambda) = |D(x,\lambda)|F(y(x,\lambda),\lambda) \quad ,$$

where $(x,\lambda)\mapsto(y(x,\lambda),\lambda)$ is the inverse mapping of $(y,\lambda)\mapsto(x(y,\lambda),\lambda)$ and $D(x,\lambda) := \det(\partial y(x,\lambda)/\partial x)$. In [15] an explicit expression for F has been obtained: Let $L_1(y,\lambda)$ be the differential operator

$$L_1(y,\lambda) = -\frac{d}{ds}\left(M_1\frac{d}{ds} + \tilde{A}_1\right) + \tilde{A}_1^T - \tilde{V}_1 \tag{3.7}$$

with

$$u = \sum_{i=1}^{m} y_i v_i(s) + \sum_{j>m} h_j(y,\lambda)v_j(s)$$

being inserted into the matrices $\tilde{A}_1=\partial A_1/\partial u$, $\tilde{V}_1=\partial^2 V_1/\partial u\partial u$. Let $U(y,\lambda;s)$ be the solution matrix of the matrix differential equation $L_1U=0$ with initial conditions $U(y,\lambda;0)=0$ and $M_1(\sigma,0)U'(y,\lambda,0)=I$, I being the (nxn)-unit matrix, and let $g^{(2)}(y,\lambda)$ be the Jacobi matrix of g with respect to y. F is then given by

$$F(y,\lambda) = |\det g^{(2)}(y,\lambda)/\det U(y,\lambda;1)|^{1/2} \tag{3.8}$$

It can be shown that F is smooth [15] . Since $\det U(y,\lambda;1) = 0$ iff $\det g^{(2)}(y,\lambda)=0$

[15], the zeros of the nominator on the r.h.s. of Eq. (3.8) are compensated by those of the denominator. For F(0,0) an expression has been found [15],[16] in terms of determinants of matrices related to the classical path $q_o(t)$.

A uniform approximation to Ψ, Eq. (3.6), can be obtained by a method developed by Duistermaat [18] and Berry [9]: From the theory of unfoldings of singularities it follows [18] that H can be written in the form

$$H(x,\lambda) = H_o(\lambda) + \sum_{j=1}^{k} H_j(\lambda)\partial f(x,d(\lambda))/\partial d_j + \sum_{i=1}^{m} \tilde{H}_i(x,\lambda)\partial f(x,d(\lambda))/\partial x_i \tag{3.9}$$

with some functions $H_o, H_j, \tilde{H}_j$, k beeing the codimension of f_o and g_o. The H_o, H_j can be calculated by inserting the k+1 stationary points of f into Eq. (3.9), so that k+1 equations are obtained determining the k+1 functions H_o, H_j [9]. At the stationary points of f the last term in Eq. (3.9) vanishes and it can therefore be neglected when integrating [9], so that we can write

$$H(x,\lambda) \simeq H_o(\lambda) + \sum_{j=1}^{k} H_j(\lambda)\partial f(x,d(\lambda))/\partial d_j = H_o(\lambda) + \sum_{j=1}^{k} H_j(\lambda) f_j(x) \quad , \tag{3.10}$$

cf. Eq. (2.16). Inserting the approximation (3.10) into Eq. (3.6) yields

$$\Psi(\lambda) \simeq H_o(\lambda)\psi(d(\lambda)) - i\hbar\Delta t \sum_{j=1}^{k} H_j(\lambda)\partial\psi(d(\lambda))/\partial d_j \quad , \tag{3.11}$$

where ψ is what we call a generalized Airy integral

$$\psi(d) := \int_{-\infty}^{\infty} dx_1 \ldots dx_m \exp\{if(x,d)/\hbar\Delta t\} \quad , \tag{3.12}$$

where the domain Ω has been replaced by an infinite integration domain what is possible because the main contribution to the integral comes from a neighborhood of the stationary points of f. Functions of the type (3.11) with (3.12) are familiar from optics, when wave functions are approximated by means of the Eikonal method in the neighborhood of a caustic [9], [19] and it is not surprising that integrals of this type occur also in quantum mechanics when a propagator is approximated semiclassically in the neighborhood of conjugate points. With the approximation (3.12) we obtain for the approximate propagator the final result

$$K(P_2,P_1) \simeq K_1(\lambda)\{H_o(\lambda)\psi(\lambda) - i\hbar\Delta t \sum_{j=1}^{k} H_j(\lambda)\partial\psi(d(\lambda))/\partial d_j\} \quad . \tag{3.13}$$

Since we have assumed that g_o is finitely determined (what generically is the case) ψ is finite in a whole neighborhood of $\lambda=0$ [20]. Consequently, the approximation formula (3.13) does not diverge on the caustics. This is the essential and physically significant difference between (3.13) and conventional formulae based on (1.2).

An approximation formula similar to Eq. (3.13) has been developed by Levit and Smilanky [21], based on an integration technique in phase space rather than in configuration space. However, these authors did not present a direct method for computing the function g resp. f by means of a splitting lemma: The catastrophe polynomial f has been introduced by means of the postulate that the number of stationary points of f and the number of classical paths coincide. Furthermore, their formula is only valid for the fixed initial and final times t_1^*, t_2^*, because they did not use a modified action functional, which is defined on a fixed function space for all times t_1, t_2.

The bifurcation set Σ,

$$\Sigma := \{\lambda : \nabla_x f(x,d(\lambda)) = 0, \det f^{(2)}(x,d(\lambda))=0\} , \qquad (3.14)$$

where $f^{(2)}$ is the Jacobi-matrix of f with respect to x, defines boundaries separating various regions in λ-space, each region giving rise to several branches of stationary points x_λ of f, determined by the equation

$$\nabla_x f(x,d(\lambda)) = 0 . \qquad (3.15)$$

At least two different branches, existing in some region, coincide on its boundary. We show now that the uniform approximation (3.13) reduces to familiar approximations for λ far away from Σ: The maximal number of possible stationary points of f is k+1. Suppose that λ is in a region where r branches $x=x_j(\lambda)$ $(1\le j\le r)$ of stationary points of f exist $(0\le r\le k+1)$. By inserting $y_\lambda=y(x_j(\lambda),\lambda)$ into Eq. (2.15), each branch determines uniquely a classical path $q_j(t;\lambda)$ connecting P_1 with P_2. On the other hand, Eq. (3.15) has, by taking **x**=z complex and continuing f analytically, k+1-r complex solutions $z=z_l(\lambda)$ $(1\le l\le k+1-r)$ with nonvanishing imaginary parts. By inserting $y(z_l(\lambda),\lambda)=y_\lambda$ into Eq.(2.13) we obtain k+1- r complex solutions $q_{Cl}(t;\lambda)$ of the Euler-Lagrange equations with the imaginary part of $q_{Cl}(t_i;\lambda)$ (i=1,2) vanishing, so that $q_{Cl}(t;\lambda)$ can be regarded as a "complex classical path connecting P_1 with P_2". We assume that the x_j and z_l are well separated so that the saddle point method [22] can be applied to approximate ψ. The resulting approximation for ψ consists of two terms ψ_1 and ψ_2 , where ψ_1 is determined by the (real) stationary points $x_j(\lambda)$. $\psi_2(\lambda)$ is determined by those complex stationary points (saddle points) which can be linked together with $z_l(\lambda)$ by an (m-dimensional) integration contour Γ (in z-space) containing the real stationary points $x_j(\lambda)$, such that (1) the real part of $(i/\hbar)f(z,d(\lambda))$ has a local maximum on Γ at $z_l(\lambda)$ (steepest descent) and (2) each point on Γ yields a local minimum of the real part of $(i/\hbar)f(z,d(\lambda))$ with respect to directions transversal to Γ. If m=1 the latter condition states that Γ lies in a valley of the topography of the real part of $(i/\hbar)f(z,d(\boldsymbol{\lambda}))$. Here, $z_1(\lambda)$ is assumed to be the saddle point with a largest value of the real part of $(i/\hbar)f(z,d(\lambda))$. We suppose that $z_l(\lambda)$ is on Γ for $1\le l\le r'$ $(r+r'\le k+1)$. Applying the saddle point method to ψ we obtain with Eq. (3.9) the following expression:

$$H_o(\lambda)\psi(d(\lambda)) - i\hbar\Delta t \sum_{j=1}^{r} H_j(\lambda)\partial\psi(d(\lambda))/\partial d_j \simeq \Psi_R(\lambda)+\Psi_C(\lambda) \qquad (3.16)$$

with

$$\Psi_R(\lambda) = (2\pi i\hbar\Delta t)^{m/2} \sum_{j=1}^{r} H(x_j(\lambda),\lambda)|\det f^{(2)}(x_j(\lambda),d(\lambda))|^{-1/2} \times \exp\{if(x_j(\lambda) ,d(\lambda))/\hbar\Delta t-m_j\pi i/2\} \qquad (3.17a)$$

$$\Psi_C(\lambda) = (2\pi i\hbar\Delta t)^{m/2} \sum_{l=1}^{r'} H(z_l(\lambda),\lambda)|\det f^{(2)}(z_l(\lambda),d(\lambda))|^{-1/2} \times \exp\{if(z_l(\lambda),d(\lambda))/\hbar\Delta t-i\beta_l(\lambda)/2\} \qquad (3.17b)$$

In Eq. (3.17a) m_j is the number of negative eigenvalues of $f^{(2)}(x_j(\lambda),d(\lambda))$ and $\beta_l(\lambda)$ in Eq. (3.17b) is given by

$$\beta_l = \sum_{i=1}^{m} \arg\alpha_{li}(\lambda)$$

with the $\alpha_{li}(\lambda)$ $(1 \leq i \leq m)$ being the eigenvalues of $f^{(2)}(z_l(\lambda),d(\lambda))$. Let $S_j=S_j(P_1,P_2)$ be the action along $q_j(t;\lambda)$ and let, by analytic continuation of the Lagrangean, $S_{Cl}=S_{Cl}(P_1,P_2)$ be the (complex) action along the complex classical path $q_{Cl}(t;\lambda)$. Then it can be shown [15] that the approximation (3.16) with (3.17), when inserted into Eq. (3.13), yields the following approximation formula for the propagator:

$$K(P_2,P_1) \simeq \sum_{j=1}^{r} K_j(P_2,P_1) + \sum_{l=1}^{r'} K_{Cl}(P_2,P_1) \quad , \tag{3.18}$$

where

$$K_j(P_2,P_1) = (2\pi i\hbar)^{-n/2}|\det\partial^2 S_j/\partial q_1\partial q_2|^{1/2}\exp\{\tfrac{i}{\hbar}S_j-(m'+m_j)\pi i/2\} \tag{3.19a}$$

$$K_{Cl}(P_2,P_1) = (2\pi i\hbar)^{-n/2}|\det\partial^2 S_{Cl}/\partial q_1\partial q_2|^{1/2}\exp\{\tfrac{i}{\hbar}S_{Cl}-i(m'\pi+\beta_l(\lambda))/2\} \tag{3.19b}$$

The imaginary part of S_{Cl} is given by $\mathrm{Im}S_{Cl}(P_1,P_2)=\mathrm{Im}f(z_l(\lambda),d(\lambda))$, so that the "partial wave amplitudes" in the second sum of Eq. (3.18) are multiplied by an exponential attenuation factor. The number $m'+m_j$ in Eq. (3.19a) is the number of focal points crossed by $q_j(t;\lambda)$ for $t_1 \leq t \leq t_2$. In those regions of λ-space where the (real) stationary points x_j cannot be linked with the saddle points z_l by a contour Γ described above, or where all solutions of Eq. (3.15) are real $(r=k+1)$, the second term in Eq. (3.18) does not occur. On the other hand, for λ in a region where no (real) stationary point of f exists (the shadow region), the first term in Eq. (3.18) is no longer present. The explicit dependence of $\tilde{f}_l(\lambda):=\mathrm{Im}f(z_l(\lambda),d(\lambda))$ on λ in the shadow region cannot be determined without further input, because it is given by both the catastrophe polynomial f and the functions $d(\lambda)$. But from a physical point of view it is obvious that $\tilde{f}_l$ increases if λ moves away from the bifurcation set since in this case the propagator must decrease.

We summarize the results: Eq. (3.13) is a uniform approximation for the path integral, valid in a whole neighborhood of the conjugate points P_1^*, P_2^*. The bifurcation set Σ divides the λ-space into various domains which are, in general, multiply covered by real and complex classical paths connecting P_1 with P_2. The bifurcation properties of f reflect exactly those of the classical paths. Outside a layer around the bifurcation set, in a given domain, Eq. (3.13) reduces to Eq. (3.18) which is the familiar approximation formula (1.3) with an additional term arising from the "relevant" complex classical paths. Both, the real and the complex classical paths are uniquely determined by the real and complex stationary points of f.

4. Examples

(1) The fold catastrophe:

Let us assume that $m'=0$, $m=1$ $(y\in R)$ and that the singularity g_0 has the form

$$g_0(y) = g_3y^3 + O(y^3) \quad , \quad g_3>0.$$

In this case the appropriate catastrophe polynomial turns out to be the fold catastrophe, i.e.,

$$f(x,d) = x^3/3 - dx,(x,d\varepsilon R) , \tag{4.1}$$

where d is a function of $\lambda=(P_1-P_1^*,P_2-P_2^*)$ with $d(0)=0$. For $d>0$ f has two stationary points given by $x_1=\sqrt{d}$, $x_2=-\sqrt{d}$, which determine real classical paths $q_i(t;\lambda)$ $(i=1,2)$ connecting P_1 with P_2. q_1 crosses a focal point at a time t' $(t_1<t'<t_2)$, whereas q_2 does not. The bifurcation set is given by d=0, thus implying an equation for the caustics which divide the λ-space in a classically allowed ("illuminated") region and a classically forbidden ("shadow") region. In the classically allowed region $(d>0)$ the two classical paths $q_i(t;\lambda)$ contribute to the propagator and coincide on the caustic. In the classically forbidden region $(d<0)$ f has two imaginary stationary points $z_i=\pm i\sqrt{(-d)}$, which determine complex classical paths $q_{Ci}(t;\lambda)$ $(i=1,2)$.

The generalized Airy integral $\psi(d)$ reduces in this case to the regular Airy function Ai,

$$\psi(d) = \int dx \exp\{i(x^3/3-dx)/\hbar\Delta t\} = 2\pi(\hbar\Delta t)^{1/3}Ai(-a), \tag{4.2}$$

where $a:=(\hbar\Delta t)^{-2/3}d$ (Fig. 2).

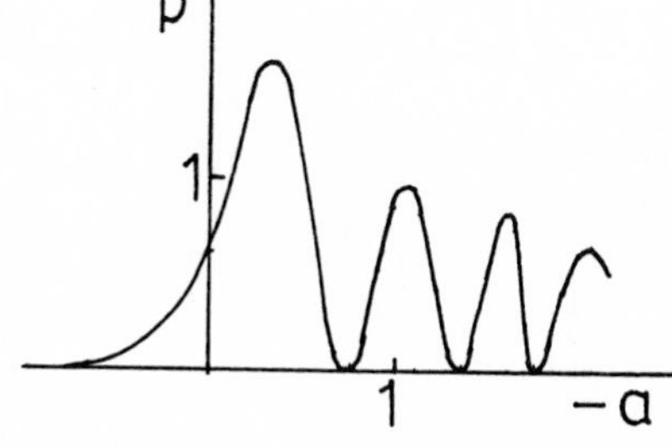

Fig. 2: Intensity $p(-a) = |Ai(-a)|^2$ of the regular Airy function

The intensity oscillations in the illuminated region result from the two contributing classical paths. In the shadow region the complex classical path, determined by $z_1=i\sqrt{(-d)}$, induces an exponentially damped intensity.

It can be shown [16] that the uniform approximation (3.13) for the propagator can be written in the form,

$$\begin{aligned} K(P_2,P_1) \simeq\ & i^{-(n+1)/2}\pi^{1/2}\{Z^{1/4}(|p_1|^{1/2}+|p_2|^{1/2})Ai(-Z) \\ & -iZ^{-1/4}(|p_1|^{1/2}-|p_2|^{1/2})Ai'(-Z)\}\exp\{i(\Phi_1+\Phi_2)/2\} , \end{aligned} \tag{4.3}$$

where

$$\Phi_j = S_j/\hbar, \quad p_j = (2\pi\hbar)^{-n}\det(\partial^2 S_j/\partial q_1\partial q_2) , \quad Z = [\tfrac{3}{4}(\Phi_2-\Phi_1)]^{2/3} ,$$

with S_j being the action along the classical path q_j $(j=1,2)$. (4.3) has been obtained by Connor and Marcus [23] using the method of Chester, Friedman and Ursell [24]. An approximation formula equivalent to (4.3) has been derived by Miller [25] by means of an a posteriori uniform approximation.

(2) The cusp catastrophe:

Suppose now that m=1 and that g_0 has the form, proposed by Schulman [26],

$$g_0(y) = -g_4 y^4 + O(y^4) \quad (g_4>0,\ y\varepsilon R) \quad .$$

The appropriate catastrophe polynomial in this case turns out to be the cusp catastrophe,

$$f(x,d) = -x^4/4 - d_2 x^2/2 - d_1 x, \ (x,d_1,d_2 \varepsilon R) \tag{4.4}$$

with $d_i=d_i(\lambda)$ (i=1,2). The bifurcation set Σ of f is determined by the standard cusp equation

$$\Sigma:\ 27 d_1^2 + 4 d_2^3 = 0 \quad , \tag{4.5}$$

which yields an equation for the caustics in λ-space. Setting

$$a = (\hbar\Delta t)^{-3/4} d_1 \ , \quad b = (\hbar\Delta t)^{-1/2} d_2 \quad ,$$

the generalized Airy integral (3.13) becomes in this case

$$\psi(d) = (\hbar\Delta t)^{1/4} \phi(a,b) \quad , \tag{4.6}$$

where

$$\phi(a,b) = \int dx \exp\{-i(ax+bx^2/2+x^4/4)\} \quad . \tag{4.7}$$

The bifurcation set of f in a-b-plane is shown in Fig. 3. In Fig. 4, the intensity $p(a,b)=|\phi(a,b)|^2$ is plotted as a function of a for various values of b. Inside the cusp there exist three real stationary points of f. Two of them coincide on Σ and become complex outside the cusp, where one has only one real stationary point. This implies that the caustics divide the λ-space into two regions, one of these covered by one and the other by three real classical paths connecting the space-time points P_1, P_2. A typical example of such a situation is shown in Fig. 5 [16].

A detailed numerical investigation of the "cusp Airy function" (Eq. (4.7)) has been carried out by Pearcey [27], who plotted curves of ϕ with constant modulus and phase in a-b-plane ("Pearcey pattern"). F.J. Wright has shown [28], that wave front dislocations [29] occur in the Pearcey pattern resulting from superpositions of real and complex stationary points of f contributing to ϕ. In terms of the path integral these superpositions can be interpreted as interferences of real and complex classical paths.

(3) The double-cusp in electron optics:

Suppose electrons, emitted by a source and accelerated by an electric field, pass through a magnetic field B, rotation symmetric to the z-axis, whose z-component B_3 is inhomogeneous. The action functional S_{12} is given up to third order in (q_1,q_2) by

are computed from f, are observed on a screen perpendicular to the z-axis in (α,β)-plane with distance γ from the origin of the initial ellipsoid.

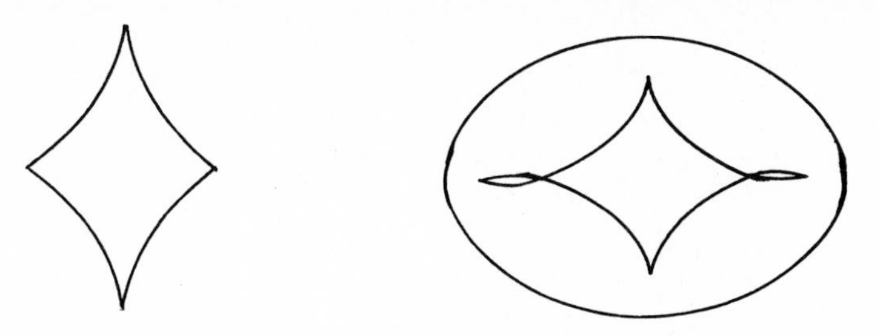

Fig. 6: Two sections of the bifurcation set of f in the (α,β)-plane.

Similar situations were discussed by Berry [9], who pointed out that in case of symmetries of the initial wavefront one might expect infinite codimensions.

Appendix

We give a formal derivation of the essential expression (2.13) which gives rise to the approximation (2.18) of S* in terms of the catastrophe polynomial f. The equation system

$$\partial W(y,a,\lambda)/\partial a_i = 0 \quad (i>m) \tag{A.1}$$

is uniquely solvable for $a=h(y,\lambda)=\{h_i(y,\lambda): i>m\}$, if (y,a,λ) is in a neighborhood of (0,0,0), since

$$\partial^2 W(0,0,0)/\partial a_i \partial a_j = \mu_i \delta_{ij} \quad (i,j>m) \quad , \tag{A.2}$$

and $\mu_i \neq 0$ for $i>m$. With

$$W_i(y,a,\lambda) := \partial W(y,a,\lambda)/\partial a_i \quad (i>m)$$

$$L_{ij}(y,a,\lambda) := \partial^2 W(y,a,\lambda)/\partial a_i \partial a_j \quad (i,j>m),$$

the following relation is easily verified:

$$\begin{aligned} W(y,a+h(y,\lambda),\lambda)-W(y,h(y,\lambda),\lambda) &= \sum_{i>m} \int_0^1 W_i(y,\Theta a+h(y,\lambda),\lambda)a_i d\Theta \\ &= \sum_{i,j>m} \int_0^1 L_{ij}(y,\Theta a+h(y,\lambda),\lambda)a_i a_j (1-\Theta)d\Theta. \end{aligned} \tag{A.3}$$

Defining

$$B_{ij}(y,a,\lambda) := 2\int_0^1 L_{ij}(y,\Theta a+h(y,\lambda),\lambda)(1-\Theta)d\Theta \quad , \tag{A.4}$$

we obtain Eq. (2.13) from Eqs.(A.3) and (A.4).

References

1. C. DeWitt-Morette, A. Maheshwari and B. Nelson "Path Integration in Non-Relativistic Quantum Mechanics", Physics Reports (in print).

2. W.H. Miller, "Classical-Limit Quantum Mechanics and the Theory of Molecular Collisions", Adv. Chem. Phys. 25 (1974), 69.

3. T. Koeling and R. Malfliet, "Semiclassical Approximations to Heavy Ion Scattering Based on the Feynman Path-Integral Method", Phys. Rep. C 22 (1975), 181.

4. R. Dashen, B. Haslacher and A. Neveu, "Non-Perturbative Methods and Extended Hadron Models in Field Theory", I: "Semi-Classical Functional Methods", Phys. Rev. D 10 (1974), 4114; II: "Two-Dimensional Models and Extended Hadrons", Phys. Rev. D 10 (1974), 4130.
   - "Particle spectrum in model field theories from semiclassical functional integral techniques", Phys. Rev. D 11 (1975), 3424.

5. R. Feynman and A. Hibbs "Quantum Mechanics and Path Integrals", McGraw Hill, 1965.

6. M. Berry and K. Mount, "Semiclassical approximations in wave mechanics", Rep.Prog.Phys. 35 (1972), 315.

7. C. DeWitt-Morette, "The Semiclassical Expansion", University of Texas at Austin, Preprint, 1976.

8. M. Gutzwiller, "Phase-Integral Approximation in Momentum Space and the Bound States of an Atom", J. Math. Phys. 8 (1967), 1979.

9. M. Berry, "Waves and Thom's Theorem", Adv. Phys. 25 (1976), 1.

10. R. Thom, "Structural Stability and Morphogenesis", Benjamin, New York, 1975.

11. J. Gelfand and A. Yaglom, "Integration in Functional Spaces and its Application in Quantum Physics", J. Math Phys. 1 (1962), 48.

12. V.I. Arnold, "Mathematical Methods of Classical Mechanics", Springer, 1978.

13. G. Dangelmayr, "Catastrophes and Bifurcations in Variational Problems", this volume.

14. R.J. Magnus, "Universal Unfoldings in Banach Spaces: Reduction and Stability", Math. Rep. 107, Battelle, Geneva 1977.

15. G. Dangelmayr, Thesis, University of Tübingen, 1978.

16. G. Dangelmayr and W. Veit, "Semiclassical Approximations of Path Integrals on and near Caustis in Terms of Catastrophes", Ann. Phys. (in print).

17. C.DeWitt-Morette, "The Small Disturbance Equation: Caustics and Catastrophes", University of Texas at Austin, Preprint, 1975.

18. J.J. Duistermaat, "Oscillatory Integrals, Lagrange Immersions and Unfoldings of Singularities", Comm. Pure Appl. Math. 27 (1974), 207.

19. R.N. Buchal and J.B. Keller, "Boundary Layer Problems in Diffraction Theory", Comm. Pure Appl. Math. 13 (1960), 85.

   D. Ludwig, "Uniform Asymptotic Expansions at a Caustic", Comm. Pure Appl. Math. 19 (1966), 215.

20. V.I. Arnold, "Critical Points of Smooth Functions", in "Proceedings of the International Congress of Mathematicians", Vol. 1, Vancouver 1974.

21. S. Levit and U. Smilansky, "The Hamiltonian Path Integrals and the Uniform Semiclassical Approximation for the Propagator", Ann. Phys. 108 (1977), 165.

22. E. Copson, "Asymptotic Expansions", Cambridge University Press, 1965.

23. J. Connor and R. Marcus, "Theory of Semiclassical Transition Probabilities for Inelastive and Reactive Collisions II: Asymptotic Evaluation of the S-Matrix", J. Chem. Phys. 55 (1971), 5636.

24. C. Chester, B. Friedman and F. Ursell, "An Extension of the Method of Steepest Descents", Proc. Cambr. Phil. Soc. 53 (1957), 599.

25. W.H. Miller, "Classical S Matrix: Numerical Application to Inelastic Collisions", J. Chem. Phys. 53 (1970), 3578.

26. L.S. Schulman, "Caustics and Multivaluedness: Two Results of Adding Path Amplitudes", in "Functional Integration and its Applications", (A.M. Arthurs, Ed.), Clarendon Press, Oxford 1975.

27. T. Pearcey, "The Structure of an Electromagnetic Field in the Neighborhood of a Caustic", Phil. Mag. Ser. 7, 37 (1946), 311.

28. F. Wright, "Wavefront Dislocations and their Analysis Using Bifurcation Theory", this volume.

29. J.F. Nye and M.V. Berry, "Dislocations in Wave Trains", Proc.R. Soc. Lond. A 336 (1974), 165.

30. W. Veit and G. Dangelmayr, in preparation

31. J. Callahan, "Special Bifurcations of the Double-Cusp", University of Warwick, Preprint, 1978.

32. E.C. Zeeman, "The Umbilic Bracelett and the Double-Cusp Catastrophe", in "Structural Stability, the Theory of Catastrophes and Applications in the Sciences", (P. Hilton, Ed.) Lecture Notes in Math. 525, Springer, 1976.

33. F. Lenz, "Theoretische Untersuchungen über die Ausbreitung von Elektronenstrahlbündeln in rotationssymmetrischen elektrischen und magnetischen Feldern", Habilitationsschrift, Aachen 1957.

34. W. Glaser, "Grundlagen der Elektronenoptik", Springer 1952.

Catastrophe and Stochasticity in Semiclassical Quantum Mechanics

M.V. Berry

H.H. Wills Physics Laboratory, University of Bristol, Tyndall Avenue
Bristol BS8 1TL, U.K.

1. Introduction

The symposium talk was an account of material already published [1,2,3,4]. Therefore this contribution will be an extended summary, in the form of a series of assertions for which justification will be found in the papers cited.

As the classical limit (Planck's constant $\hbar \to 0$) is approached, quantum bound states bifurcate into two universality classes distinguished by the morphologies of their wave functions $\psi(\underline{q})$. PERCIVAL [5] calls these 'regular' and 'irregular' states; they are associated with different sorts of classical motion as foreseen by EINSTEIN [6]. In what follows, the classical motion is Hamiltonian, with N degrees of freedom and hence a 2N-dimensional phase space.

Regular states are associated with integrable motion [7,8,9], that is motion restricted by the existence of N constants of motion to N-dimensional tori in phase space. The projections of these tori onto the coordinate space $\underline{q}$ are singular on caustics that generically take the form of the catastrophes classified by Professor THOM [10].

Irregular states are associated with ergodic classical motion, that is motion where only the Hamiltonian is conserved, so that the system explores the whole 2N-1-dimensional energy surface in phase space. Neighbouring trajectories separate exponentially fast (if $N > 1$) and this implies that ergodic motion is stochastic. The projections of stochastic regions onto $\underline{q}$ are bounded by 'anti-caustics' [1] whose strength is not infinite and which need not be catastrophes. The distinction between caustics and anticaustics is analogous to the distinction between the boundaries of projections from 3 space to 2 space of surfaces and volumes respectively.

These classical motions are connected with quantum mechanics via the Wigner function $W(\underline{q}, \underline{p})$, defined for a state $\psi(\underline{q})$ by

$$W(\underline{q}, \underline{p}) \equiv \left(\frac{2\pi}{\hbar}\right)^N \int d^N X\, e^{-i\underline{p}\cdot\underline{X}/\hbar}\; \psi(\underline{q} - \tfrac{1}{2}\underline{X})\; \psi^*(\underline{q} + \tfrac{1}{2}\underline{X}). \tag{1}$$

Its projection 'down' $\underline{p}$ onto $\underline{q}$ gives the coordinate probability density:

$$\int d^N p\, W(\underline{q}, \underline{p}) = |\psi(\underline{q})|^2, \tag{2}$$

and its projection 'along' q gives the momentum probability density.

The assumption is made that for typical semiclassical quantum energy eigenstates $W(\underline{q}, \underline{p})$ is localised on the manifolds in phase space that are explored by typical classical orbits, that is tori for integrable systems and whole energy surfaces for ergodic systems. Each quantum state has its own manifold, selected by a quantum condition. For tori the quantum condition [11,12,8] is that the action $\oint \underline{p} \cdot d\underline{q}$ round each irreducible cycle must be a half-integer multiple of $2\pi\hbar$. For energy surfaces the quantum condition is unknown. (Conjecture: the phase space volume enclosed by the energy surface E_n is quantised by

$$\int_{E_n > \text{Hamiltonian}} d^N q \int d^N p = (n + \mu)\, h^N, \tag{3}$$

where n is an integer and μ a constant. This would lead for these ergodic systems to an asymptotically regular sequence of levels in sharp contrast to the clustering [13] generic in integrable systems.)

The status of the assumption that $W(\underline{q}, \underline{p})$ is localised on the typical classical orbits is as follows: for integrable systems W is known to condense onto a delta function on a torus when $\hbar = 0$,[2,14]. Moreover, when $\hbar$ is small but not zero more refined asymptotic results [2,3,15] show W to take the form of 'Airy' fringes decorating the torus. For ergodic systems it is still a conjecture that the W corresponding to an eigenstate condenses onto the energy surface [1,16]. (An alternative conjecture, due to GUTZWILLER [17,18], is that quantum states in ergodic systems correspond to the exceptional (but dense) unstable closed orbits. My opinion is that it is implausible for quantum conditions to select orbits that are both untypical and unstable, and indeed GUTZWILLER's own analysis yields not delta-function energy levels, but a spectrum of Lorentzian resonances of width $\hbar$, strongly suggesting that what he has found are not true bound states but 'quasi-modes' [19,20] which decay after times of order $\hbar^{-1}$.)

Making the assumption about W leads easily [1,16] to strikingly different predictions about the morphology of regular and irregular states. The probability density $|\psi(\underline{q})|^2$, smoothed over the (semiclassically very rapid) oscillations near $\underline{q}$, is given by (2) simply by projecting the classical torus or energy surface onto $\underline{q}$. The pattern of oscillations of ψ near $\underline{q}$, as embodied in the autocorrelation function [1] of ψ, is obtained from the momenta $\underline{p}$ of intersections of the fibre through $\underline{q}$ with the classical manifold W (by de Broglie's rule, each $\underline{p}$ gives an oscillatory contribution to ψ, with wave vector $\underline{k} = \underline{p}/\hbar$). The predictions are:

Regular states have probability densities rising to high values on caustics at classical boundaries, and are decorated with vivid patterns of highly anisotropic interference oscillations with a discrete spectrum of wave vectors.

Irregular states have probability densities falling to zero (or, when N = 2, remaining constant) on anticaustics at classical boundaries, and are decorated with more moderate oscillations (like those of a Gaussian random function) which for ergodic systems are statistically isotropic and which have a continuous spectrum of wave vectors.

A generic Hamiltonian classical system with $N > 1$ is neither fully integrable nor fully ergodic [7,8,9]. Some orbits lie on N-tori while others explore 2N-1-dimensional stochastic regions of the energy surface. In the corresponding quantum systems this should lead to a mixed spectrum, with some states regular and some irregular. If the system is quasi-integrable, that is if its departure from integrability depends on a perturbation parameter ε, then I have speculated [2,8] that there should be three semiclassical régimes depending on the value of $\hbar$ in comparison with ε (both quantities being small).

It is not easy to test these predictions by actual or numerical experiment. In nature, irregular states (of, say, vibrating asymmetric molecules) would be vulnerable to perturbation [5]. In the computer, the small value of $\hbar$ means that ψ oscillates rapidly and is vulnerable to numerical noise. The regular and irregular states are emergent morphologies; they appear as $\hbar \to 0$ and for finite $\hbar$ are encoded in the Schrödinger equation in a form too economical to be easily reconstructed.

To get over this computational problem we have invented quantum maps [4, see also 21]. Instead of quantising the simplest stationary Hamiltonian system with generic properties, which has N =2, we have directly quantised a class of area-preserving maps of the plane. These are obtained by taking a system with N =1 but with Hamiltonian periodic in time with period T, and taking snapshots of the phase plane q,p at times nT where n is an integer. Such maps can be made to exhibit both integrable and stochastic behaviour. The corresponding quantum systems evolve under the action of a unitary operator and the mapping of wave functions ψ is described by an integral equation.

To see a continuous transition from a regular to an irregular ψ, an initial state (not an eigenfunction of the map) was chosen to correspond to a curve C in qp whose points map stochastically. Under the classical map, C soon develops complication by throwing off 'tendrils' snaking violently back and forth. In the projection of C onto q, caustics proliferate rapidly and when they are closer together than the de Broglie wavelength they no longer contribute individually to ψ. In the computer-generated pictures [4] it is easy to see the transition from regular states, with intense caustics and spectrally pure oscillations, to irregular states, with anti-caustics and oscillations with multiple scales. Fig. 1 shows one example taken from several presented in [4], where a detailed discussion is given of the relation between C and ψ. (Actually, the tendril is one of two morphological elements in C's developing complication. The other is the 'whorl,' a wrapping of C around itself, associated not with stochasticity but with the smooth invariant curves (tori) associated with stable regions of the map. Again details are given in [4].)

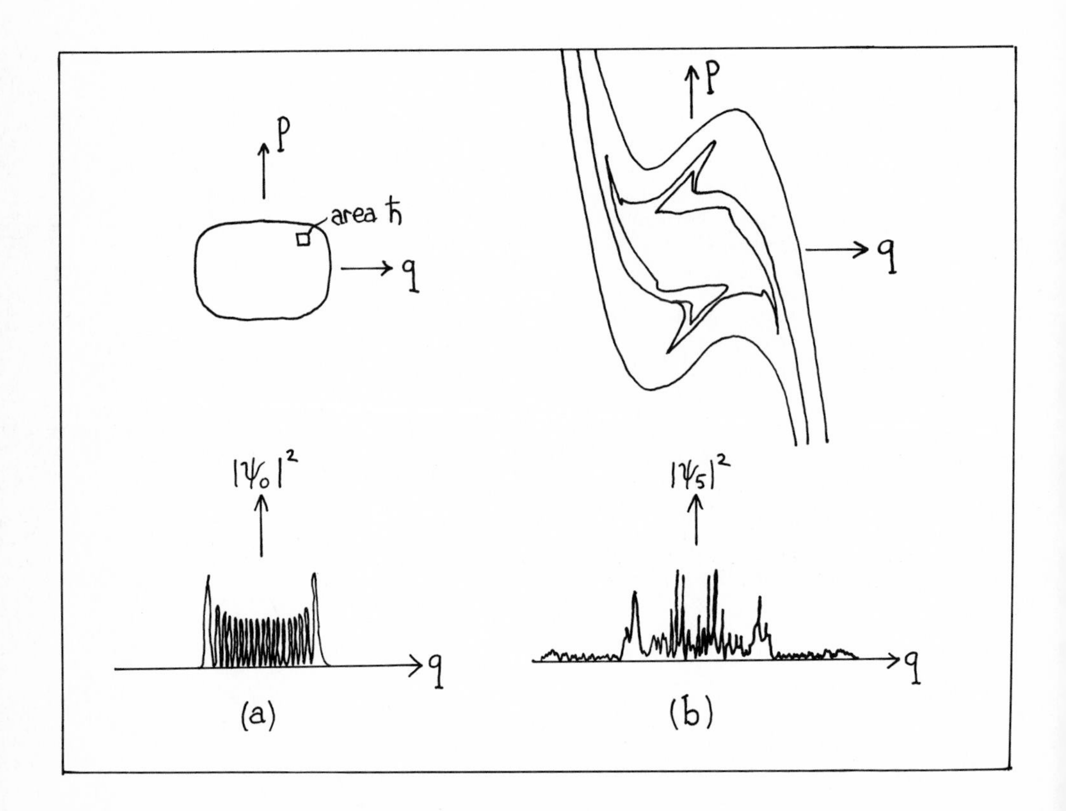

Fig. 1 Transition of state ψ from regular (a) to irregular (b) after five iterations of a quantum map, as computed by Dr. M. Tabor. The state is represented by the classical curve C shown above, which develops 'tendrils' under iteration.

In an extension of the quantum map formalism to phase spaces with the topology of a 2-torus, we [22] have succeeded in quantising 'Arnol'd's cat' [7,8,9], a map that is completely ergodic. To achieve this it is necessary to quantise Planck's constant by h = area of torus/M, where M is an integer. The quantum cat map has M eigenstates and in the semiclassical limit $M \to \infty$ neither the distribution of eigenvalues nor the eigenfunctions $\psi(q)$ themselves can be described by smooth function but instead seem to have a 'number-theoretic' dependence on M.

The two sorts of quantum state discussed here show how Planck's constant plays a much more subtle role than previously supposed. For regular states, $\hbar$ imposes extra detail, in the form of quantum oscillations, on a classical background that is smooth apart from isolated caustic singularities. For irregular states, however, $\hbar$ acts as a quantum smoothing parameter on a classical orbit structure which because of stochasticity has detail down to arbitrarily fine scales.

References

1. M.V. Berry: J. Phys. A. 10, 2083-2091 (1977)
2. M.V. Berry: Phil. Trans. Roy. Soc. Lond. A 287, 231-271 (1977)
3. M.V. Berry, N.L. Balazs: J. Phys. A., in press (1979)
4. M.V. Berry, N.L. Balazs, M. Tabor, A. Voros: submitted to Annals of Physics
5. I.C. Percival: J. Phys. B. 6, L229-323 (1973)
6. A. Einstein: Verh. Dt. Phys. Ges. 19, 82-92 (1917)
7. J. Ford: Fundamental Problems in Statistical Mechanics III, 215-255 (ed.: E.G.D. Cohen; North-Holland, Amsterdam, 1975)
8. M.V. Berry: Am. Inst. Phys. Conf. Ser. 44, Chap. 2 (Nonlinear dynamics, ed.: S. Jorna; A.I.P., New York 1978)
9. V.I. Arnol'd, A. Avez: Ergodic Problems of Classical Mechanics (Benjamin, U.S.A. 1968)
10. R. Thom: Structural stability and morphogenesis (Benjamin, U.S.A. 1975) (original French edition 1972)
11. V.P. Maslov: Théorie des Perturbations et des Méthodes Asymptotiques (Dunod, Paris 1972) (original Russian edition 1965)
12. I.C. Percival: Adv. Chem. Phys. 36, 1-61 (1977)
13. M.V. Berry, M. Tabor: Proc. Roy. Soc. Lond. A 356, 375-394 (1977)
14. A. Voros: Ann. Inst. H. Poincaré A 24, 31-90 (1976)
15. N.L. Balazs, G.G. Zipfel Jr: Ann. Phys. N. Y. 77, 139-156 (1973)
16. A. Voros: in: Stochastic Behaviour in Classical and Quantum Hamiltonian Systems (G. Casati, J. Ford, eds.) Springer notes in physics (1978, to be published)
17. M.C. Gutzwiller: J. Math. Phys, 12, 343-358 (1971)
18. M.C. Gutzwiller: Proc. Nato Conf. on Path Integrals (Antwerp 1977) (to be published)
19. V.I. Arnol'd: Funkt. Anal. Ego. Pril. 6, 12-20 (1972) (English translation: Funct. Anal. Applic. 6, 94-101)
20. M.V. Berry: J. Phys. A. 10, L193-194 (1977)
21. G. Casati, B.V. Chirikov, F.M. Izraelev, J.Ford: in: Stochastic Behaviour in Classical and Quantum Hamiltonian Systems (G. Casati, J. Ford, eds.) Springer notes in physics (1978, to be published)
22. M.V. Berry, J.H. Hannay: to be published

Part IV

Defects and Dislocations

Stable Defects in Ordered Media

René Thom

Institut des Hautes Etudes Scientifiques, 35, route de Chartres
F-91440 Bures-sur-Yvette, France

1. The definition of an ordered medium

Nature provides us with many examples of so-called "ordered media." The best-known are liquid crystals (nematics, smectics, cholesterics, etc.), but a precise definition of such an ordered structure leads to some conceptual difficulties. Physicists are now so well acquainted with the notion of symmetries--and the corresponding mathematical notion of groups--that it is quite difficult for them to give up this idea in the case of objects which locally exhibit such a symmetry but in a variable manner, as in the case of a crystal in which the mesh of the lattice varies according to a global law of the entire medium. Mathematicians, on the other hand, have at their disposal a notion which at first glance seems to be very well suited to describing such objects, namely the notion of pseudo-group. Let us recall that a pseudo-group Γ in a domain V of Euclidean space is defined by a family of local homeomorphisms h_{ji} having as source a neighborhood U_i in V, and as target an open set V_j in V. If the target V_j of h_{ji} is contained in the source U_j of h_{kj}, where h_{kj} is another morphism of the pseudo-group Γ, then the composed map $h_{kj} \circ h_{ji}$ also belongs to the family Γ. Identity homeomorphisms $U_i \simeq U_i$ also belong to Γ. Intuitively, a pseudo-group differs from a group by the fact that the composition of two maps of the pseudo-group may not be defined. As a result, pseudo-groups no longer have that rich mathematical structure associated with groups, which are so interesting to physicists (representations, invariants, etc.). But the transformations of a pseudo-group Γ which leave a point x fixed form a group Γ_x, called the isotropy subgroup of Γ at the point x. However, contrary to what happens for perfect crystals, this isotropy subgroup may vary with the position of x. There are two basic types of variations for Γ_x: Either the dimension of Γ_x (and its algebraic type as a Lie group) varies discontinuously when x belongs to some subset Y of U. This corresponds to a special case of the notion of defect, to be discussed later. Or the group Γ_x varies continuously among the full group of local Euclidean equivalences. Mathematically speaking, the Lie subalgebras of a given dimension k in the Lie algebra A of an n-dimensional Lie group form an algebraically "constructible" set; there exists an algebraic set K in U, with a "stratification" in the Grassman variety $G^k(U)$ of k-planes, such that the k-dimensional Lie subalgebras can be continuously deformed into one another by a global isotopy in U, iff they belong to the same stratum of K. If the global group of local equivalences is compact (O(3), for instance), then any two such isotopic subgroups are conjugate through an inner automorphism of the ambient compact group: This results from Palais' rigidity theorem [1].

For any domain V of Euclidean space carrying an ordered structure, it is convenient to consider the totality of diffeomorphisms of V which leave the ordered structure invariant. They form a group--in general, of infinite dimension--the group I(V) of automorphisms of V. If I(V) acts transitively on V, then all isotropy subgroups Γ_x for x in V belong to the same continuous family. The concept of pseudo-group appears as a very natural tool for describing media endowed with local symmetry. Hence, physicists should show less reluctance to give up global symmetries. It is obvious that global symmetries are perfectly unrealistic. But the need for using known mathematical theories keeps physicists stubbornly attached to these hypotheses: in statistical mechanics, they claim that water cannot freeze unless it fills the whole universe, and many workers in high energy physics still believe that the universe is a Minkowski space. Why not believe that the earth is flat?

The pseudo-group notion, nevertheless, has some shortcomings: It works fairly well when the global isomorphism group of the structure is transitive (no defects). In fact, as a parenthetical remark, let me add that Elie Cartan's profound and still little understood work on primitive pseudo-groups could have a bearing on the problem of phase transitions: There are, according to Elie Cartan, four types of "primitive pseudo-groups." Among them, three seem to have a fairly immediate interpretation as "phases": All diffeomorphisms (gaseous phase), conservative diffeomorphisms (keeping the volume invariant, i.e. liquid phases), symplectomorphisms (describing the motion of solid bodies).

The main shortcoming of the pseudo-group notion is that it does not allow the description of crystal structures--in particular of slowly varying crystal structures. I propose here a strengthening of this notion, namely the concept of semi-local structure.

2. Semi-local structure

A semi-local structure on an open set U of Euclidean space R^n is defined by a family of distinguished open sets U_i in U together with maps g_i: $U_i \to M$ onto a fixed compact manifold M of dimension $n-k$, with k a positive integer. This manifold M is acted on by a subgroup G of diffeomorphisms and for any non-void intersection $U_i \cap U_j$ there is a map h_{ji}: $U_i \cap U_j \to G$, constant in G, such that for any $x \in U_i \cap U_j$ we have $g_j(x) = h_{ji} \cdot g_i(x)$.

The gluing elements h_{ji} satisfy the obvious transitivity relations for any triple of indices (i, j, k). Moreover, all maps g_i are submersions (surjective maps of maximal rank).

As a consequence of these assumptions, at each point $x \in U_i$ the level varieties of the maps g_i are k-dimensional regular submanifolds independent of the local distinguished chart U_i containing x. Hence, any domain carrying a semi-local structure also carries a Haefliger foliation of dimension k.

A semi-local structure is said to be trivial if the "structure group" G acting on M can be taken as the identity map of M onto itself. If the structure defined on U is trivial, there is a global submersion $U \to M$, defining that structure.

The standard crystal structure is an example: If D is a subgroup of the translation group T in R^n, such that the D-action has a bounded fundamental domain, then the quotient T/D is an n-dimensional torus, and the map $T \to T/D$ defines a trivial semi-local structure on R^n. In this example, the space T is the universal covering of the torus T/D. For a trivial semi-local structure g: $U \to M$ there always exists a "connection," i.e., a field of n-planes in U transversal to the fibers $g^{-1}(m)$ of the map g. If U is non-compact, this connection can be made integrable (Gromov's theorem). Integrable means that the field of n-planes is tangent to a family of n-manifolds transversal to the fibers of the map. A leaf of such a transversal foliation is mapped into M by a map locally surjective. If such a leaf can be embedded into a compact set in U, then it is a covering of the "base manifold" M. In many cases, it is possible to deform the connection in such a way that a leaf L of the connection becomes the universal covering $\tilde{M}$ of M. Hence, in a trivial semi-local structure, there are--in general--distinguished neighborhoods diffeomorphic to the product of the universal covering $\tilde{M}$ of M by some k-dimensional Euclidean space.

Can ordinary three-dimensional Euclidean space R^3 appear as such a product? Obviously, there exists a trivial semi-local structure of R^3 over the circle S^1, as $R^3 \simeq R^1 \times R^2$, where R^1 is the universal covering of S^1. In the same way, R^3 is a trivial semi-local structure over any surface of strictly positive genus, because such a surface admits the plane R^2 as its universal covering. For $k = 0$, R^3 is the universal covering of the so-called Raumformen of dimension three.

We may abstractly define an ordered structure in a domain U of Euclidean space as the data of a semi-local structure on U with base manifold M. The local associated foliation defines a pseudo-group Γ, with local isotropy subgroups Γ_x, all isomorphic to the subgroup of local equivalences leaving a k-dimensional linear subvariety, or more generally, a flag of linear spaces contained in this k-dimensional variety, fixed. There is also the possibility of building a hierarchy of semi-local structures: Suppose the base manifold M is fibered over a manifold B with fiber F, and suppose also that the structure group G acts equivariantly in M and in B. Then any semi-local structure over M is also a semi-local structure over B. The fibers over B are bundles having F as base space, and the fiber over M as fiber. In such a case, the local isotropy group Γ_x leaves the flag defined by the composed foliation invariant.

3. Examples of ordered structures

(a) Nematics. A nematic is a fluid composed of thin, elongated rodlike molecules. Due to the interaction between neighboring molecules, the axes of the molecules tend to orient themselves parallel to a locally given direction. Hence the structure is defined in R^3 by a field of local, non-orientable directions. Here we have only a foliated structure, and no semi-local structure. Therefore, the structure is entirely defined by the field of its isotropy subgroups.
(b) Smectics. Here the fluid is a nematic, but its molecules tend to arrange themselves in such a way that their extremities form layers normal to the direction of the molecules. As a result, the fluid carries a semi-local structure over the circle S^1: The fibers over S^1 are the two-dimensional layers which are in fact "parallel" (in the differential-geometric sense of "parallel surfaces") to one another.
(c) Cholesterics. A cholesteric is a nematic where the molecules arrange themselves into locally plane layers, and the orientation of the molecules is rotated by an angle α when one passes from one layer to the next.

In case α is a divisor of 2π, one has an example of a hierarchical semi-local structure over $S^1 \times R \to S^1$. Note here that, contrary to our axiomatic requirement, the base manifold $S^1 \times R$ is not compact.

4. Defects

Let D be a closed subset of a domain U in Euclidean space. If the complementary set U - D carries an ordered structure, then D can be said to be a defect of that structure. Of course, such a definition is rather imprecise. Let us say that a defect D is stable, if by subjecting the medium to a small perturbation of its boundary data (or its global physical or thermodynamical parameters) the new medium exhibits a defect D' which is the image of D under an ε-homeomorphism of U onto itself. Such a definition, of course, is akin to the standard definition of structural stability for a dynamical system. In the same way as stable singularities of smooth maps are extremely restricted in topological types, one might hope that a given ordered structure may develop only stable defects belonging to a finite catalog of topological types. I do not know to what extent such a conjecture is borne out by experiment. But the principle stated recently by M. Kléman and G. Toulouse goes quite a way in this direction.

To understand the origin of such a principle, it suffices to appeal to the elementary catastrophe scheme--or, what amounts to almost the same thing--to Landau's mean field theory. Let us assume that the local configurations of locally interacting molecules form an "internal" space S. The energy of such a configuration is defined by a potential function $V: S \to R$. Now, in general, V in invariant under the action of the global Euclidean group E. Hence, the problem of minimizing the energy V and applying bifurcation theory to the singularities (minima) of the potential V is no longer solved by elementary catastrophe theory. But generalizing this theory to functions invariant under a Lie group action is now mathematically accessible. The main difference lies in the fact that in such a case the minima of V are no longer points, but orbits or strata of orbits. In standard catastrophe theory, catastrophe points arise because of the conflict between stable régimes (and sometimes, as in the case of the cusp catastrophe, because of the conflict of a régime with itself). In the same way, defects in ordered structures arise because the local symmetry prescribed locally by the minimum of V may not be compatible with the topological constraints defined by the boundary data. In a sense, defects can be considered as degenerate "phases" of the medium and arise from a generalized "phase diagram" defined by the bifurcation of the minima of V.

Now, a defect could be only spurious: If in a regular ordered structure defined on an open set Y of R^3, we substract a closed set A and consider it as a defect in Y (because Y-A bears a regular structure), then obviously, such a defect may not be considered as existent (and a fortiori not stable). Hence, if for a defect D in U the regular structure defined in U-D cannot be smoothly extended over D, then one has to expect that D as a defect will be stable, in the sense that no slight perturbation of the medium will destroy it.

On a defect point x, the singularity of V (on S) has in general a larger symmetry (isotropy) subgroup than the nearby regular points. A typical example is given by the potential function $V(x,y)$ on $R^2(x,y)$ invariant under rotation around 0, defined by $V = r^4/4 - r^2/2$, $r^2 = x^2 + y^2$.

The function V reaches its minimum along the circle $r = 1$, $z = -1/4$. Suppose there is a loop g in U, formed by regular points such that the local state at a point u of g is a point $x(u)$, $y(u)$ which describes the circle $r = 1$ when u describes g . Then I may claim that any disk in U having g as boundary has to contain at least one defect point. For no disk can be smoothly mapped into its boundary. Hence, such a loop g is topologically linked to the defect set D. Such a disk, to minimize the integral of the energy, has to be mapped homeomorphically onto the unit disk of the Oxy plane: The counterimage of the origin, which is a maximum of $V(x,y)$ will be a defect point. On such a point, V admits the full rotation group as isotropy subgroup, whereas on a point of the circle $r = 1$ the isotropy subgroup is the identity group. Now the gradient trajectories arriving at $r = 0$, $z = 0$ on the potential well $z - V(x,y) = 0$ form a two-dimensional disk (as in standard Morse theory) which has the minimum circle $r = 1$ as boundary.

Due to the topological constraints the maximum $r = 0$ becomes "stabilized." This is a case of "threshold stabilization." In fact, it could occur that a local minimum appears around 0, inside a sink limited by a small circle of radius $r = \varepsilon$. This would correspond to the formation of a "core" inside the defect: the precise molecular configuration on the defect is then associated with this core. (See Fig. 1).

To state the Kléman-Toulouse principle, we still have to introduce the notion of "manifolds of regular structures." It is the set of minima for the local potential V. But this potential has to be taken semi-locally, to allow for the possibility of semi-local structure of equilibrium.

The stratum of minima for V is then a fiber bundle Φ having as basis a space of moduli M, and as fiber a homogeneous space of type E/K (Euclidean group modulo closed subgroup). The fiber corresponds to the "punctual" parametrization of regular structures $(E(x)/\Gamma(x))$, the base M to the semi-local parametrization. Then we shall admit that any stable defect D has a stratified structure (like a semi-algebraic variety). Let Z be its stratum of maximal dimension (lowest codimension). Let z be a point in Z, H_z the normal plane to Z at z. Denote by s a sphere of small radius in Z with center z (such a sphere is locally linked to the stratum Z.) There exists a map k which associates to any point $t \in s$ the corresponding local regular structure defined at that regular point. Then the Kléman-Toulouse principle states: If the defect D is stable, then the map $k: s \to \Phi$ is non-homotopic to zero [2, 3].

This implies that the regular structure cannot be extended smoothly across the defect D , which implies that the defect has to stay under slight perturbations (but does not prove that its topological type is stable).

As stated here, the Kléman-Toulouse principle also takes into account the defects of crystal structures (dislocations). The space of local structures is a torus, and a dislocation of dimension one has a linked circle s of dimension one. The map k of s into M has a degree, which in the one-dimensional homology group $H_1(M)$ is the classical Burgers vector of the dislocation. The Volterra process corresponds to defining the map k of s into M.

The application of the Kléman-Toulouse principle may lead to difficulties. As I have shown in [4], the example of smectic A is particularly interesting in this respect. The principle implies that if the manifold of regular structures Φ is connected, then there are no stable walls (any map of S^0 in a connected space is null-homotopic). Hence the surface layers of a smectic A should have the property that their normals have no two-dimensional envelope. Hence the appearance, for such surfaces, of Dupin cyclids for which the focal surface degenerates into two focal conics (the product of these conics is in fact the base space M of a semi-local structure which completes the one described above). This configuration was found by Friedel and is now classic. But around these conics, the Kléman-Toulouse principle does not strictly apply (at least to the "punctual part of the manifold Φ"). Moreover, between several semi-local structures (even of the same type) there must be walls. There is no doubt that the Kléman-Toulouse principle has far-reaching validity (in many instances, it can be strengthened, as it appears as a consequence of the transversality lemma). But it may lead to contradictions to itself. The case of smectic A reflects the fact that the field of two-planes defined by the layers of the fluid must satisfy an integrability condition (of the non-linear type $\theta \wedge d\theta = 0$ of the Frobenius condition). Such a condition creates the appearance of a semi-local structure for the field of directions, and this fact modifies the manifold Φ of regular structures.

As a conclusion, I would like to point out that the Kléman-Toulouse principle is apparently the first example of the introduction of rather sophisticated concepts of algebraic topology to physics. Despite its difficulties, I am inclined to believe that it will not remain unique.

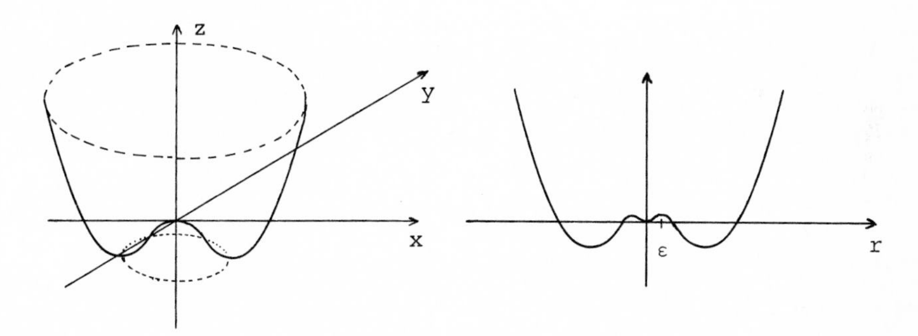

Fig. 1

References

1 H. Stewart, R. Palais: Deformations of compact differentiable transformation group, Amer. J. Math. 82 (1960), 935 - 37

2 M. Kléman, G. Toulouse: Principles of a classification of defects in ordered media, Journal de Physique Lettres 37, L149, 1976

3 G.E. Volovik, E.P. Minnev: Pi'sma v.ZhFIF 24, 1976, 605

4 R. Thom: Sur certains aspects qualitatifs de la théorie des équations aux dérivées partielles, Colloque IRIA, Décembre 1977 (Ed. J.J. Lions)

Structural Stability in Evolving Flow Fields

J.F. Nye

H.H. Wills Physics Laboratory, University of Bristol, Tyndall Avenue
Bristol BS8 1TL, U.K.

Abstract

The changing topology of an evolving flow field can be studied by mapping it into velocity space and observing the singularities. In two dimensions the structurally stable singularities are folds, which are lines, and cusps, which are isolated points on the folds [2]. In three dimensions they are fold surfaces, which contain ribs, and the ribs themselves may contain isolated singular points. As the field changes with time these singularities interact to produce events. Between events the basic structure of the flow field remains the same. How many different kinds of event will occur generically depends on whether or not the flow field is constrained (for instance by being irrotational); the catalogue of events can be deduced from known results on stable mappings and from catastrophe theory. The paper is mainly a summary of previously published results [1,6].

1. Introduction

The work on general vector fields which this paper describes arose from an attempt to solve a problem in glaciology. The floating ice that covers the surface of the Arctic Ocean has a pattern of motion which, when averaged on a scale of say 100 km, can be modelled as the motion of a two-dimensional continuum. The Arctic Ice Dynamics Joint Experiment was designed to measure this changing velocity field and to produce a mathematical model. Thus part of the problem was, in essence, to compare a series of observed flow fields with the corresponding computed fields. It became apparent that, in additon to detailed numerical description of an evolving flow field, some sort of qualitative description was needed to specify its basic structure and how it changed. One could then ask "Are the two flow fields similar, and are they evolving in a similar way?" in the same way that one might ask the same structural question about two animals. This type of problem will arise in any investigation in which "generic" vector fields, without special symmetry, have to be compared.

2. Events in Evolving Two-dimensional Flows

A detailed account of the work summarised in this and the following section is given in [1]. Consider the continuous two-dimensional velocity field with components $u = u(x,y)$ and $v = v(x,y)$, and map each point (x,y) in physical space into the associated point (u,v) in velocity space. The singular points of the mapping, where the Jacobian $J = \partial(u,v)/\partial(x,y) = 0$, form smooth lines in (x,y) space called fold-lines, which contain special points called cusp-points. In (u,v) space the fold-lines map into folds which have cusps at the images of the cusp-points [2]. These folds and cusps can be taken to characterise the topological structure of the flow field. For example, the flow field in Fig. 1a has the fold-lines shown in Fig. 1b (heavy lines) and cusp-points α and β. They map into the folds and cusps in velocity space shown in Fig. 1c. To visualise the mapping take the (x,y) plane and fold it along CD, with appropriate stretchings, to bring its various points to the appropriate places in (u,v) (thus pairs of points on either side of CD have the same velocity). Then in the bottom sheet of the fold make a pleat at α and another at β, thus bringing yet

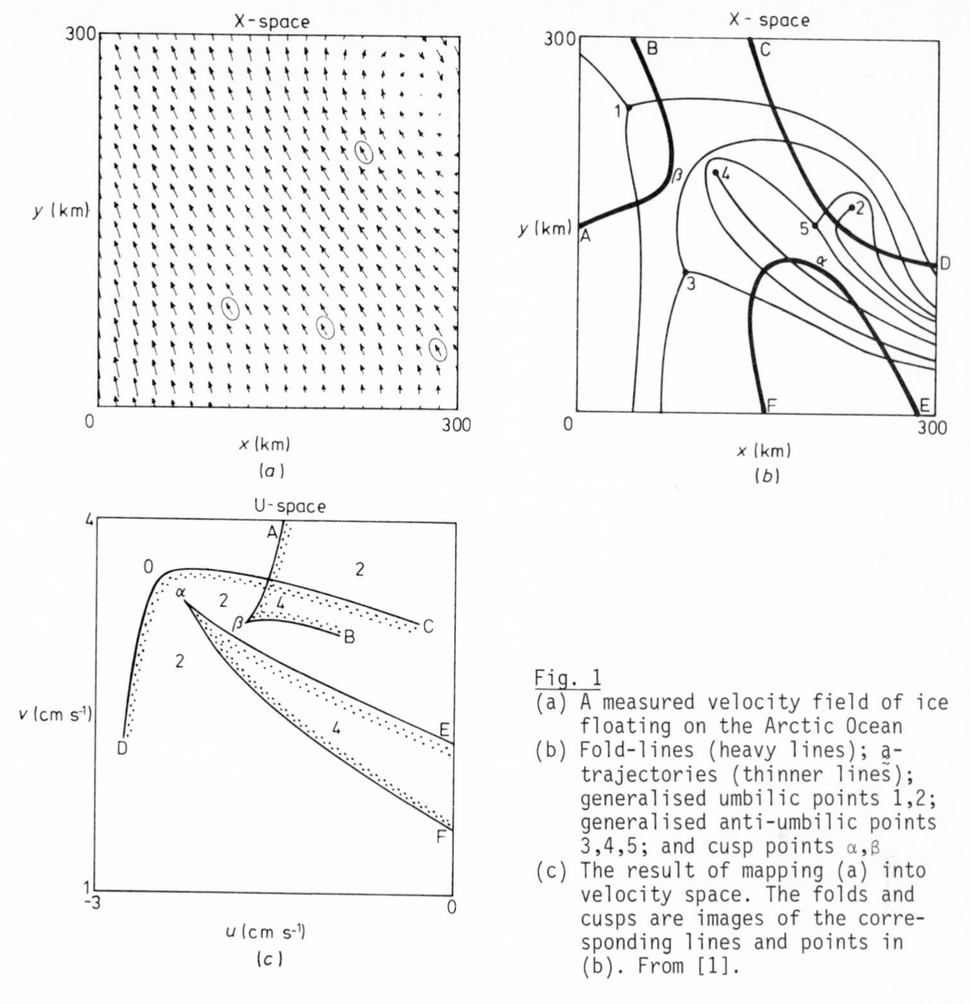

Fig. 1
(a) A measured velocity field of ice floating on the Arctic Ocean
(b) Fold-lines (heavy lines); $\underset{\sim}{a}$-trajectories (thinner lines); generalised umbilic points 1,2; generalised anti-umbilic points 3,4,5; and cusp points α,β
(c) The result of mapping (a) into velocity space. The folds and cusps are images of the corresponding lines and points in (b). From [1].

more points in (x,y) to the same point in (u,v). This is the basic geometry of the mapping.

The folds and cusps denote places where the spatial variations in velocity have a particular character. They have the important property that they are stable features not destroyed by small perturbations of the flow. Another way in which they are significant is the following. Imagine that the velocity (u,v) is measured at grid points in (x,y) (as the wind might be measured in an ideal meteorological experiment) and the velocity at each grid point is plotted as a dot in (u,v). The density of dots would be high at the folds and higher still at the cusps, becoming infinite at these places as the grid size is reduced to zero. The velocities are "focused" at the singularities.

Now allow the velocity field to evolve with time. The pattern of folds and cusps will change in a continuous way, except that at certain moments the folds and cusps will interact to produce what will be called an event. To analyse this consider the mapping $(x,y,t)\to(u,v,\tau)$ where $t=\tau$. τ is simply another time label. In maps from R^3 to R^3 the singular sets that are generic and stable are the fold, cusp and swallowtail; but because this particular mapping has the special feature that $t = \tau$ it turns out that there are certain restrictions on the orientation of the singular sets in (u,v,τ) space. The successive planes τ = constant slice through the singular sets to produce two sorts of events as seen in (u,v): (a) when τ = constant cuts a singular point, and (b) when τ = constant is tangent to a rib (the locus of a cusp in (u,v,τ)). (a) gives swallowtail events, while (b) gives beak-to-beak and lips events. These are the only generic events that will occur in a freely evolving flow field.

Figure 2 shows an example of the flow field of sea ice. Beak-to-beak events occur between the first and second and between the second and third time sections. Three swallowtail events occur between the third and fourth sections.

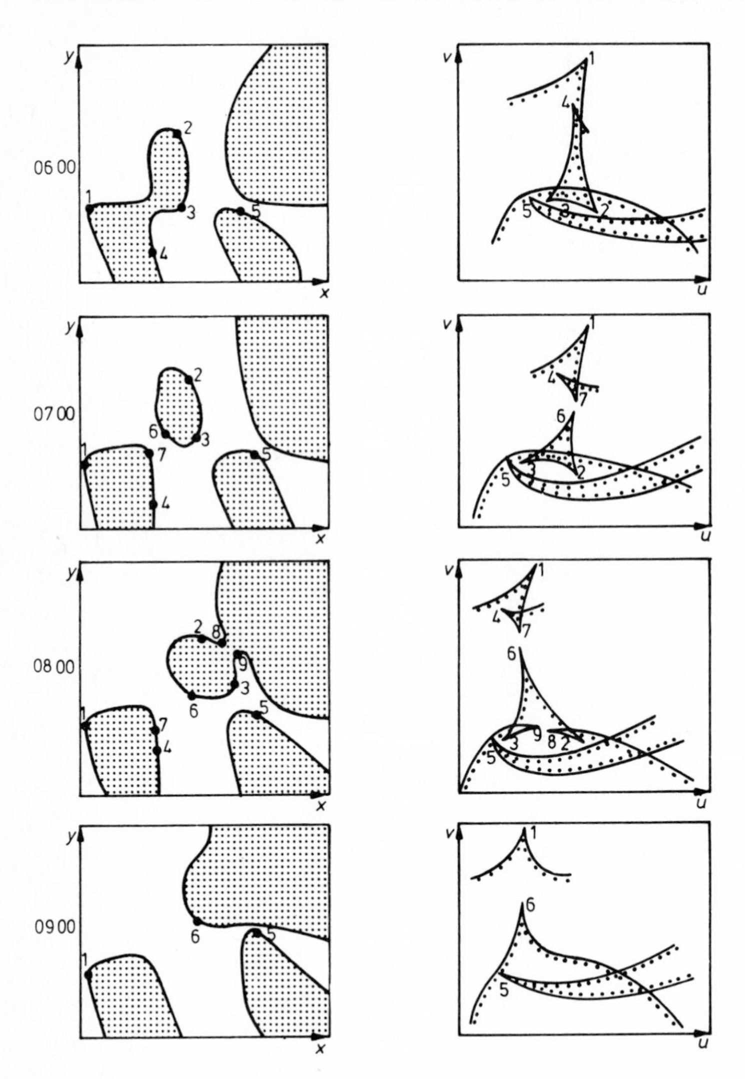

Fig. 2 A flow field of sea ice at hourly intervals. At each time the singular set (fold-lines and cusp-points) is shown in (x,y) and its image (folds and cusps) in (u,v). The (x,y) space is 600 km square and the (u,v) space is roughly 10 $cm\,s^{-1}$ square. Figure 1 is an expanded version of part of the first field. From [1]

However, the list of events is longer if the field is subject to the contstraint of being incompressible or irrotational. We can deal with these cases by using catastrophe theory. If we have a time dependent irrotational flow derived from a flow potential $\phi(x,y,t)$, we can construct a function Φ

$$\Phi(x,y,t;u,v,\tau) = \phi(x,y;\tau) + ux + vy + \frac{1}{2}t^2 - \tau t, \tag{1}$$

where (x,y,t) are state variables and (u,v,τ) are controls. Then the stationary condition

$$\Phi_x = \Phi_y = \Phi_t = 0 \tag{2}$$

(subscripts denote differentiation) gives at once

$$u = -\phi_x, \qquad v = -\phi_y, \qquad \tau = t,$$

which is the required flow. Thus the singular set in control space (u,v,τ) of the gradient mapping defined by (2) is precisely what we need for the problem of irrotational flow. A similar procedure can be used for incompressible flow by using the stream function ψ in place of ϕ. For both these constrained flows the stable singularities in (u,v,τ) are those on THOM's list [3] of codimension up to 3: fold, cusp, swallowtail, elliptic umbilic and hyperbolic umbilic. When we slice these singular sets with planes τ = constant, two new events appear, namely the elliptic and hyperbolic umbilics (Table 1a).

Constraint does not always lengthen the list. A two-dimensional field that is both irrotational and incompressible can contain only one sort of singularity (points) and as it evolves there are no events at all. A little constraint enriches behaviour; too much impoverishes it.

Figure 3 shows an incompressible time-evolving flow field: a geostrophic wind field computed from measured pressure values. A hyperbolic umbilic event occurs at the second time section: fold A is inside D in the first section and is outside D in the third section.

There is a perfect mathematical analogy [1] between patterns like Fig. 3 and the optical caustics produced at infinity by slightly perturbed plane wavefronts -- for example the image of a point made by a sheet of uneven glass [4].

3. Generalised Umbilic and Antiumbilic Points

Let L be the velocity gradient matrix

$$L = \begin{pmatrix} u_x & u_y \\ v_x & v_y \end{pmatrix} .$$

Then at a given point P we can find the vector $L\underset{\sim}{n}$, where $\underset{\sim}{n}$ is a unit vector in an arbitrary direction. As $\underset{\sim}{n}$ varies in direction at P there will be one direction of $\underset{\sim}{n}$ (and its opposite) that minimises $|L\underset{\sim}{n}|$. Following WHITNEY [2] we call this special direction $\underset{\sim}{a}$. The thinner lines in Figs. 1b and 3 are drawn parallel to $\underset{\sim}{a}$ at each point. These $\underset{\sim}{a}$-trajectories are significant in two respects. First, a point of tangency in the (x,y) space between an $\underset{\sim}{a}$-trajectory and a fold-line, as at α and β in Fig. 1b, determines a cusp in (u,v) space (α and β in Fig. 1c). Second, notice that there are discrete points in Fig. 1b where the $\underset{\sim}{a}$ direction becomes indeterminate. At these points L takes either the form

$$\begin{pmatrix} a & b \\ -b & a \end{pmatrix} \quad \text{or} \quad \begin{pmatrix} a & b \\ b & -a \end{pmatrix} \tag{3}$$

(generalised umbilic point u) (generalised antiumbilic point $\bar{u}$).

Points 1,2 are u, lying in regions where J > 0, while points 3,4, 5 are $\bar{u}$, lying in the region where J < 0.

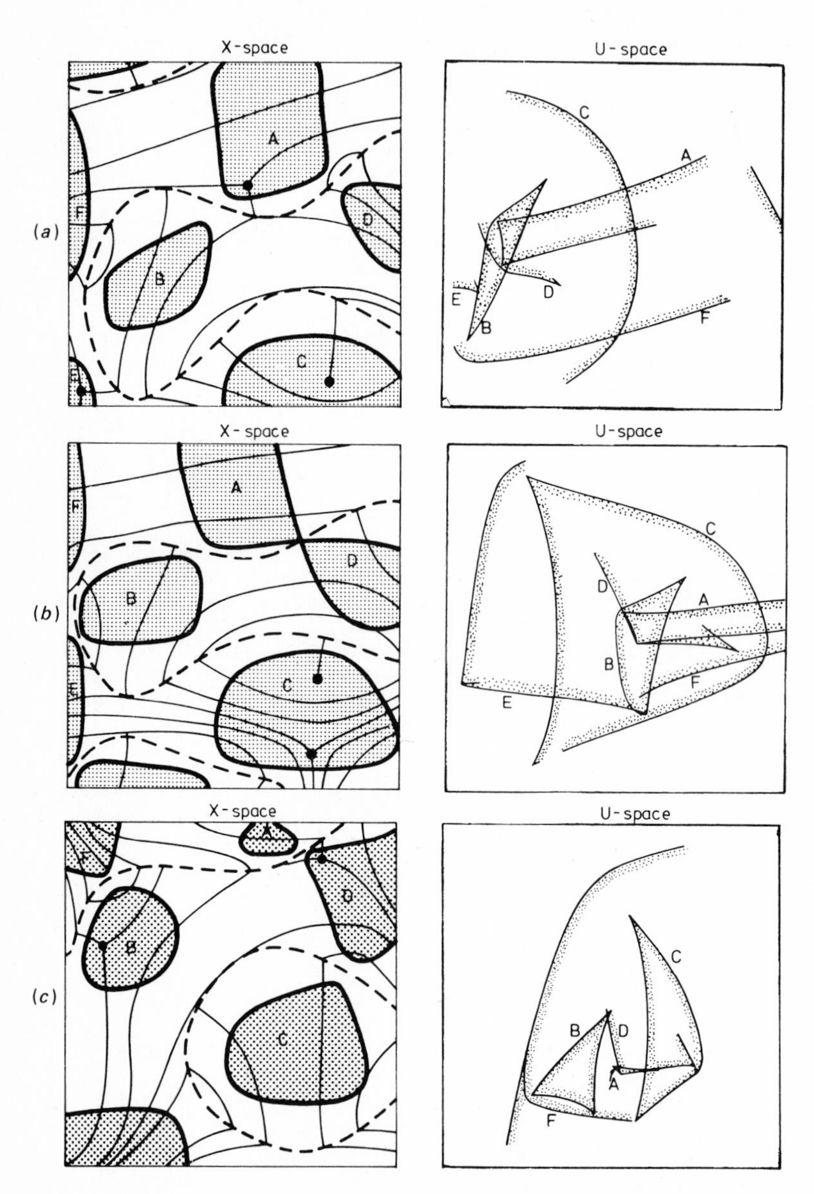

Fig. 3 A geostrophic wind field at 6 hourly intervals mapped as in Fig. 1. The (x,y) space is 1125 km square and the (u,v) space is about 20 m s^{-1} square. From [1]

If the field is irrotational or incompressible, one of the two conditions for an antiumbilic point ($u_x+v_y = 0$, $u_y-v_x = 0$) is satisfied everywhere. The remaining condition defines a <u>line</u> of antiumbilic points. Thus in Fig. 3 the broken line is an antiumbilic line, while the heavy dots are umbilic points. If an umbilic point collides with a fold-line we have an <u>umbilic event</u>, either elliptic or hyperbolic as previously described.

For these constrained flows (irrotational or incompressible) the velocity field is derived from the gradient of a scalar (ϕ or a stream function ψ). The scalar can be thought of as defining a surface of small slope whose height at each point (x,y) is ϕ or ψ (in Fig. 3 ψ is in fact proportional to the atmospheric pressure).

The $\underset{\sim}{a}$ direction then denotes one of the two directions of principal curvature of this surface. At an umbilic point of a surface the two principal curvatures, C_1 and C_2, are equal and the two principal directions of curvature are indeterminate. On the line where $C_1=-C_2$ the points are defined as antiumbilic.

A general vector field, on the other hand, is not derivable from a scalar; there is no underlying surface and so there are no umbilic or antiumbilic points. However, their place is taken by generalised umbilic and antiumbilic points, in the sense defined by (3).

BERRY and HANNAY [5] have shown that there are 4 different kinds of umbilic points of a surface depending on the local pattern of lines of curvature and whether they are elliptic or hyperbolic. In a similar way there turn out to be 6 different kinds of generalised umbilic points in a two-dimensional vector field and 6 different kinds of generalised antiumbilic points. This same classification applies to the isotropic points of any symmetric tensor, a stress field for example. Thus the positive and negative isotropic points of photoelasticity are generalised umbilic points and there are 6 different kinds.

4. Events in Evolving Three-dimensional Flows

An extension of the ideas in section 2 from two to three space dimensions, plus time, is given in [6]. The general approach is the same. At given time the singular set is defined by the vanishing of the Jacobian of the mapping from (x,y,z) to (u,v,w), where (u,v,w) are velocity components:

$$J = \partial(u,v,w)/\partial(x,y,z) = 0$$

The set S of points in (u,v,w) space thus defined form a surface consisting of fold sheets, which are sharply creased at ribs (lines of cusps). As the flow field evolves with time the three-dimensional figure S changes and we can recognise discrete events. To list the generic events in unconstrained flow fields (Table 1b) we first consider the stable singular sets in maps from R^4 to R^4: $(x,y,z,t)\rightarrow(u,v,w,\tau)$ with $t = \tau$. Then we slice them with τ = constant. As before, events can arise in two ways, either by τ = constant cutting a singular point in R^4 or from some kind of tangency. If the field is constrained to be irrotational we can again use catastrophe theory to list the singular sets, this time in R^4, and once again the list of generic events is different (Table 1b).

<u>Table 1</u> Events in evolving flow fields. (a) Two space dimensions [1]

Unconstrained	Irrotational or incompressible
Beak-to-beak	Beak-to-beak
Lips	Lips
Swallowtail	Swallowtail
	Elliptic umbilic
	Hyperbolic umbilic

Table 1 (b) Three space dimensions [6]

Unconstrained	Irrotational
Cusp events:	Cusp events:
Elliptic rib collapse (2 kinds)	Elliptic rib collapse (2 kinds)
Hyperbolic rib interchange	Hyperbolic rib interchange
Swallowtail pairs:	Swallowtail pars:
Head to head	Head to head
Tail to tail	Tail to tail
Side to side	Side to side
Elliptic umbilic	Elliptic umbilic pair
Hyperbolic umbilic	Hyperbolic umbilic pair
Butterfly (3 kinds)	Butterfly (3 kinds)
	Parabolic umbilic (2 kinds)

5. Discussion

We have described a way of looking at flow fields, or vector fields in general, which is natural to a mathematician who thinks of functions as mappings, but which is not the conventional way for a physicist. The most important property of the singularities in velocity space that we consider is that they are structurally stable: they are not destroyed by small perturbations of the flow. In fact they continue in existence from one event to the next. Thus the topological structure of the flow remains the same between events. In this sense the singularities in velocity space and the events which alter them can be said to characterise the evolution of the flow.

In conclusion I should like to say explicitly what is already evident from the references, that the work described here was done in collaboration with my colleagues Alan Thorndike and Clifford Cooley.

References

1. A.S. Thorndike, C.R. Cooley, J.F. Nye: J.Phys.A: Math. Gen. 11, No. 8, 1455-1490 (1978)
2. H. Whitney: Ann. Math., N.Y. 62, 374-410 (1955)
3. R. Thom: Structural Stability and Morphogenesis (Reading, Mass.: Benjamin) (Translation of Stabilité Structurelle et Morphogênèse 1972 (Reading, Mass.: Benjamin)) (1975)
4. J.F. Nye: This volume (1979)
5. M.V. Berry, J.H. Hannay: J.Phys.A: Math.Gen. 10, 1809-21 (1977)
6. J.F. Nye, A.S. Thorndike: Submitted to J.Phys.A: Math. Gen. (1979)

Wavefront Dislocations and Their Analysis Using Catastrophe Theory

F.J. Wright

H.H. Wills Physics Laboratory, University of Bristol, Tyndall Avenue
Bristol BS8 1TL, U.K.

Abstract

We aim to shed some new light on the concept of wavefront dislocations by using ideas from catastrophe theory. Wavefront dislocations were originally defined as phase singularities in a complex wavefunction, which provides a very convenient way to analyse their properties. The reason why wavefront dislocations are common features of wavefields is explained by considering some typical global behaviour of wavefronts. Catastrophe theory tells us that in two dimensions we should consider the wavefronts around a cusp caustic. Consideration of the way dislocations are likely to be observed experimentally suggests that they may be regarded as wavefront catastrophes of a real wavefunction. We can therefore use catastrophe theory to analyse the local forms of structurally stable dislocations.

1. Introduction

Wavefront dislocations were introduced into wave theory in 1974 by NYE & BERRY [1], to explain the experimentally observed appearance and disappearance of crest/trough pairs in a wavefield. Each crest/trough pair may be associated with a wavefront, and a wavefront dislocation is a line (in xyz-space) constituting the edge of a wavefront. Wavefront dislocations are so called by analogy with crystal defects.

To analyse wavefront dislocations Nye & Berry worked with a complex wavefunction, and defined the dislocation lines as the lines on which the phase of the complex wavefunction is indeterminate, which requires that the amplitude be zero. When working with a quantum mechanical wavefunction, which is essentially complex, this is an ideal definition. Several authors have studied quantum mechanical dislocations in the context of quantum whirlpools [2], quantized magnetic flux lines [3] and magnetic monopoles [4]. They play an essential role in the Aharonov-Bohm effect, in which a quantum wavefunction is affected by a zero magnetic field with a non-zero vector potential [5].

All physical wavefunctions are real, however, and any imaginary part is added purely for computational convenience. The above definition of wavefront dislocations has nevertheless proved extremely convenient. NYE [6] and BERRY, NYE & WRIGHT [7] have emphasized the way that dislocation lines underpin diffraction patterns and provide a useful description of their structure. The shortcomings of this theory of wavefront dislocations, which we will call the 'complex theory', are that it is somewhat removed from the way in which dislocations manifest themselves experimentally, and it requires more information than just the real physical wavefunction.

Before venturing into new theories of wavefront dislocations, we review the complex theory of Nye & Berry, making a few generalizations. After introducing some catastrophe theory, we use it to gain some insight into why wavefront dislocations must occur in a general wavefield.

If wavefront dislocations are fundamental features of all wavefields, as we believe they are, it should be possible to construct a theory using only a real wavefunction, based upon the way that dislocations are actually observed. The main aim of this paper is to suggest a possible way of doing this, using catastrophe theory. This approach has the advantage that it should facilitate analysis of the structural stability of wavefront dislocations, although here I present no more than a basis for such analysis.

2. Complex Theory of Wavefront Dislocations

We shall represent a general time dependent scalar wave by the complex wavefunction

$$\psi(\underline{r},t) = \rho(\underline{r},t)\, e^{i\chi(\underline{r},t)} \tag{1}$$

at a point $\underline{r}$ and time t, and we call the real valued functions ρ and χ respectively the amplitude and phase of the wave. It is understood that for most physical applications we take the real part of ψ.

We wish to know what singularities may occur in the wavefronts, so our first task is to decide what we mean by wavefronts. Intuitively we probably think of the crests or troughs of a real wave as the wavefronts, but there are crests in space at a fixed time, and crests in time at a fixed space point. It is also common theoretically to regard the wavefronts as surfaces of constant phase, so what is the relationship among these different definitions?

Let us begin by considering the simplest possible case: the monochromatic plane wave

$$\psi(\underline{r},t) = A\, e^{i(\underline{k}.\underline{r} - \omega t)}.$$

The amplitude A and phase $(\underline{k}.\underline{r} - \omega t)$ are uniquely defined by the real part of ψ alone, so we can construct the complex wavefunction from the real wave. The crests in space and the crests in time both correspond to a phase of $2n\pi$, therefore our three possible definitions of wave fronts are identical.

Now consider a general monchromatic wave:

$$\psi(\underline{r},t) = A(\underline{r})\, e^{i(B(\underline{r}) - \omega t)}.$$

Again the amplitude and phase are uniquely defined by the real part of ψ alone, because we know the time variation. The crests in time still correspond to a phase of $2n\pi$, but the crests in space correspond to phases varying with position. Perhaps we should regard crests in time or equiphase surfaces as wavefronts?

When we return to our general time dependent wavefunction (1) we find that both crests in time and crests in space correspond to different phases, which vary with space and time. Clearly, we must make an arbitrary choice of our definition of wavefronts, and when using a complex wavefunction the most convenient choice is to define wavefronts to be surfaces of constant phase. In practice, we only consider a quasimonochromatic wave (i.e. a wave whose amplitude varies with time much more slowly than its phase) and then the difference between the different notions of wavefront is small.

The problem in using phase to define wavefronts is that in the general case it is not possible to tell from a real wavefunction what the phase is, because the amplitude and phase of (1) are not uniquely defined by the real part of ψ alone. We also need the imaginary part. For a monochromatic wave we can compute the imaginary part from the real part by taking the time derivative, or by taking the real part at time $t \pm \pi/2\omega$, but for a general wave we cannot rigorously do this. (For a quasimonchromatic wave we can construct a plausible imaginary part, but it is not unique, unless the unmodulated carrier is also available. Using the latter, WALFORD, HOLDORF & OAKBERG [8] were able to produce a very useful amplitude-phase display for quasimonochromatic waves.)

NYE & BERRY [1] explain that the complex wavefunction (1) really corresponds to performing the following experiment: observe a real wavefunction (Re ψ), then shift the carrier phase by $\pi/2$ and observe the resulting real wavefunction (Im ψ). The significance of the amplitude $\rho(\underline{r},t)$ is that it is the envelope of the real wavefunction as the carrier phase is varied. The fact that this procedure is rarely followed in practice provides part of the motivation for this paper.

We consider wavefunctions of the form (1), whose wavefronts satisfy $\chi(\underline{r},t)$ = constant. Since the wavefunction $\psi(\underline{r},t)$ is assumed to be analytic everywhere, the phase $\chi(\underline{r},t)$ is well defined (modulo 2π) everywhere except where $\psi(\underline{r},t) = 0$ (i.e. $\rho(\underline{r},t) = 0$). We define a wavefront dislocation to be a connected set of points in $(\underline{r},t)$ satisfying $\rho(\underline{r},t) = 0$, which requires two conditions (Re $\psi = 0$ and Im $\psi = 0$) to be satisfied. Therefore, wavefront dislocations are singularities of codimension 2, i.e. in a two-dimensional (x,y) space at fixed t they are points, in (x,y,t) space-time they are lines and in (x,y,z) space at fixed t they are lines (the dislocation lines usually referred to), etc.

The nature of the singularity is that if we follow a closed path encircling the dislocation once, the phase changes by a total of $2n\pi$. n is an integer which defines the strength of the dislocation, relative to some specified direction round the path, and we admit the degenerate case n = 0. To see this, consider $\psi(\underline{r},t)$ as a mapping from space-time into the complex plane; the dislocation maps into the origin. A closed circuit C in space-time maps into a closed circuit C' in the complex plane (since $\psi(\underline{r},t)$ is single valued). C' will encircle the origin an integer number of times, which is the number of units of 2π by which the phase has changed. A continuous change in C will produce a continuous change in C'. The number of times that C' encircles the origin can only change if C' crosses the origin, i.e. if C crosses the dislocation. If we forbid circuits from intersecting the dislocation, then all topologically equivalent circuits C produce the same value of n, so the dislocation strength n is a topological invariant.

We can similarly prove that a dislocation cannot end, because if it did the space-time circuit C could be shrunk to a point not on the dislocation without passing through it. This implies that the image circuit C' could be shrunk to a point other than the origin without passing through the origin, which is clearly impossible. Therefore, dislocations must either form closed loops, or pass out of the domain of the wavefield (i.e. 'end' on boundaries such as reflecting or radiating surfaces). They may interact with each other such that the total strength is conserved.

An alternative view of dislocation strength results from writing it as

$$\frac{1}{2\pi} \oint_C d\,\chi(\underline{r},t) = \mathrm{Re}\left\{ \frac{1}{2\pi i} \oint_{C'} \frac{d\psi}{\psi} \right\} .$$

But by residue calculus

$$\frac{1}{2\pi i} \oint_{C'} \frac{d\psi}{\psi} = n \text{ (which is real)}$$

if C' encircles the pole (of $1/\psi$) at the origin n times as C encircles the dislocation once. Therefore, dislocation strength is given by

$$\frac{1}{2\pi i} \oint_C \frac{d\psi(\underline{r},t)}{\psi(\underline{r},t)} .$$

The phase of a wavefunction increases by 2π between successive wavefronts, therefore if we move along some path P such that the phase changes by 2m , then we must have passed through m different wavefronts (each one may have been traversed any odd number of times) as in Fig. 1, where m = 3.

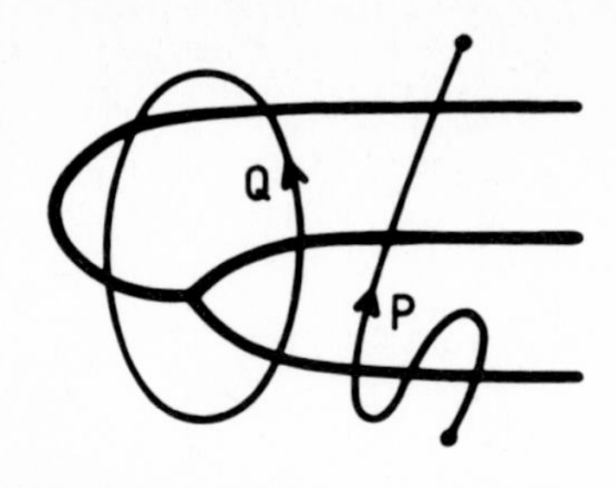

Fig. 1 Paths through a set of wavefronts

If the path is closed, such as path Q in Fig. 1, then we must have passed around the ends of m wavefronts. Therefore, a dislocation of strength m constitutes the coincident edges of m wavefronts. Figure 2 shows the full phase structure around a typical single strength dislocation in two dimensions, both locally and globally. Phase 0 is emphasized to represent wavefronts. Note the essential phase saddle (here at phase π).

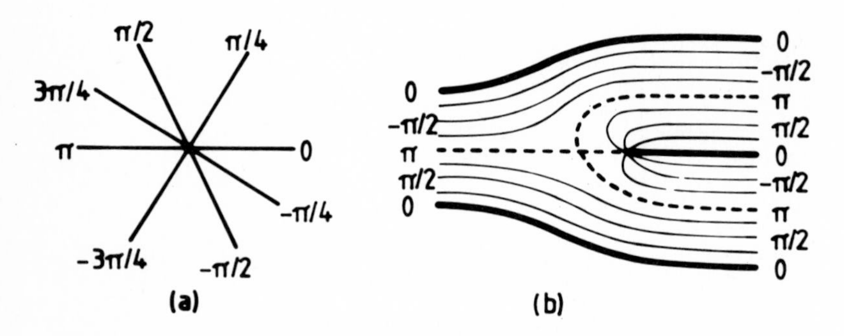

Fig. 2 Local (a) and global (b) equiphase lines around a single strength dislocation in two dimensions

3. Local Complex Models

We have been discussing the properties of a hypothetical wavefunction, without showing that solutions of the wave equation exist which exhibit such properties. To avoid the problem of specifying and satisfying appropriate boundary conditions, we shall construct the simplest possible local wavefunction in the neighborhood of a dislocation. Higher order terms could in principle be added to satisfy global constraints, without changing the local dislocation structure. These local models allow us to study the possible configurations of wavefronts around a dislocation, and the behaviour of the dislocation as time varies.

Let us consider what dislocation structure we expect in a three-dimensional wavefield, by analogy with crystal defects. We shall regard the dislocation as a local perturbation of a plane wave. If the dislocation line lies in the plane of the asymptotic wavefronts, we say it is of pure edge type; if it is perpendicular we say it is of pure screw type. Anything else is of mixed screw-edge type, and may be characterized by the angle of the dislocation line to the plane of the asymptotic wavefronts. For example, a single strength straight pure edge dislocation is essentially Fig. 2b translated along the dislocation, as shown in Fig. 3a, whereas a double straight pure screw dislocation is a right helicoid, as shown in Fig. 3b. All dislocations except a pure edge link sets of wavefronts together, e.g. that in Fig. 3b links alternate wavefronts into two sets, since it has strength 2.

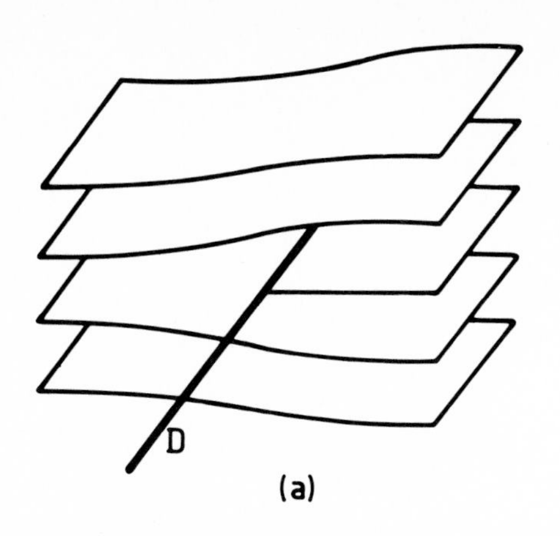

Fig.3 Wavefronts propagating upwards around two straight dislocation lines labelled D:
(a) single strength pure edge; (b) double strength pure screw

Motion of a dislocation perpendicular to the asymptotic planes is called glide and motion in the asymptotic planes is called climb; general motion is a combination of the two. A pure screw dislocation can only climb. A general dislocation is a twisted space curve, whose type and motion change along its length.

Following closely the methods of NYE & BERRY [1], we consider wavefunctions which are polynomial modulations of plane waves travelling along the z-axis. The simplest, which we shall call static, have z and t appearing only as $\zeta = z - ct$, and therefore travel rigidly along the z-axis at the wave velocity c. This case is the closest analogy to a fixed crystal dislocation. It is to be contrasted with dislocations in monochromatic waves, which are fixed in space while the wavefronts sweep round them. The dislocations are really gliding backwards with glide velocity c relative to the asymptotic wavefronts.

Since we are setting up a local model, it should suffice to consider straight dislocation lines, if they exist. It can be shown [9] that the most general straight static dislocation has the local form

$$\psi(x,y,\zeta) = (x + \beta_s y + \beta_e \zeta)^n \, e^{ik\zeta} \tag{2}$$

where β_s and β_e are not both real, and $\beta_s = \pm i$ unless $n = 1$. The strength of the dislocation is n, as is readily verified by considering the mapping of some simple circuit in (x,y,ζ). The equation of the dislocation line is

$$\left.\begin{aligned} \mathrm{Re}\,(x + \beta_s y + \beta_e \zeta) &= 0 \\ \mathrm{Im}\,(\beta_s y + \beta_e \zeta) &= 0 \end{aligned}\right\}$$

and the asymptotic wavefronts satisfy ζ = constant.

$\beta_e = 0$ gives the equation of the dislocation line as $x = y = 0$, which is pure screw. If β_s is pure real, the dislocation line is given by

$$\left.\begin{aligned} x + \beta_s y &= 0 \\ \zeta &= 0 \end{aligned}\right\}$$

which is pure edge, but this satisfies the wave equation only if $n = 1$. In fact, it can be proved [9] that static dislocations which are globally pure edge can only have single strength. The general form of (2) represents a multiple mixed dislocation.

One could go on in this vein to generate hosts of more complicated sets of curved dislocations. More interesting is to consider simple models of moving dislocations, for which z and t do not appear only as $\zeta = z - ct$. This requires more complicated modulation than the essentially linear functions used above, but it is possible [1 & 9] to generate models of sets of dislocations which glide and climb, pairs of equal and opposite strength which annihilate each other, or are born together, etc. However, we do not need to consider the equations of these for our present purposes.

To conclude our discussion of complex models of dislocations, we present a model satisfying the global constraint that the wavefunction tends to zero at infinity. With the assistance of Dr. John Hannay, the following wavefunction in spherical polar coordinates was constructed:

$$\psi(r,\theta,\phi,t) = \left(\frac{1}{kr} + \frac{i}{(kr)^2}\right) \sin\theta\, e^{i(kr + \phi - \omega t)}.$$

This is a radiating dipole field with a straight single strength pure screw dislocation along the z-axis, which should be physically realizable.

4. The Cuspoid Catastrophes

The essence of catastrophe theory [10, 11, 12] is an answer to the question [11, p. 27]: in a k-parameter family of functions, which local types do we typically meet? We only need to consider functions ϕ of a single variable t, which restricts our interest to the cuspoid catastrophes. We wish to know how the form of our function can vary in real space, so we consider three-parameter families. But it may be that not all three parameters play a significant role (because of symmetries, for example). One significant parameter gives us the fold catastrophe, for which the form of the function $\phi(t)$ varies with x as

$$\phi(t;x) = t^3/3 + xt. \qquad (3)$$

Two significant parameters give the cusp catastrophe, for which $\phi(t)$ varies with x and y as

$$\phi(t;x,y) = t^4/4 + xt^2/2 + yt, \qquad (4)$$

and all three parameters give the swallowtail catstrophe, for which

$$\phi(t;x,y,z) = t^5/5 + zt^3/3 + yt^2/2 + xt.$$

t is called the state variable, and the parameters x,y,z are called control variables, implying that the controls are set, and then behaviour with respect to t is analysed.

The important property of these families of functions is that they are structurally stable. This means that a small perturbation of the form of one member of a family produces (after a smooth change of coordinates) just another member of the same family. It is this stability which makes them typical, because the occurrence of a particular family does not depend on any special circumstance occurring.

The critical points of these so-called potential functions satisfy $\partial\phi/\partial t = 0$, and for certain values of the control parameters two or more critical points coalesce, when $\partial^2\phi/\partial t^2 = 0$. The set of control parameters on which this occurs is called the bifurcation set, because it bifurcates the control space (locally) into tow regions in which the number of critical points differs by 2. The names of the catastrophes (other than the fold) are due to the shapes of the bifurcation sets, as sketched in Fig. 4.

Also of interest are the graphs of the critical values of ϕ, which we call ϕ_c, as a function of the control parameters. Let us write the cuspoid potential functions in the general form

$$\phi(t;\underline{x}) = t^n/n + \sum_{r=1}^{n-2} x_r\, t^r/r,$$

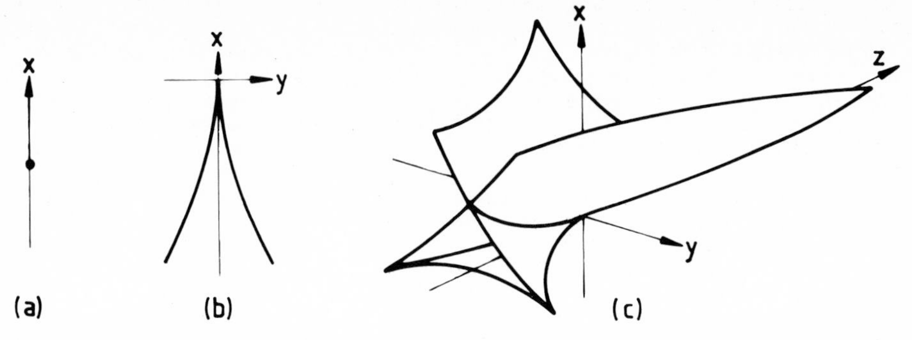

Fig.4 Bifurcation sets of the first three cuspoid catastrophes: (a) fold; (b) cusp; (c) swallowtail

then it can be shown that the graph of the critical values of a cuspoid potential function of degree n is identical to the bifurcation set of the potential function of degree $(n+1)$ with its control parameter x_1 replaced by $-\phi_c$, and x_r replaced by $nx_{r-1}/(r-1)$ for $r = 2$ to $(n-1)$. The cusp-line of the critical value graph of ϕ projects down the ϕ_c axis into its own bifurcation set.

5. Why Do Wavefields Need Dislocations?

We have considered the local structure of wavefront dislocations, and proved that they can occur. But why should they occur? This is an intuitive question, and we shall suggest an intuitive answer. The fact is that they do occur. To understand why, we must study the global wavefront structure of some typical wavefield.

For simplicity we restrict ourselves to a two-dimensional space, in which the wavefronts are lines. We will generate a monochromatic wavefield geometrically by specifying an initial wavefront, and constructing its subsequent form as time progresses, in the spirit of BERRY [13]. A wavefront evolves geometrically such that after a time T every point of the wavefront has travalled a distance $D = cT$ along its local normal or ray, where c is the wave velocity. If a finite length of wavefront evolves into a point, then that point is a local focus, which is generally part of a caustic. The caustic is the envelope of all the rays, that is, the evolute of all the wavefronts, and conversely the wavefronts are all different involutes of the caustic. Our problem is to construct the whole family of involutes from one given number.

Consider the point P at (X,Y) a distance $\rho(s;X,Y)$ from a point Q on the initial wavefront, parametrized by s. PQ is a ray or wave normal if $\partial\rho/\partial s = 0$, i.e. if ρ takes a critical value $\rho_c(X,Y)$. The wavefront after a time T is the set of points (X,Y) for which ρ is critical and equal to D, i.e. it is the section $\rho_c(X,Y) = D$ through the critical value graph of ρ. On the caustic, two solutions of the ray equation $\partial\rho/\partial s = 0$ coalesce, so that $\partial^2\rho/\partial s^2 = 0$ also.

We know that a plane wavefront will only evolve into another plane wavefront, so let us consider the next simplest case of a parabolic initial wavefront having equation

$$Y = as^2$$

where s is the X coordinate of a point on the initial wavefront. Then

$$\rho(s;X,y) = \sqrt{(X-s)^2 + (Y-as^2)^2}\,.$$

We can solve $\partial\rho/\partial s = \partial^2\rho/\partial s^2 = 0$ to give the equation of the caustic:

$$(Y - 1/2a)^3 = 27X^2/16a. \tag{5}$$

This is a cusp of precisely the same form as the bifurcation set of the cusp catastrophe, so catastrophe theory tells us that a smooth change of coordinates will map $\rho(s;X,Y)$ onto the cusp catastrophe potential function (4). It also tells us that ρ has the most complicated local structure which will typically occur in a two-dimensional parameter space, so we know that we have captured the most complicated local wavefront structure typical of two dimensions. Near the cusp point the mapping is approximately

$$\rho(s;X,Y) \simeq a^3\ \phi(s;x(Y),\ y(X)) + Y$$

where $x(Y) \approx (1 - 2aY)/a^3$ and $y(X) \approx -2X/a^2$.

Each wavefront is given by

$$\rho_c(X,Y) \approx a^3\phi_c(x(Y),\ y(X)) + Y = D$$

for a particular value of D. In (x,y) coordinates the wavefronts are given locally by the intersection of the critical value graph $\phi_c(x,y)$ with the plane

$$\phi_c = (D - 1/2a)/a^3 + x/2a^2.$$

(The mapping from ρ to ϕ is such that the sections through the critical value graph are not given by ϕ_c = constant, which we might initially have expected.) Therefore, we expect the geometrical wavefronts around the cusped caustic to look like particular sections through the critical value graph of the cusp catastrophe, which is the bifurcation set of the swallowtail catastrophe, sketched in Fig. 4c. (The generalization to higher dimensions must similarly be true.)

To actually plot the wavefronts it is just as easy to work with $\rho(s;X,Y)$ rather than $\phi(s;x,y)$ and so solve the problem exactly. In Fig. 5 we plot the cusped caustic (5) with $a = 1/2$, and a set of exact wavefronts with D increasing in steps of 1/2 from 0. They have precisely the swallowtail form which we predicted.

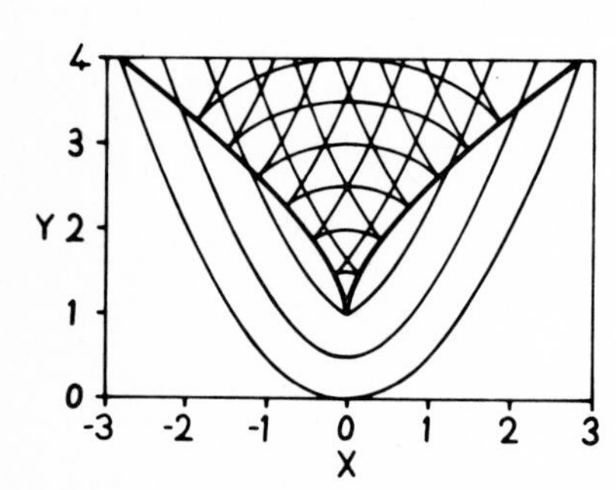

Fig.5 Geometrical wavefronts around a cusped caustic

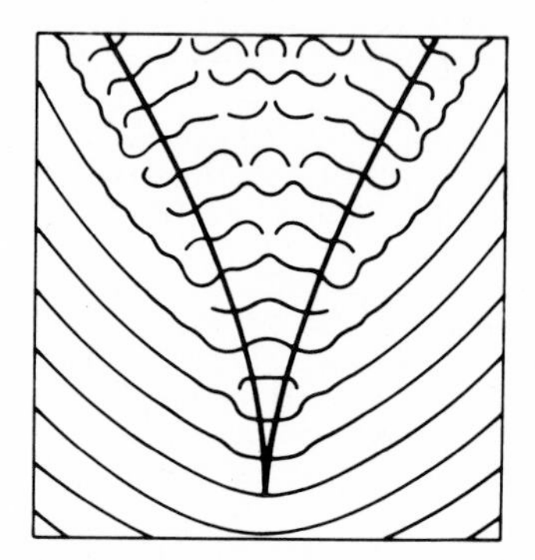

Fig.6 Diffraction wavefronts around a cusped caustic (enlarged)

One might imagine that the peculiar form of these wavefronts results from having considered a special case (a parabolic wavefront), but our invocation of catastrophe theory assures us that these are a typical set (locally) of two-dimensional geometrical wavefronts with wavelength D. The wavefronts below the caustic are simple, but above the caustic each wavefront has three branches, corresponding to the three rays present in this region. Those parts of the wavefronts which are concave upwards have yet to focus on the caustic; those parts (above the caustic) which are convex upwards have already focused on the lower part of the caustic, and are now diverging.

The point we wish to illustrate is the way that the wavefronts get tangled up above the caustic, becoming littered with singularities in the form of cusps and intersections. These wavefronts are geometrical; we have made no reference to wavefunctions or phase. Diffraction causes the three rays above the caustic to interfere and produce a single-valued complex wavefunction having a well defined phase everywhere, except at dislocations. The diffraction wavefronts are defined as equiphase lines at a particular value of phase, which will not typically coincide with the value at a phase saddle. It is even less likely to coincide with a phase value at which a saddle and extremum have coalesced. Therefore, diffraction wavefronts will not typically display intersections or cusps.

As we 'turn on' the diffraction, the geometrical wavefronts must disentangle themselves as they turn into diffraction wavefronts. In doing so, they tear and shed sections of wavefront which are left as disconnected pieces. The tears on which the sections of wavefront end are wavefront dislocations. An enlarged view of a typical set of diffraction wavefronts is shown in Fig. 6, taken from computations in [9]. Analysis of the dislocation structure around a caustic requires a study of the appropriate diffraction catastrophe, which is currently being pursued at Bristol [6,7,9,13].

We see from Fig. 6 that the cusp diffraction catastrophe has a row of dislocations just below the caustic, to cater for the geometrical wavefront cusps, and a triangular array of dislocation pairs above the caustic, to cater for the intersections of the geometrical wavefronts. After looking at Fig. 5 it is not surprising that the dislocations above the caustic fall into a triangular lattice, nor that the lattice planes are distorted to be convex upwards like the diverging geometrical wavefronts. We conjecture that use of the critical value graphs combined with stationary phase methods could provide a powerful tool for studying the dislocation structure of diffraction catastrophes.

6. Catastrophe Theory of Wavefront Dislocations

Catastrophe theory can be used to analyse the local structure of wavefront dislocations in a way closely related to their experimental observation. Consider a real wavefunction ψ, and imagine observing it on an oscilloscope by placing a receiver at some point $\underline{r}$. The oscilloscope will display ψ as a function of time t, and the form of the function which we observe will vary with $\underline{r}$, i.e. we are observing a family of functions of t parametrized by $\underline{r}$. We wish to know what local form this family can typically have, and catastrophe theory gives us the answer. To fit our previous notation we will write our wavefunction as $\psi(t;\underline{r})$.

It is convenient to introduce some terminology for moving dislocations which is more specific than previously. To distinguish the form of a dislocation in space-time from its spatial manifestation at some fixed time, we shall call the former the dislocation world set (borrowing relativistic nomenclature), and the latter the dislocation set. The projection of the world set into space gives the dislocation trajectory, the point set traced out by the dislocation set as it moves with time. The dislocation set itself is a section at constant time through the world set.

The closest we can get to our previous definition of wavefronts as equiphase lines of a complex wavefunction is to define them as alternate zeros, say every zero following a crest. This corresponds exactly to choosing a phase $\pm\pi/2$ for the complex wavefunction. Therefore, the wavefronts are given by alternate solutions of

$$\psi(t;\underline{r}) = 0.$$

A wavefront ends where two consecutive zeros coalesce, i.e. where

$$\partial\psi(t;\underline{r})/\partial t = 0,$$

so the dislocation is given by the condition that ψ is critical and zero. The dislocation trajectory is the section

$$\psi_c(\underline{r}) = 0$$

through the critical value graph of ψ. When we apply catastrophe theory and map our wavefunction ψ onto a catastrophe polynomial ϕ, we can add any function of $\underline{r}$ to ϕ, just as happened in the previous section. Catastrophe theory allows the dislocation trajectory to look like any section through the critical value graph $\phi_c(\underline{r})$, and the wavefronts to look like any section through the graph of $\phi(t;\underline{r})$. All it tells us is what the whole family of such dislocations looks like. We conclude that we have not chosen a suitable definition of wavefronts to which to apply catastrophe theory alone.

The alternate definition of wavefronts is as crests (or troughs) on the oscilloscope display, given by

$$\partial\psi(t;\underline{r})/\partial t = 0.$$

The surface in $(t,\underline{r})$ space satisfying this equation is called the critical (or catastrophe) manifold, and in our case a section t = constant through this critical manifold gives the wavefronts at time t, plus the complementary troughs (or crests). A wavefront ends where a crest and trough coalesce, i.e. where

$$\partial^2\psi(t;\underline{r})/\partial t^2 = 0.$$

The dislocation trajectory is precisely the bifurcation set, and the world set is the fold on the critical manifold, which projects down the t-axis into the bifurcation set. Notice that now the actual values of ψ are not relevant.

Let us begin with the simplest case of one significant space variable x, so that the wavefunction typically has the fold catastrophe form (3) locally:

$$\psi(t;x) = t^3/3 + xt.$$

The crests and troughs have equation $t^2 + x = 0$, and the dislocation occurs at $t = x = 0$. Fig. 7 shows the crest wavefront as a thick line, the trough as a thinner line. Superimposed on this is the wavefunction at three particular values of x. Some global structure, outside the scope of catastrophe theory, has been added as dashed lines to show how the local structure might arise in a realistic wavefield.

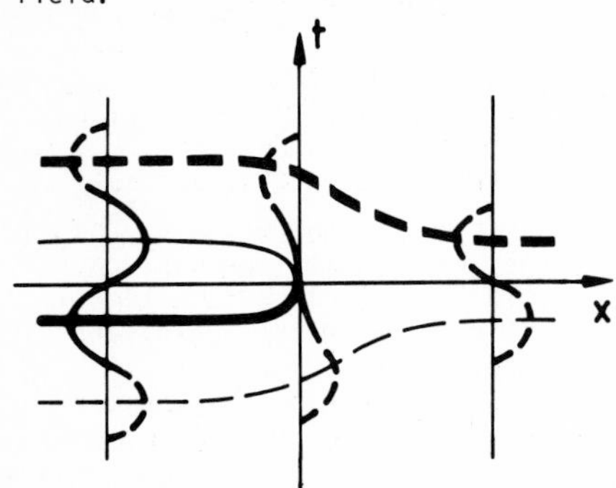

Fig.7 Wavefronts (thick) and wavefunctions around a 'fold dislocation'. Global structure is shown dashed

The behaviour of this wavefunction is precisely the type of observation which led Nye & Berry to the idea of wavefront dislocations.

We could perhaps realize this one-dimensional wave on a string, but it is more interesting to consider waves in two dimensions. The easiest two-dimensional dislocation to generate is the static dislocation obtained by replacing t by $\zeta = z - ct$. This gives essentially the same wavefronts in xz space as those in xt space shown in Fig. 7, but sweeping up the z-axis with velocity c.

Suppose instead we include another control variable y by making a smooth coordinate change such that

$$\psi(t;x,y) = t^3/3 - yt^2 + (x + y^2)t.$$

Then the critical manifold becomes

$$t = \pm\sqrt{-x} + y$$

as shown in Fig. 8.

The dislocation trajectory (bifurcation set) is the y-axis and the dislocation world line (fold line) has equation $x = 0$, $y = t$. The local wavefront is the half-parabola in which the plane t = constant cuts the critical manifold above the fold. As time increases the wavefront and dislocation point move together in the y-direction with uniform speed, but we cannot really distinguish glide and climb in this local theory, because we have no asymptotic wavefronts.

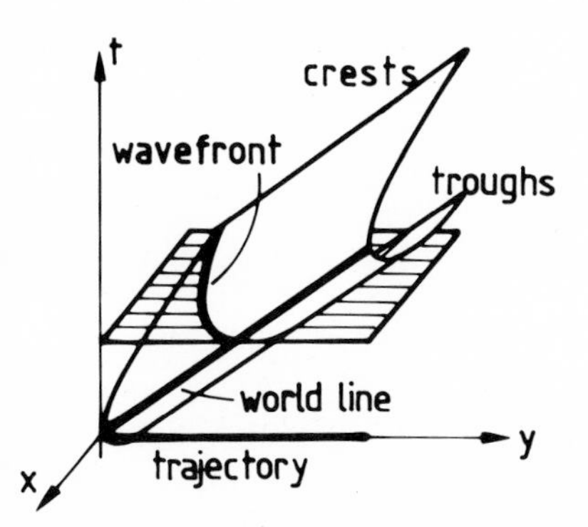

Fig.8 A two-dimensional 'fold dislocation'

By making more complicated coordinate changes we can distort the global shape of the critical manifold almost anyhow, and with it the dislocation world line and wavefronts. The velocity of the dislocation along its trajectory depends on the angle the world line makes with the t-axis, which can be anything other than zero for a fold catastrophe. In particular, the world line cannot be a straight line parallel to the t-axis, would give the trajectory as a point. We shall see later that this would be the case for a monochromatic dislocation. This is not a catastrophe and is not structurally stable, because a small perturbation tilting the world line would turn the point trajectory into a line. If the world line is perpendicular to the t-axis, a wavefront appears or disappears abruptly; this is the limit of a dislocation moving with infinite velocity, which is permissible [1].

If we give the world line a local minimum or maximum with respect to time, then we will produce the birth or death of a pair of dislocations. By giving the world line a cubic-type kink we can produce dislocation 'skip', in which a dislocation moves along a trajectory and a pair is born some way in front of it. One member of this pair moves backwards and annihilates the original trajectory as though it were the original which had 'skipped over' part of its trajectory. A multiply kinked trajectory produces multiple skip. Catastrophe theory tells us that all these events are typical: for example, dislocations are born, move at varying speeds and skip, in the soundfield of a pulsed piston radiator [9].

Once again we can make this model into a static dislocation in three dimensions by replacing t by $\zeta = z - ct$. Then at fixed t the whole critical manifold above or below the fold becomes the wavefront and the fold becomes the dislocation line.

As time progresses, this whole structure sweeps along the z-axis, and the dislocatic line sweeps out a trajectory surface. A static dislocation line in three dimensions can have any shape allowed for a fold line, which is always a regular curve; singularities only occur in the trajectory surface, as we shall see.

Finally let us look briefly at the cusp catastrophe, whose critical manifold is sketched in Fig. 9. The fold line twists around such that at one point its tangent is vertical, and it is this point which projects into the cusp point of the bifurcation set.

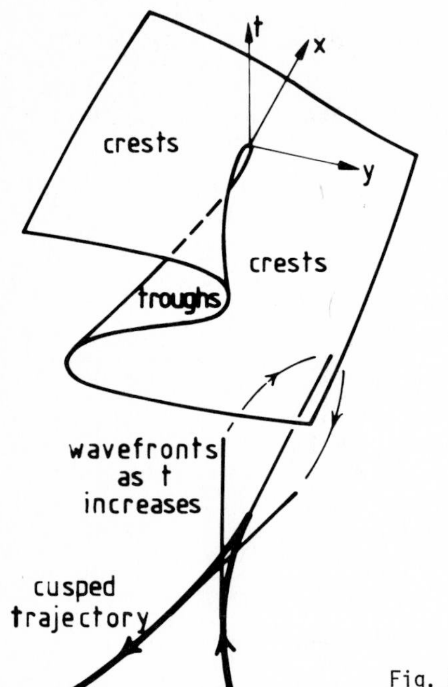

Fig. 9 A two-dimensional 'cusp dislocation'

Planes t = constant cut the critical manifold of the standard cusp in straight line and as t varies these straight lines roll around the cusp. Each such wavefront lin ends on a dislocation where it touches the cusp, and as the wavefronts swing round, the dislocation traces out the cusped trajectory. At the cusp point itself, the dislocation stops its motion and doubles back. A distorted critical manifold would produce curved wavefronts, as for the fold in Fig. 8, but their local behaviour wou be the same.

If we make this model into a static three-dimensional model by replacing t by ζ, the dislocation line is the regular fold line, but the trajectory surface has a cusp line in it. The wavefunction is always locally symmetrical in the time-like variable at the cusp point, whereas at a fold point it is locally antisymmetrical, so a symmetrical dislocation is much less likely than an antisymmetrical one.

The most complicated typical behaviour in three dimensions is given by the swallowtail catastrophe. The standard form will give the dislocation line as a straight line, and the wavefronts will be planes rolling round the swallowtail trajectory surface sketched in Fig. 4c. Since the swallowtail critical manifold has four sheets 'inside the tail', we have the possibility of two dislocations interacting. However, a double dislocation requires that the first four derivative of ψ vanish (two per dislocation), and this only happens at a single point (the origin) for the swallowtail. Thus a double dislocation line corresponds to a line of swallowtail points, which requires at least four control variables. Therefore, we will not typically encounter double dislocation lines in three dimensions unless we can vary at least one extra control variable to look for them. Similarly a triple dislocation line would require six control variables to occur typically.

We have found that motion, birth and death of dislocations are typical occurrences, depending on the global form of the wavefield. Distinctions between glide and climb, edge and screw cannot effectively be made in this local theory. In fact, such distinctions can be very difficult to make in practice for anything but the simplest (i.e. nearly plane) wavefield! What catastrophe theory really applies to is the structure of a dislocation in the neighbourhood of a singularity of its trajectory. Rather than pursue the general theory into the wilderness of higher dimensions, let us see how some of the basic theory works in practice, when applied to the simple complex model wavefunction we analysed earlier.

7. A Simple Example

In section 3 we showed that a suitable local complex wavefunction for a static straight single strength mixed screw-edge dislocation is, from (2) with $n = 1$,

$$\psi(t;x,y,z) = (x + \beta_s y + \beta_e(z - ct))\, e^{ik(z - ct)}$$

where β_s and β_e are not both real. To apply catastrophe theory we require a real wavefunction, therefore to preserve generality we include a constant phase ϕ before taking the real part. We define $T = \omega t - kz$, factor out β_e for convenience (assuming $\beta_e \neq 0$), and take our general real wavefunction to be

$$\psi = \mathrm{Re}\ (\alpha x + \beta y + T)\ e^{i(\phi - T)}$$

$$= (X + T) \cos (T - \phi) + Y \sin (T - \phi)$$

where $X + iY \equiv \alpha x + \beta y$.

If we shift T and X by $T \to T + \phi$, $X \to X - \phi$ we get

$$\psi = (X + T) \cos T + Y \sin T,$$

i.e. by suitable coordinate changes we have absorbed all the arbitrary constants. In particular, changing the projection of our complex wavefunction into a real wavefunction (e.g. changing from taking the real part to taking the imaginary part) only produces a translation in the XT plane

The dislocation trajectory in catastrophe theory is found by solving

$$\partial\psi/\partial T = \partial^2\Psi/\partial T^2 = 0$$

to be

$$X = \pi/2 \pm \{\mathrm{atan} \sqrt{-(Y + 2)/(Y + 1)} - \sqrt{-(Y + 2)\ (Y + 1)}\}.$$

Requiring also $\partial^3\psi/\partial t^3 = 0$ gives the cusp points as

$$X = -T = -(n + 1/2)\pi, \quad Y = -2.$$

The trajectory is a periodically cusped line as shown in Fig. 10, and is essentially a z = constant section through the trajectory surface in (x,y,z) space. It looks like a projection of a helix, which it is because the dislocation world line (the fold line) is a distorted helix. To see this we define

$$\xi = X + T, \quad \eta = Y + 3/2$$

giving the solution of (8) as

$$\xi^2 + \eta^2 = (\tfrac{1}{2})^2, \ \tan T = -(1/2 - \eta)/\xi$$

which is a regular circular helix of radius 1/2 and pitch 45^0 in (ξ,η,T). The mapping back to (X,Y,T) shears the helix so that its axis lies at 45^0 to the (X,Y) plane. Then at Y = -1 the helix is parallel to the (X,Y) plane, whereas at Y = -2 it is perpendicular to it. These perpendicular points project into the cusp points of the trajectory, as indicated on Fig. 10.

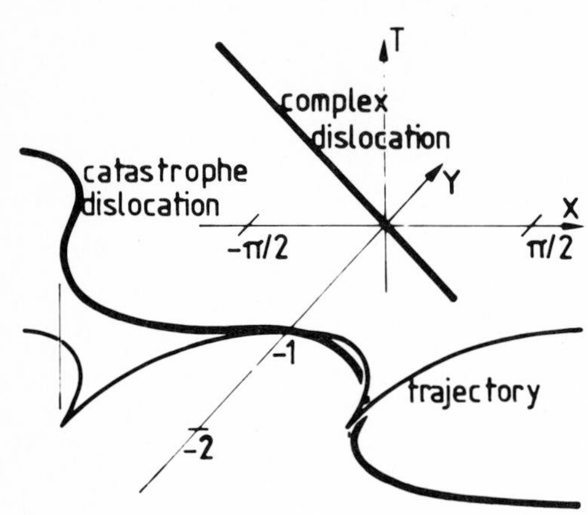

Fig. 10 The static straight single strength mixed screw-edge dislocation

The original complex wavefunction in the new coordinates is

$$\psi = (X + iY + T)\, e^{-iT}.$$

The 'complex' dislocation line, defined as a zero of amplitude, is given by

$$X = -T, \ Y = 0.$$

This is a straight line parallel to the axis of the helical catastrophe dislocation line, and a small distance (3 helix radii) away from it.

In the pure edge limit of the complex dislocation α and β are pure imaginary, so that X is replaced by ϕ which is then not removable. In the pure screw limit $|\alpha|$ and $|\beta| \to \infty$ so that the scale of the mapping of xy into XY increases. In xy space the catastrophe dislocation line aproaches the T-axis, and in the limit coalesces with it. Then the fold line is a straight line parallel to the T-axis, which is not a catastrophe, and is not structurally stable, as we mentioned before. But screw dislocations are monochromatic; in fact, we could write a general monochromatic wavefunction as

$$\psi = X \cos T + Y \sin T,$$

which would give the fold line as

$$X = Y = 0.$$

Therefore, monochromatic dislocations are not structurally stable. This is not surprising, because their trajectories would be points (in two dimensions, say), but the slightest perturbation would turn these into lines.

Pure edge dislocations are structurally stable in space-time, except for values of ϕ which occur on a YT section through a cusp in the trajectory. In this case the dislocation will be symmetric, so a symmetric pure edge dislocation is not structurally stable in space-time alone; only the family around it with varying ϕ is stable. Therefore, we will not typically see symmetric pure edge dislocations.

Before leaving this example, let us examine the wavefronts. The critical manifold has equation

$$Y + 1 = \xi \tan T$$

which is a right helicoid, as sketched in Fig. 3b, with a T-repeat of π and centred on $\xi = 0$, $Y = -1$. The mapping back to (X,Y,T) shears this helicoid along X just as it did the fold line. The sheet of this sheared helicoid which cuts the plane $T = 0$ in the half-line $Y = -1$, $X > 0$ is the crest wavefront. The surfaces of phase 0 and π of the complex wavefunction

$$\psi = (X + iY + T)\, e^{-iT}$$

have the equation

$$Y = \xi \tan T$$

and the sheet with phase 0 passes through $T = 0$ along the half-line $Y = 0, X > 0$. Thus the complex wavefront with phase 0 is identical to the crest wavefront, shifted one unit along the Y axis!

8. Conclusion

We have seen that wavefront dislocations can exist and why they should exist, and we have considered two theories of the local structure of wavefront dislocations. The complex theory consists essentially of thinking up ad hoc wavefunctions, and then analysing them. The catastrophe theory has more direct connection with experiment, and should be able to provide a rigorous classification of the local structure of stable dislocations. Our example has shown that the geometry of catastrophe dislocations is much more complicated than the equivalent complex dislocation. In fact, catastrophe analysis of anything but the simplest complex dislocation has so far proved intractable, and is probably not the best way to proceed.

Perhaps the way forward lies in an amalgamation of the two approaches which combines the simplicity and tractability of the complex theory with the power of catastrophe theory to assess structural stability, by using catastrophe theory to suggest suitable wavefunctions for complex analysis. We have tried to show that such a goal is worth pursuing, and to set up a basis from which to pursue it.

Acknowledgement

I should like to thank Dr.M.V.Berry and Prof.J.F.Nye for suggesting aspects of this work to me, and for many valuable discussions. I should also like to acknowledge the financial support of the Science Research Council.

References

1. J.F.Nye, M.V.Berry: Proc. Roy. Soc. Lond. A 336, 165 - 190 1974
2. J.O.Hirschfelder, A.C.Christoph: J. Chem. Phys. 61, 5435 - 55 (1974)
   J.O.Hirschfelder, C.J.Goebel, L.W.Bruch: J. Chem. Phys. 61, 5456 - 59 (1974)
   J.O.Hirschfelder, K.T.Tang: J. Chem. Phys 64, 760 - 785 (1976)
   J.O.Hirschfelder, K.T.Tang: J. Chem. Phys. 65, 470 - 486 (1976)
3. J.Reiss: Ann. Phys. 57, 301 - 321 (1970)
   J.Reiss: Phys. Rev. D2, 647 - 653 (1970)
   J.Reiss: Phys. Rev. B13, 3862 - 69 (1976)
4. P.A.M.Dirac: Proc. Roy. Soc. A133, 60 (1931)

5. Y.Aharonov, D.Bohm: Phys. Rev. 115, 485 (1959)
R.G.Chambers: Phys. Rev. Let. 5, 3 (1960)
H.Erlichson: Am. J. Phys. 38, 162 - 173 (1970)
M.V.Berry, R.G.Chambers, C.Upstill: in preparation (1979)
6. J.F.Nye: Optical Cuastics and Diffraction Catastrophes - this volume (1979)
7. M.V.Berry, J.F.Nye, F.J.Wright: Phil. Trans. Roy. Soc. Lond., in press (1979)
8. M.E.R.Walford, P.C.Holdorf, R.G.Oakberg: J. Glaciol. 18, 217 - 229 (1977)
9. F.J.Wright: Wavefield Singularities (Ph.D. Thesis, Univerity of Bristol) (1977)
10. R.Thom: Structural Stability and Morphogenesis (Reading, Mass: Benjamin) (1975)
11. T.Poston, I.N.Stewart: Taylor Expansions and Catastrophes (London, Pitman) (1976)
12. T.Poston, I.N.Stewart: Catastrophe Theory and its Applications (London, Pitman) (1978)
13. M.V.Berry: Adv. Phys. 25, 1 - 26 (1976)

Part V

Statistical Mechanics and Phase Transitions

Structural Stability in Statistical Mechanics

Mario Rasetti

Istituto di Fisica, Politecnico di Torino, I-10100 Torino, Italy

1. Introduction

In spite of the fact that phase transition theory presents one of the most formidable problems of mathematical physics, a phase transition is--in an intuitive way--an elementary phenomenon. It is characterized by the fact that a system undergoing a phase transition is "stable" both below and above the critical temperature (or, more generally, in some subregions of the thermodynamic state space) in the sense that variations of the temperature (or of the whole set of parameters characterizing the state of the system) within a phase do not change the gross properties of the state itself. Such a kind of stability is lost at the temperature of phase transition (or on the set of critical points which separate different stability domains, in the above control parameter space).

It is therefore quite appealing to try to build up a sort of analogous dynamical system, whose set of equilibrium configurations--continuously defined or parametrized over the state space P (in most cases either a Euclidean space of a differentiable manifold of finite dimension)--identifies, through the pattern of its structural stability picture, both the nature and the localization of the possible phase transitions [1].

More precisely, we are led to study the set χ of all such dynamical systems at once, defining a topology over it and hence an equivalence relation (connected for instance to the existence of some homeomorphism identifying different orbits) between arbitrarily close systems. In this significance structural stability is obviously a global requirement, implying that the perturbation of the whole system leaves its quality preserved no matter what the initial conditions are.

Thus, just the major problem of structural stability theory [2], namely deciding whether stable systems are open-dense in the space χ of flows, and classifying them accordingly would rephrase our problem.

In particular, if we write

$$\chi = S \cup \Sigma, \tag{1.1}$$

where S is the open-dense set of stable systems, the determination of Σ--the complementary set corresponding to unstable systems--and of its structure would then not only lead to isolating the thermodynamic states at the phase transitions but to recognizing classes of equivalence of different systems.

The natural way of doing this is moving a generic point $p \in P$ through P: correspondingly, the state of the system parametrized by p will move within χ describing some orbit. As the latter crosses Σ at $p = p_c$, the system abruptly changes its quality.

In general Σ is stratified into strata of various codimension and a trajectory in χ parametrized by a point p of dimension n_p will meet only strata of codimension

$\leq n_p$: the analysis of such strata will indicate the generic ways in which a system can bifurcate.

The previous intuitive procedure--appealing as it may sound at first--is not as simple to realize in practice, and in some instances it might be incorrect in rigorous terms.

There are, though, examples in which it is indeed implicitly applied.

The first part of this paper, heuristic in character, will first briefly deal with a couple of such examples, successively showing how their intrinsic nature of approximate scheme or mean-field theory devoids them of the proper universal character.

The second part will then try a different pattern of attack, still exploiting globality as the leading characteristic of the problem, and discuss the point of view on which work is in progress [22].

2. Phase Transitions as Structural Instabilities

2.1 The Renormalization Group

There are several instances in which the logical scheme discussed in the Introduction is--more or less declaredly--applied: most famous among them the renormalization group method [3].

The latter is an approximation scheme which has produced plenty of interesting estimates and a somewhat deeper insight into many questions connected with critical phenomena. It is based on two different concepts: first, the scale invariance of a system near a phase transition, second, the description of the system itself in terms of an effective probability distribution u_p, embodying the totality of its statistical properties at the point $p \in P$.

Let us denote by Θ the space of the u_p's.

The rescaling procedure induces orbits on Θ, which may be thought of as the result of the action of some transformation group G. A dynamical system can then be associated to this situation by requiring that its flow in Θ for a given p be just the one whose orbits coincide with the orbits induced by the operations of G:

$$du_p/dL \equiv \dot{u}_p = X_p(u_p), \tag{2.1}$$

where the variable L parametrizing the orbit is, e.g., the scaling length.

It should be noted here that the differential equation (2.1) is indeed a limiting representation of a discrete recursive scheme. It is known [4] that recursion relations of this type possess a rich structure, characterized by successive bifurcations and by the appearance of stability sets which asymptotically approach universality. The latter are in fact independent on the explicit form of X_p, but are controlled only by its local structure in the neighborhood of its stationary points.

Let us denote by u_p^* an attractor, namely a fixed point under G-action

$$G\, u_p^* = u_p^*. \tag{2.2}$$

The feature of the attractor is implied by the matrix

$$J_{j,k}(p) = \partial X_p^j(u_p^*)/\partial u_p^{*k} \tag{2.3}$$

having only eigenvalues with strictly negative real part.

On the other hand, the Jacobian determinant being non-zero, by the implicit function theorem there exists a continuous function u_p^* such that

$$X_p(u_p^*) = 0, \tag{2.4}$$

which just defines u_p^*--the fixed point under rescaling--as an equilibrium state for our fictitious dynamical system.

As p moves in P, different "accidents" may occur. The crossing of the boundary between two basins of attraction for instance may correspond to the condition

$$\partial X_p(u_p^*)/\partial u_p^* = 0. \tag{2.5}$$

In other words, events like the coalescing and disappearing of two distinct orbits may take place, or a limit point may turn into a limit cycle, etc. We can consider the set Σ defined by Eq. (2.5) as the set of singular points of a vector field over P, formally getting from Eq. (2.4)

$$\partial p/\partial u_p^* = -[(\partial X_p/\partial u_p^*)/(\partial X_p/\partial p)] \equiv Y(p) \tag{2.6}$$

whereby the singular points are just $p = p_c$ such that

$$Y(p_c) = 0, \tag{2.7}$$

obviously equivalent to (2.5).

The latter is consistent with Wilson's theory [3,5], and Eq. (2.6) is in fact Wilson's renormalization group equation.

Notice that the group referred to as renormalization group is not G, but actually the group generating Eq. (2.6).

The system of Eqs. (2.4) and (2.5) is characteristic of catastrophe theory [6] which provides a tool for unfolding at once the manifold variety of possible situations.

I will not focus my attention here on this aspect of the picture because it is clear that its suggestions, however nontrivial and helpful in a heuristic sense, rely heavily on the basic approximations of the scheme, which are to be further discussed and clarified. Therefore, in order to do so, and before trying to attack the problem in more general terms, let me discuss in some detail another simple example.

It will require some approximations itself, but it is more closely in the perspective of gaining an insight into the structure of exact solutions.

2.2 A One-dimensional Crystal

The system is an N-site one-dimensional lattice. On each of its sites a random configuration "spin" variable s_i, $i = 1, \ldots, N$, assuming--with equal a priori probabilities--the values ± 1 (whereby $s_i^2 = 1, \forall i$).

The energy is given by

$$H = \frac{1}{2} \sum_{i=1}^{N} \sum_{j=1}^{N} s_i J_{ij} s_j + \mathcal{B} \sum_{i=1}^{N} s_i \tag{2.8}$$

where the interaction matrix $J = (J_{ij})(i = 1, \ldots, N;\ j = 1, \ldots, N)$ is required to have strictly positive determinant and $\mathcal{B}$ is an external magnetic field.

The thermodynamic equilibrium properties of the system are completely determined by the canonical partition function

$$Z_N = \sum_{\{s\}} \exp(-\beta H), \qquad \beta = 1/K_B T \tag{2.9}$$

where {s} denotes the set of all possible configurations of the spin variables, T is the temperature, and K_B Boltzmann's constant; together with the set of all n-point functionals.

Let us focus our attention on Eq. (2.9). The first step in computing Z_N is to map the stochastic process in the discrete variables s_i into another stochastic process in a set of N variables x_i, i = 1, . . ., N; each $x_i \in \mathcal{R}$ ranging continuously from $-\infty$ to $+\infty$. This can be done by using Schlömilch's [7] integral representation for the Gaussian form obtained upon insertion of (2.8) into (2.9), and performing explicitly the summation--now trivial--over {s}.

One obtains

$$Z_N = (2/\pi)^{N/2} \, 1/\sqrt{\det J} \int_{-\infty}^{+\infty} \cdots \int \prod_{i=1}^{N} \{dx_i \ \cosh \left[\sqrt{\beta}\, x_i - \beta \mathcal{B}\right]\}$$

$$\times \exp \{-\frac{1}{2} \sum_{i=1}^{N} \sum_{j=1}^{N} x_i \ (J^{-1})_{ij} \ x_j\}. \tag{2.10}$$

The reason for performing such a map is worth some further analysis, and I make a short digression here to devote a little attention to it, trying to state things in general terms, the meaning of which will hopefully be clear in the further manipulations of Eq. (2.10).

It is known [8] that an axiomatic quantum field theory in the sense of Wightman can be characterized in terms of a set of monotone continuous linear functionals--the set of all n-point functionals or n dynamical variable equal time expectation values--on the Borchers algebra.

The test-function space may be thought of as an abstract complete Hausdorff locally convex vector space E over the field $\mathcal{C}$ of complex numbers. The topology τ on E induces the projective topology $\tau^{(n)}$--itself Hausdorff and locally convex--with respect to which the completion $E^{(n)}$ of the n-fold tensor product

$$E^{\otimes n} = E \otimes E \otimes \ldots \otimes E \ (n \text{ times}) \tag{2.11}$$

can be performed.

The tensor algebra of E is

$$\mathcal{T}(E) = \bigoplus_{n=0}^{\infty} (E^{\otimes n}, \tau^{(n)}) \tag{2.12}$$

giving by completion

$$\overline{E} = \bigoplus_{n=0}^{\infty} E^{(n)}; \quad E^{(0)} = \mathcal{C} \tag{2.13}$$

We assume that all $E^{(n)}$ (and therefore $\overline{E}$) are barreled and that E (and hence all $E^{(n)}$ and $\overline{E}$) is nuclear [9].

If $\mathcal{M}^{(2)}$ denotes the set of all square summable infinite matrices $A = \{A_{ij}\}_{ij \in \mathcal{Z}}$ of complex numbers:

$$A_{ij} \in \mathcal{M}^{(2)} \ \text{iff} \ \sum_{i=1}^{\infty} |A_{ij}|^2 < \infty, \ \sum_{j=1}^{\infty} |A_{ij}|^2 < \infty \tag{2.14}$$

then let us define $\mathcal{M}^{(2)}(E)$ to be the set of all *-homeomorphisms m from E into $\mathcal{M}^{(2)}$ such that

$$x \longrightarrow \left[\sum_{i=1}^{\infty} |m_{ij} \ (x)|^2\right]^{\frac{1}{2}} = q_j(x), \qquad x \in E \tag{2.15}$$

for all $j \in \mathcal{Z}$ is a continuous seminorm.

$\mathcal{M}^{(2)}(E)$ has a proper subset $\mathcal{M}^{(2)}_{\infty}(E)$ defined in the following way: (i) all $m \in \mathcal{M}^{(2)}_{\infty}(E)$ generate an associative *-algebra $\mathcal{A}(m)$ in $\mathcal{M}^{(2)}$, (ii)

$$\bigwedge_{n \in \mathcal{Z}} x_1 \otimes \ldots \otimes x_n \longrightarrow \left[\sum_{i=1}^{\infty} |(m(x_1)\ldots m(x_n))_{ij}|^2 \right]^{\frac{1}{2}} = q_j^{(n)}$$

is a continuous seminorm on $E^{\otimes n}$ for all $j \in \mathcal{Z}$. We think of the matrices m as acting in some Hilbert space l^2 of square summable sequences $Z = (Z_i)_{i \in \mathcal{Z}}$ of complex numbers. We will denote by $\{e_n;\ n = 0,1,2,\ldots\}$ an orthonormal basis of l^2.

Now any continuous linear monotone functional

$$W = \{1, W_1, W_2, \ldots\} \tag{2.16}$$

on $\overline{E}$ corresponds to some $m = m_W \in \mathcal{M}^{(2)}_{\infty}(E)$ such that the restriction of W to $\mathcal{T}(E)$ has the following representation

$$W_n(x_1 \otimes \ldots \otimes x_n) = \sum_{j_1=0}^{\infty} \cdots \sum_{j_{n-1}=0}^{\infty} m_{oj_1}(x_1)\, m_{j_1 j_2}(x_2) \ldots m_{j_{n-1}o}(x_n) =$$

$$= \langle e_o, m(x_1) \ldots m(x_n) e_o \rangle \tag{2.17}$$

where all $x_i \in E$. Of course the matrix m is determined by (2.17) only up to a unitary transformation (in the sense of von Neumann [10]).

Moreover, each state W on $\overline{E}$ is the limit of a sequence of finite order matrix states

$$\{W^{(\mathcal{N})}\}_{\mathcal{N} \in \mathcal{Z}} = \{W_n^{(\mathcal{N})};\ n = 0,1,2,\ldots\}$$

such that the convergence

$$W_n^{(\mathcal{N})}(x_1 \otimes \ldots \otimes x_n) \xrightarrow[\mathcal{N} \to \infty]{} W_n(x_1 \otimes \ldots \otimes x_n) \tag{2.18}$$

is weak on $\mathcal{T}(E)$.

All this structure takes an especially manageable form in the case when E is one-dimensional, say $E = \mathcal{C}\, e_1$.

Then $E^{(n)} = \mathcal{C}\, e_1 \otimes \ldots \otimes e_1$ and the topology of the locally convex sum (2.13) is defined by all sequences $\{c_n \geq 0\}_{n \in \mathcal{Z}}$. With the definition of inner product

$$(\zeta, z) = \sum_i \zeta_i\, z_{n-i} \tag{2.19}$$

and involution

$$(z)^{\dagger} = \left(\sum_{n=0}^{\infty} z_n e_n \right)^{\dagger} = \sum_{n=0}^{\infty} z_n^* e_n, \tag{2.20}$$

$\overline{E}$ becomes a barreled nuclear locally convex *-algebra with unit.

We are interested in the sequences of positive type, namely those sequences $c = \{1, c_1, c_2, \ldots\}$ such that for all $\mathcal{N} \in \mathcal{Z}$ and all $\{z_0, z_1, \ldots, z_{\mathcal{N}}\} \in \mathcal{C}^{\mathcal{N}+1}$ we have

$$\sum_{n=0}^{\mathcal{N}} \sum_{m=0}^{\mathcal{N}} z_n^* c_{n+m}\, z_m \geq 0. \tag{2.21}$$

They are in one-to-one correspondence with the normalized linear functionals (2.16).

Finally theorem (2.17) generates the following characterization: a sequence c is of positive type iff there is an element $m = m_c \in \mathcal{M}^{(2)}_{\infty}(\mathcal{C})$ such that

$$c_n = \sum_{j_1=0}^{\infty} \cdots \sum_{j_{n-1}=0}^{\infty} m_{oj_1} m_{j_1 j_2} \cdots m_{j_{n-1}o};\quad c_1 = m_{oo}, \tag{2.22}$$

in other words iff there exists a symmetric operator A, in some separable Hilbert space $\mathcal{X}$ containing a cyclic unit vector e_o for $\{A^n \mid n = 0,1,2,\dots\}$ such that

$$c_n = (e_o, A^n e_o) \qquad \forall\, n \geqslant 0. \tag{2.23}$$

Then it has been shown by Brüning [11] that each $m_c \in \mathcal{M}^{(2)}_{\infty}(E)$ satisfying (2.22) and (2.23) is unitarily equivalent to a standard form element of $\mathcal{M}^{(2)}_{\infty}(E)$

$$a = \begin{vmatrix} a_{oo} & a_{o1} & o & o & o & \dots \\ a_{1o} & a_{11} & a_{12} & o & o & \dots \\ o & a_{21} & a_{22} & a_{23} & o & \dots \\ o & o & a_{32} & a_{33} & a_{34} & \dots \\ \cdot & \cdot & \cdot & \cdot & \cdot & \\ \cdot & \cdot & \cdot & \cdot & \cdot & \\ \cdot & \cdot & \cdot & \cdot & \cdot & \end{vmatrix} \tag{2.24}$$

Eq. (2.24) suggests--in view of the thermodynamic limit to be eventually performed on our system--that the most general form for J^{-1} in Eq. (2.10) be of the standard type (2.24).

Actually, with no real loss of generality and in order to simplify the notation, I restrict the attention to the case

$$(J^{-1})_{ij} = -\rho\,\delta_{ij} + \zeta^{(+)}\,\delta_{i,j-1} + \zeta^{(-)}\,\delta_{i,j+1}. \tag{2.25}$$

This choice on the one hand gives a more transparent structure of differential operators to the density matrix (in the x representation) entering Eq. (2.10) and guarantees that J_{ij} has long enough range to make a phase transition possible according to van Hove's theorem. On the other hand it has interesting connections also with the general question of the integrability conditions and stochastization time of the system.

In a system with a finite number N of degrees of freedom, by virtue of the Liouville theorem, the system itself is fully integrable (i.e., it is possible to separate the variables and to introduce action-angle variables) if there exist N commuting integrals of motion. The integrable systems are not completely randomized, since there is no exchange of energy between the degrees of freedom.

The only possible cause of stochastization is then the deviation of the discrete system from a fully integrable continuous limit.

The choice (2.25) gives rise to a completely integrable system when periodic boundary conditions are assumed.

Define now the integral (Hilbert-Schmidt) kernel

$$K(x|y) = \langle x|\,\mathbb{T}\,|y\rangle = \sqrt{2/\pi}\,\left[\det J\right]^{-N/2} \sqrt{\cosh\{\sqrt{\beta}x - \beta\mathcal{B}\}}\,\cdot$$
$$\cdot\sqrt{\cosh\{\sqrt{\beta}y - \beta\mathcal{B}\}}\,\exp\{-\tfrac{1}{2}\,[(\zeta^{(+)}+\zeta^{(-)})(y-x)^2$$
$$-\,(\zeta^{(+)}+\zeta^{(-)}-\rho)(x^2+y^2)]\}. \tag{2.26}$$

One has obviously

$$Z_N = \int_{-\infty}^{+\infty}\!\!\dots\int \prod_{i=1}^{N} (dx_i \langle x_i|\,\mathbb{T}\,|x_{i+1}\rangle) = \mathrm{Tr}(\mathbb{T}^N) \tag{2.27}$$

if one imposes the periodic boundary condition $x_{N+1} \equiv x_1$. In other words, Z_N is the N-th iterate of $K(x|y)$ and may be written

$$Z_N = \sum_{K=1}^{\infty} \lambda_K^N, \tag{2.28}$$

where λ_K are the eigenvalues of the integral equation

$$\int dy\ K(x|y)\ \Phi(y) = \lambda \Phi(x) \tag{2.29}$$

The larger N (the "thermodynamic limit"), the stronger is the relevance in Z_N of the maximum eigenvalue

$$\lambda_M = \max_K \{\lambda_K\}, \tag{2.30}$$

so that the thermodynamic potentials, e.g. the free energy,

$$F = -(1/\beta N)\ \ell n\ Z_N \underset{N\to\infty}{\approx} -(1/\beta)\ \ell n\ \lambda_M \tag{2.31}$$

are essentially controlled by its behavior.

Obviously, any pathologic situation is then determined by a bifurcation of λ_M as the control parameters (in this simple example only two, β and $\mathcal{B}$) vary. Let us simplify further the notation, setting

$$t = (\zeta^{(+)} + \zeta^{(-)})^{-1} \tag{2.32}$$

$$\Omega = \rho - \frac{1}{t} \tag{2.33}$$

$$\Lambda_K = \frac{1}{2}\sqrt{\pi/2t}\ (\det J)^{1/2N}\ \lambda_K \qquad \forall K \geqslant 0 \tag{2.34}$$

$$V(x) = (2/t)\ \{\Omega x^2 - \ell n\ \cosh\ (\sqrt{\beta}x - \beta\mathcal{B})\} \tag{2.35}$$

so that Eq. (2.29) reads

$$1/\sqrt{2\pi t} \int_{-\infty}^{+\infty} dy\ \exp\{-\frac{t}{4}\ V(y)\}\ \exp\{(-1/2t)(x-y)^2\}\ \exp\{-\frac{t}{4}\ V(x)\}\ \Phi(y) = \Lambda\ \Phi(x). \tag{2.36}$$

By the well known formal identity

$$(1/\sqrt{2\pi t}) \int_{-\infty}^{+\infty} dy\ \exp\{(-1/2t)(x-y)^2\}\ f(y) = \exp\{\frac{t}{2}\ d^2/dx^2\}\ f(x), \tag{2.37}$$

Eq. (2.36) can be cast into the differential operator form

$$e^{-\frac{t}{4}V(x)}\ \exp\{\frac{t}{2}\ d^2/dx^2\}\ \{e^{-\frac{t}{4}V(x)}\ \Phi(x)\} = \Lambda\ \Phi(x). \tag{2.38}$$

In order to tackle the latter eigenvalue problem, a straightforward approximation procedure, well familiar in the theory of stochastic processes, suggests looking first at the exponential functions. If t is small enough, since V(x) is obviously a smooth function not too rapidly varying over a domain of the order $\sqrt{t}$, a series expansion of the left-hand side of Eq. (2.38) truncated at the first order in t contains most of the interesting features of the model.

Indeed such an expansion generates the Schrödinger-like equation

$$d^2\Phi/dx^2 + [E - V(x)]\ \Phi(x) = 0, \tag{2.39}$$

whose eigenvalues are connected to those of (2.36) by

$$E_K = -(2/t)\ \ell n\ \Lambda_K \qquad \forall K \geqslant 0. \tag{2.40}$$

Obviously, the singular part, if any, of the free energy (2.31) is then given by

$$F_{sing.} = -\beta t E_m/2 \tag{2.41}$$

where

$$E_m = \min_K \{E_K\}. \tag{2.42}$$

The relevance of the "ground state" eigensolution of Eq. (2.39) expressed by Eq. (2.42) makes the former especially easy to handle. In fact, only the neighborhoods of the minima of $V(x)$ are relevant [12], and the behavior of $V(x)$ in proximity of these.

If we write, locally,

$$V(x) = V_{om} + \frac{1}{2}\,\omega(x-x_m)^2 + O((x-x_m)^3), \tag{2.43}$$

where $x = x_m$ is a stationary point of the potential, namely one of the solutions of the equation

$$\partial V/\partial x = 0, \tag{2.44}$$

the points where

$$\omega = \partial^2 V/\partial x^2 \Big|_{x=x_m} \tag{2.45}$$

is positive give the physically interesting minima. When one neglects the terms $O((x-x_m)^3)$ in (2.43), all of a sudden Eq. (2.39) is the most familiar object to physicists: the harmonic oscillator.

Where ω is large enough [13], the Φ's may then be safely thought of as the well known eigenfunctions of the harmonic oscillator, centered at $x=x_m$, and the corresponding eigenvalues are given by

$$E_K = V_{om} + \sqrt{2\omega}\,(K+1/2) \tag{2.46}$$

whereby

$$E_m = V_{om} + \sqrt{\omega/2}. \tag{2.47}$$

The larger ω, the closer the result (2.47) is to being exact. A much greater deal of attention is to be paid in the situation when ω, which is a function of β and $\mathcal{B}$, is vanishingly small. A lengthier calculation, amounting essentially to a perturbation expansion, shows that in this case the result (2.47) is still essentially correct in a very small neighborhood--in the space of parameters--of the set of values of β and $\mathcal{B}$ corresponding to $\omega \approx 0$. The perturbation series, though, becomes non-Borel-summable everywhere in P when $\omega = 0$ exactly. Due to the heuristic character of the present discussion, I will not enter here into any further details concerning the latter point and I proceed with the analysis of the gross properties of the model referring to Eq. (2.47), which for such a purpose is to be regarded as correct.

By insertion in (2.44) of the explicit form (2.35) of $V(x)$, x_m is seen to satisfy the transcendental equation

$$2\Omega x_m/\sqrt{\beta} = \tanh\ (\sqrt{\beta}x_m - \beta\mathcal{B}). \tag{2.48}$$

It is straightforward to check that this equation exhibits a bifurcating pattern of solutions when one varies the "control" parameters β and $\mathcal{B}$, showing different basins of attraction and different global behavior--corresponding to states or phases--of the eigensolutions.

The set Σ of critical parameters, corresponding to a catastrophe for the potential $V(x)$ is determined by the "tangence condition"

$$\sqrt{1-2\Omega/\beta} = \tanh\ \{\beta[1/2\Omega\ \sqrt{1-2\Omega/\beta} - \mathcal{B}]\}. \tag{2.49}$$

In absence of external magnetic field, $\mathcal{B} = 0$, the latter gives $\beta = \beta_c = 2\Omega$. Expanding in a small neighborhood of the point $(0,2\Omega)$ in the $(\mathcal{B},\beta)$-plane, one gets from (2.49)

$$\mathcal{B}^2 = (2\Omega)^{-5}(\beta - 2\Omega)^3 \tag{2.50}$$

whereby $\omega \propto \mathcal{B}^{2/3}$.

The catastrophe is cuspidal [14], as could be expected from a simple inspection of the potential V(x) which, upon varying β and $\mathcal{B}$, shows two attractors (minima) coalescing into a single one at the critical points and by considering V as a classical potential.

3. A Tentative Rigorous Approach

The previous examples, while confirming the intuitive belief that, indeed, a phase transition can be thought of as a structural instability of some kind, and that an answer can be expected in terms of structural stability theory to the problem of Gibbs phase rule [15], yet point out the complexity of recognizing a unique structure whose instability determines the transition, whereby a classification of the interactions might possibly be achieved (mainly in the perspective of constructing new solvable models). The first delicate question concerns the definition of equilibrium thermodynamic states.

3.1 The Thermodynamic States and Their Evolution

In the two above examples, we viewed the states in turn as a set of suitable thermodynamic effective measures, or as the set of eigenstates of a virtual associate quantum system. We have to bypass this sort of ambiguity, aiming at a unique object to deal with. The key open question is: the laws of thermodynamics provide a direct mathematical relationship between the finely structured microscopic state space on the one hand and a much coarser thermodynamical phase space on the other. We must pick up as clearly and rigorously as possible in our definition such a connection, in that our final goal--a classification of phase transitions as structural instabilities--will find its relevance as much as it will allow to state "a priori" that a given (microscopic) interaction is able to produce a desired phase portrait.

The two above questions find a common answer in the fact that it is possible to realize a thermodynamic state as a cross-section of a fiber bundle, $\mathcal{E}$. Without entering into a detailed proof of such a statement, (for which the reader is referred to [22]) I simply point out here that the ingredients are:(i) the infinite-dimensional Lie group generated by the Hamiltonian H together with the set of projection operators onto the eigenvector subspaces corresponding to all the eigenvalues of H, whereby a standard fiber is induced, (ii) the (differentiable) manifold of the physical variables describing the thermodynamic state (the choice of such variables is dictated by the thermodynamic potential function which is minimized at equilibrium, and it might possibly not be the whole set of parameters in P) acting as base space of the bundle, (iii) the manifold of microscopic states, direct product of the Hilbert state spaces for each constituent of the system, serving as total space. Within such a scheme the relationship between the microscopic states and the thermodynamic states is then naturally provided by the canonical projection map $\pi_{\mathcal{E}}$ of the total space of the fiber bundle $\mathcal{E}$ on the base space (which I denote by P for the sake of clarity).

A justification for such a construction is first of all the following. The fiber $\pi_{\mathcal{E}}^{-1}(p)$, $p \in P$ of which the thermodynamic state identified by p is a cross-section, is homeomorphic to the group G constituting it. The latter in general does not have a vector space structure, and, therefore, the space of thermodynamic states defined above is not a vector space. This is consistent with the well known "superselection rule" according to which no temperature state can be obtained as a linear superposition of two or more temperature states. On the other hand, just this construction emphasizes a subtle difficulty inherent in the implicit program sketched in the introduction of this paper: our virtual dynamical systems χ, whereby we move a state into another, may be describable not in terms of vector fields but of generalized dynamical systems determined in terms of the usual morphisms of the fiber bundle.

There are different--though essentially equivalent--ways of constructing the thermodynamic states defined before.

An especially interesting one from the point of view of present discussion is due to Takahashi and Umezawa [16].

It is based on the observation--already utilized and partly discussed in Sec. 2--that the statistical average $<\mathcal{O}>$ of a dynamical observable represented by the operator $\mathcal{O}$ in the theory, has properties similar to the vacuum expectation value of $\mathcal{O}$ in a quantum field theory. Hence, by Wightman's theorem [8] asserting the existence of a unique quantum field theory when the (vacuum) expectation values of all the possible n-field operator products are given, a representation is obtained for a finite temperature quantum field theory in which temperature dependent "vacuum states" (here denoted simply as $|\beta>$) in a suitable space are constructed, realizing the requirement that the statistical average of $\mathcal{O}$ over the canonical ensemble, namely

$$<\mathcal{O}> = \mathrm{Tr}\ \{\mathcal{O}\ e^{-\beta H}\}/\mathrm{Tr}\ \{e^{-\beta H}\} = Z^{-1}\ \mathrm{Tr}\ \{\mathcal{O}\, e^{-\beta H}\} \tag{3.1}$$

be expressible as an expectation value

$$<\mathcal{O}> = <\beta|\ \mathcal{O}\ |\beta>/<\beta|\beta>. \tag{3.2}$$

The $|\beta>$'s are indeed some sort of coherent states [17] parametrized in terms of temperature and other thermodynamic variables (such as chemical potential, external fields, etc.), and the above choice has several advantages which I merely list here. It allows handling from the beginning systems with an infinite number of degrees of freedom, conveniently leading both to a definition of a natural measure (e.g., a generalized Gaussian measure; in infinite-dimensional analysis the Lebesgue measure does not exist) and to the construction of a Fock space F in which, by the Schwartz-Grothendieck kernel theory [18], the operators are simply characterized in terms of their symbols. Moreover the theory of kernels and symbols of operators and sesquilinear forms in F allows giving a full characterization of any operator in terms of integral representations. Finally there is a natural way of transforming a temperature state into another, strictly reminiscent--at least formally--of Bloch's equation:

$$\partial|\beta>/\partial\beta = -\mathcal{G}(\beta)\,|\beta> \tag{3.3}$$

where--in the realization of $|\beta>$ of Ref. [16]--

$$\mathcal{G}(\beta) = (H - U_\beta)/2 \tag{3.4}$$

namely $\mathcal{G}$ describes the distribution of energy eigenvalues over the internal energy $U_\beta = <H> = -\partial(\ell n\ Z\ (\beta))\partial\beta$ of the system.

The formal solution of Eq. (3.3) is

$$|\beta> = {}^{-\beta\mathcal{J}}\ |0> \tag{3.5}$$

where

$$\mathcal{J} = \beta^{-1}\int_0^\beta d\theta\ \ \mathcal{G}(\theta) \tag{3.6}$$

in the representation referred to above is given by

$$\mathcal{J} = (H + \beta^{-1}\ \ell n\ \frac{Z(\beta)}{Z(0)})/2, \tag{3.7}$$

$Z(\beta)$ being the canonical ensemble partition function.

Eqs. (3.3) and (3.5) have the drawback that neither $\mathcal{G}$ nor $\mathcal{J}$ are positive definite operators.

The fiber bundle morphism thus is more conveniently handled in the realization of the thermodynamic state by

$$|\beth>_\beta = \sqrt{Z(\beta)/Z(o)}\ |\beta> \tag{3.8}$$

$|\beth>_\beta$ is not normalized to 1 but has a temperature dependent norm (notice, however, that $|\beth>_\beta$ is always well defined, even when $Z(\beta)$ is not).

Indeed, the equations equivalent to (3.3) and (3.5) read

$$\partial|\beth>_\beta/\partial\beta = -H|\beth>_\beta/2 \tag{3.9}$$

--formally a heat equation--and

$$|\beth>_\beta = e^{-\beta H/2}|0>, \tag{3.10}$$

respectively, where

$$|0> \equiv |\beth>_o \cdot \tag{3.11}$$

is the ∞-temperature reference vacuum in the Hilbert space of $|\beta>$. I will now focus my attention on the operator τ such that

$$\exp(-\beta\tau^\dagger\tau) = \exp(-\tfrac{1}{2}\beta H)/Z(\beta)^{\frac{1}{2}} \tag{3.12}$$

It will turn out that just τ, or, more precisely, its closure $\mathcal{T}$ over the Hilbert space $\mathcal{H}$ defined by τ when some boundary conditions are specified, is the universal object characterizing the entire structure.

In general $\mathcal{T}$ is defined over some field $\mathcal{M}$, endowed with a topology induced both by the physical boundary conditions and by the structure of the Lie algebra of the invariance group G of the Hamiltonian, which is usually a compact connected real manifold.

In general we have

$$G(\Lambda) = \bigotimes_{s\in\mathcal{M}} G_s \tag{3.13}$$

where G_s is a copy of G associated to each site $s \in \mathcal{M}$, and Λ = volume $(\mathcal{M})$, is the volume occupied by the system. In typical physical applications, G is either simply an abelian group (or a free product of abelian groups), or the semi-direct product of SO(n) by an abelian group ("spin" systems).

The measure induced as Gibbs measure in such a case is the Haar measure on the homogeneous space G/K, where K is the isotropy subgroup of a point in the unit n-sphere. $\mathcal{T}$ is then given in terms of elements of the group ring.

3.2 The Atiyah-Singer Index

If $\mathcal{M}$ is compact and C^∞, $\mathcal{T}$ has the transmission property, namely it maps $C^\infty(\mathcal{E}')$ onto $C^\infty(\mathcal{E}'')$, where $\mathcal{E}'$ and $\mathcal{E}''$ denote smooth (vector) bundles over $\mathcal{M}$, i.e. it maps smooth sections of the bundle $\mathcal{E}'$ into smooth sections of $\mathcal{E}''$. Thus in general it is a pseudo-differential operator. Moreover, dim $\mathcal{E}'$ = dim $\mathcal{E}''$ and for any local representation its symbol is invertible (i.e., it is a non-singular matrix). Hence $\mathcal{T}$ is elliptic.

The theory of elliptic operators [19] guarantees that both the kernel and the cokernel of $\mathcal{T}$ are finite-dimensional, so that its Atiyah-Singer index [20]

$$\text{index}\ \mathcal{T} = \dim\ker\mathcal{T} - \dim\ \text{coker}\ \mathcal{T}$$
$$= \dim\ker\mathcal{T} - \dim\ker\mathcal{T}^{\dagger}, \qquad (3.14)$$

where $\mathcal{T}^{\dagger}$ denotes the formal adjoint of $\mathcal{T}$ with respect to some global hermitian inner product in $\mathcal{E}$, say $(\cdot,\cdot)_{\mathcal{E}}$, is well defined.

When $\partial\mathcal{M}$ is empty, ellipticity implies the Fredholm property, hence index $\mathcal{T}$ is essentially the Euler characteristic.

When $\partial\mathcal{M} \neq 0$, ellipticity does not imply the Fredholm property, which can be ensured only by a suitable choice of the boundary conditions.

Notice that $\mathcal{T}^{\dagger}$ carries global boundary conditions in a way complementary to those characterizing $\mathcal{T}$.

Let $\Gamma_{\lambda}(\mathcal{E})$ be the eigenspace of $T_{\mathcal{E}} = \mathcal{T}^{\dagger}\mathcal{T}$(essentially, by Eqs. (3.9) and (3.10) the operator describing (one-half of) the Hamiltonian, equipped, though, with the proper boundary conditions) on $\mathcal{E}$ associated with the real eigenvalue λ

$$\Gamma_{\lambda}(\mathcal{E}) = \{t \in \mathcal{E}\ |T_{\mathcal{E}}\ t = \lambda t\} \qquad (3.15)$$

The countable sequence of such subspaces ($\Gamma_{\lambda}(\mathcal{E}) = 0$ except for a discrete set of non-negative λ's), gives an orthogonal direct sum decomposition of the Hilbert space $\mathcal{H}(\mathcal{E})$ obtained from $\mathcal{E}$ by completion relative to $(\cdot,\cdot)_{\mathcal{E}}$,

$$\mathcal{H}(\mathcal{E}) = \oplus\Gamma_{\lambda}(\mathcal{E}). \qquad (3.16)$$

If $\mathcal{T}: \mathcal{E}' \longrightarrow \mathcal{E}''$, the Hodge theorem [21] states two important properties (isomorphisms):

i) for $\lambda > 0$, $\mathcal{T}:\Gamma_{\lambda}(\mathcal{E}') \rightarrow \Gamma_{\lambda}(\mathcal{E}'')$ is an isomorphism

ii) $\Gamma_{o}(\mathcal{E}') \approx \ker\mathcal{T}$; $\Gamma_{o}(\mathcal{E}'') \approx \text{coker}\ \mathcal{T} \equiv \ker\mathcal{T}^{\dagger}$.

It follows that Eq. (3.14) can be written

$$\text{index}\ \mathcal{T} = \dim\Gamma_{o}(\mathcal{E}') - \dim\Gamma_{o}(\mathcal{E}''). \qquad (3.17)$$

Now, for any $\beta > 0$ we can construct the bounded operators $\exp(-\beta\mathcal{T}^{\dagger}\mathcal{T})$ and $\exp(-\beta\mathcal{T}\mathcal{T}^{\dagger})$ which are the formal fundamental solution of the Bloch equation (Eq. (3.9) with suitable boundary conditions)

$$\partial/\partial\beta + T_{\mathcal{E}'} = 0 \qquad (3.18)$$

(or equivalently, a "propagator" such that according the the global version of (3.10)

$$|\beth>_{\beta} = e^{-\beta\mathcal{T}^{\dagger}\mathcal{T}}|0> \qquad (3.19)$$

and its analog for $T^{*}_{\mathcal{E}} = \mathcal{T}\mathcal{T}^{\dagger} = T_{\mathcal{E}''}$, namely for the operator obtained by self-adjoint continuation of $T_{\mathcal{E}}$ over the double of $\mathcal{M}$.

In a neighborhood of $\partial\mathcal{M}$, $T_{\mathcal{E}}$ and $T^{*}_{\mathcal{E}}$ are isomorphic via the symbol σ (such an isomorphism can actually be lifted along the conormal to $\partial\mathcal{M}$, to the entire manifold $\mathcal{M}$). Besides, having a discrete spectrum with finite multiplicities, their non-zero eigenvalues coincide (almost trivially, indeed: if $\mathcal{T}^{\dagger}\mathcal{T}|\psi> = \mu|\psi>$, then $\mathcal{T}\mathcal{T}^{\dagger}\times(\mathcal{T}|\psi>) = \mu(\mathcal{T}|\psi>)$, and thus $|\psi> \longrightarrow \mathcal{T}|\psi>$ defines an isomorphism of the μ-eigenspace $\Gamma_{\mu}(\mathcal{E})$ of $T_{\mathcal{E}}$ into that of $T^{*}_{\mathcal{E}}$, with inverse $|\psi> \longrightarrow \mu^{-1}\mathcal{T}^{\dagger}|\psi>$). Also, the null space of $T^{*}_{\mathcal{E}}$ coincides with that of $\mathcal{T}^{\dagger}$.

Finally, the series

$$Z_{\beta}(T_{\mathcal{E}}) = \sum_{\lambda} e^{-\beta\lambda}\dim\Gamma_{\lambda}(\mathcal{E}) = \text{Tr}\ (e^{-\beta T_{\mathcal{E}}}) \qquad (3.20)$$

and the corresponding one for $T_{\mathcal{E}}^*$ converge for almost all $\beta > 0$.

The Atiyah-Bott-Patodi theorem [21] states then that (3.14) or (3.17) can be rewritten:

$$\text{index}\, \mathcal{T} = \text{Tr}\, e^{-\beta \mathcal{T}^\dagger \mathcal{T}} - \text{Tr}\, e^{-\beta \mathcal{T}\mathcal{T}^\dagger}. \tag{3.21}$$

Let us analyze briefly what information the integer-valued functor index $\mathcal{T}$ contributes to our physical picture.

First of all, the index $\mathcal{T}$ is well defined whenever the partition function $Z(\beta')$, where $\beta' = \beta/2$, is convergent. The breaking of such a convergence for a discrete set of "critical" values $\{\beta_c\}$ which is usually connected in statistical mechanics with the appearance of a phase transition, in the present picture implies that index $\mathcal{T}$ is either non-uniquely determined or undefined. In other words the set Σ is identified by the property that index $\mathcal{T}$ abruptly jumps when the control parameter point crosses it. Thus, index $\mathcal{T}$ acts as an indicator of the occurrence of a phase transition [22]. The index, however, carries a finer information, viz., its absolute value in each stability domain and the amount of the above-mentioned jump. These qualities point out to it as the relevant object for a classification of phase transitions. But there is more physics to be learned from it.

Index $\mathcal{T}$, on the one hand, is a measure of the asymmetry of the spectrum of $\mathcal{T}$ and it accounts for the dimension of ker $\mathcal{T}$, namely the multiplicity of zero eigenvalues of $\mathcal{T}$ itself (Eq. (3.17)), through the difference in the asymptotic behavior of the kernels $\exp(-\beta\tau^\dagger\tau)$ and $\exp(-\beta\tau\tau^\dagger)$ on the double of $\mathcal{M}$ (the latter can in fact be identified as a suitable Pontrjagin form of the metric on $\mathcal{M}$, hence also the topology of $\mathcal{M}$ is accounted for). In other words, index $\mathcal{T}$ measures the stability of the partition function with respect to global changes in the boundary conditions.

The latter property is just a restatement of Peierl's criterion for the appearance of a phase transition [23].

On the other hand it can be shown [16] that up to a trivial factor, $\exp(-\beta\mathcal{S})$ is a Bogolubov transformation connecting states with different temperature. A highly nontrivial analysis [22] in terms of K-theory shows that index $\mathcal{T} > 0$ implies the global existence of a transformation $\mathcal{S}_B$ (generalized Bogolubov automorphism) such that $\mathcal{S}_B^{-1}\mathcal{T}\mathcal{S}_B$ is stably represented as a direct sum of operators of lower order.

A jump in index $\mathcal{T}$ then has the following meaning. Varying the parameters by moving p in P the operator $\mathcal{T}$ moves within the bundle of all operators having the same Jordan form--guaranteed by the existence of $\mathcal{S}_B$--realizing an endomorphism of the state space. Corresponding to the decomposition of the space generated by the $\mathcal{T}$'s into bundles the parameter space decomposes into submanifolds.

The exceptional parameters for which the global conditions for the existence of $\mathcal{S}_B$ fail (and which correspond to a bifurcation of the eigenvalues--in particular the null eigenvalue of $\mathcal{T}$--as a function of p) define the set--which is a finite union of smooth manifolds, each corresponding to a bundle--over which index $\mathcal{T}$ is undefined.

If index $\mathcal{T} = 0$, then $\mathcal{T}$ has locally distinct eigenvalues for every $s \in \mathcal{M}$.

The latter general situation is just what we encountered in the simple example discussed in Sec. 2.2; and it is indeed a characterization in global terms of the profound mechanism whereby a phase transition is generated, known as Kac's mechanism [24].

The analysis just concluded has thus picked up an object by which we are able, in principle, to perform a classification of phase transitions (note: a classification actually exploiting the microscopic properties of the system). Obviously, the next question is: what is the practical advantage of going through all this, besides heading toward a different but difficult classification of phase transitions?

The answer is quite easy: the practical advantage remains within the Atiyah-Singer index theorem [25] which states that

$$\text{index } \mathcal{T} = \gamma(\mathcal{T}) \tag{3.22}$$

where $\gamma(\mathcal{T})$ is the topological index of $\mathcal{T}$ (as opposed to index $\mathcal{T}$, often referred to as the analytical index of $\mathcal{T}$).

Now, the topological index can be computed in ways which do not require the solution of the whole statistical mechanical problem. Indeed it can be written as an integral over the fundamental cycle of the cotangent bundle of $\mathcal{M}$, and may be expressed in terms of the symbol of the operator $\mathcal{T}$ and the curvature tensor of the manifold as

$$\gamma(\mathcal{T}) = \{\text{ch } \mathcal{T} \cdot \text{Td } \mathcal{M}\}, \tag{3.23}$$

where the curly brackets denote the evaluation of the product of cohomologies over the entire manifold, Td is the Todd cohomology class of the tangent bundle of $\mathcal{M}$, and ch the relative Chern class.

The problem of classification is then reduced to the determination of the Pontrjagin characteristic classes of the fiber bundle, and the entire structure exhibits its intrinsic purely geometric character quite explicitly.

3.3 An Example: The Ising Model

To conclude, let me describe briefly one of the applications of the above scheme to the two-dimensional Ising model. As is well known, the latter consists of a two-dimensional lattice (e.g., a square lattice) in each site of which is a spin variable σ, assuming the values ± 1.

The interaction is bilinear and nearest-neighbor:

$$H = (-J_1/2) \sum_{(i,j)\in\Lambda} \sigma_{ij}\,\sigma_{i+1,j} - (J_2/2) \sum_{(i,j)\in\Lambda} \sigma_{ij}\,\sigma_{i,j+1} \tag{3.24}$$

where Λ denotes the "volume" occupied by the system endowed by the periodic boundary conditions with the topology of a torus.

Due to just this topology and to the differential character of the interaction, the operator $\mathcal{T}$ for the model in question turns out to be a Toeplitz operator [26, 27] as can be easily checked by exploiting--according to Onsager's [28] classical analysis--the structure of the algebra of G, and performing the Fourier analysis in the representation of the transfer matrix. This makes life particularly easy: indeed for a Toeplitz operator $\mathcal{T}_{[s]}$--which is a convolution product operator by a smooth function s(z) defined over the unit circle S' and generates therefore a closed path in the complex plane, $s: S' \longrightarrow \mathbb{C} - \{0\}$--the Atiyah-Singer index reduces to the winding number W, which is the one and only topological invariant under continuous deformation.

$\mathcal{T}_{[s]}$ is a Fredholm operator if $s(z) \neq 0$ on S' and in such a case

$$W \equiv \text{index } \mathcal{T}_{[s]} = (2\pi)^{-1} \left[\arg\{s(e^{i\pi})\} - \arg\{s(e^{-i\pi})\}\right]. \tag{3.25}$$

If, moreover, s is differentiable, then

$$W = (2\pi i)^{-1} \oint ds/s. \tag{3.26}$$

Notice that $\mathcal{T}_{[s]}$ is globally invertible if and only if index $\mathcal{T}_{[s]} = 0$.

The computation of W, upon introduction into (3.25) of the explicit expression for s(z), namely [28,29]

$$s(z) = e^{-i\pi/2} z\sqrt{(\tau_1\tau_2 z-1)(\tau_2 z-\tau_1) / (z-\tau_1\tau_2)(\tau_1 z-\tau_2)}, \tag{3.27}$$

where

$$\tau_1 = \exp(-2\beta J_1) \tag{3.28}$$

$$\tau_2 = \tanh(\beta J_2) \tag{3.29}$$

gives

$$W = \begin{cases} 1 & \text{for } \beta > \beta_c \\ 0 & \text{for } \beta < \beta_c \end{cases}, \tag{3.30}$$

β_c being just Onsager's critical inverse temperature, and thus the solution of the equation

$$\tau_1 = \tau_2 \tag{3.31}$$

In correspondence with β_c, W is undefined.

Acknowledgement:

It is my pleasant duty to dedicate this paper to Prof. René Thom to whom I owe so much in the way of inspiration with gratitude and admiration.

References

1 V. de Alfaro and M. Rasetti, Fortschritte der Physik 26, 143 (1978)

2 Widely exhaustive references are:
R. Thom, Stabilité Structurelle et Morphogénèse, W.A.Benjamin, Inc., 1972
W. Güttinger, in: Lecture Notes in Biomathematics, Vol. 4, (M. Conrad et al., Eds.), Springer 1974
E.C. Zeeman, Catastrophe Theory, Addison Wesley Publishing Co., 1977
T. Poston and I. Stewart, Catastrophe Theory and its Applications, Pitman 1978

3 K.G.Wilson and J. Kogut, Phys. Reports 12C, 75 (1974)

4 M.J.Feigenbaum, Quantitative Universality for a Class of Nonlinear Transformations, Los Alamos Scientific Laboratory preprint LA-UR-77-1063, 1977, to be published in SIAM Journal.
N. Metropolis, M.L.Stein, and P.R.Stein, J.Comb.Theory, 15, 25 (1973)
J. Guckenheimer, Inventiones Math. 39, 165 (1977)

5 K.G.Wilson, Phys. Rev. B4, 3174 (1974); ibid. 3184

6 R. Thom, Bull. Amer. Math. Soc. 75, 2 (1969)

7 O. Schlömilch, Compendium der Höheren Analysis, Braunschweig, 1789; vol. 2, page 497

8 A.S.Wightman, Phys. Rev. 101, 860 (1956)
H.J.Borchers, "Algebraic Aspects of Wightman Field Theory," in Statistical Mechanics and Field Theory, R.N.Sen and C. Weil (eds.), Israel University Press, Jerusalem, Israel, 1972

9 F. Treves, Topological Vector Spaces, Distributions and Kernels, Academic Press, New York, N.Y., 1967

10 J. von Neumann, Zur Theorie der Unbeschränkten Matrizen, in Collected Works, Pergamon Press, London, 1961; vol. 2, page 144

11 E. Brüning, On the Characterization of Relativistic Quantum Field Theories in Terms of Finitely Many Vacuum Expectation Values, Bielefeld preprints BJ-TP 77/06 and BI-TP 77/23, 1977

12 C. Bender, to be published in Advances in Mathematics
C. Bender and T.T.Wu, Phys. Rev. D7, 1620 (1973)
E. Brézin, J.C.Le Guillou, and J. Zinn-Justin, Phys. Rev. D15, 1544 (1977); ibid. 1558

13 B. Simon, Ann. Phys. (N.Y.) 97, 279 (1976)
E.M.Harrell, Comm. Math. Phys. 60, 73 (1978)

14 R. Thom, Topology 8, 313 (1969)

15 D. Ruelle, Commun. Math. Phys. 53, 195 (1977)
In this beautiful paper the Gibbs phase rule problem in statistical mechanics is concisely stated in the following very clear way. Let $\mathcal{H}_0$ be some interaction for which n phases coexist, and denote by Θ a suitable (Banach) space of interactions, $\mathcal{H}_0 \in \Theta$. The Gibbs rule consists of the three following requirements:
i) $\mathcal{H}_0$ contains a manifold $\mathcal{M}$ of codimension n-1, of coexistence of at least n phases.
ii) In proximity of $\mathcal{M}$ there is a manifold with boundary M_K for each subset $K = \{k_1, k_2, \ldots, k_{|K|}\}$ of $\{1, \ldots, n\}$, of coexistence of at least $|K|$ phases. M_K has codimension $|K|-1$, and its boundary is $\partial M_K = \bigcup_{\substack{J \supset K \\ J \neq K}} M_J$.
iii) There exists a homomorphism ω-tangent to the identity at $\mathcal{H}_0$-mapping a neighborhood of $\mathcal{H}_0$ into itself and such that $\omega^{-1} M_K$ is a locally convex polyhedral cone.
See also: L. Mistura, J. Phys. A, 9, 2139 (1976)

16 Y. Takahashi and H. Umezawa, Collective Phenomena 2, 55 (1975)

17 M. Rasetti, Int. J. Theor. Phys. 13, 425 (1975); ibid. 14, 1 (1975)
A. Lanke, J. Math. Phys. 19, 1110 (1978)

18 A. Grothendieck, "Produits Topologiques et Espaces Nucléaires," Mem. Amer. Math. Soc. N. 16, Providence, R.I., 1955

19 L. Hörmander, Comm. Pure Appl. Math. 18, 501 (1965)
R.T.Seeley, Trans. Amer. Math. Soc. 117, 167 (1965)
R.T.Seeley, Proc. Sympos. Pure Math. 10, 288 (1971)

20 M.F.Atiyah and I.M.Singer, Bull. Amer. Math. Soc. 69, 422 (1963)
M.F.Atiyah and I.M.Singer, Ann. of Math. 87, 484, (1968); ibid, 87 546 (1968); ibid. 93, 199 (1971); ibid. 93, 139 (1971);
M.F.Atiyah and I.M.Singer, Publ. Math. Inst. Hautes Etudes Sci. (Paris) 37 (1969)

21 M.F.Atiyah, R.Bott, and V.K.Patodi, Inventiones Math. 19, 279 (1973)

22 M. Rasetti, "Global Approach to Phase Transitions," in Fundamental Problems in Statistical Mechanics, N.N.Bogolubov, jr. and A.M.Kurbatov (eds.), J.I.N.R. Publ., Dubna, U.S.S.R., 1977
M. Rasetti, The Atiyah-Singer Index in Phase Transition Theory, Istituto di Fisica del Politecnico di Torino, preprint IFPT-TH-3-77, to be published Comm. Math. Phys.

23 R. Peierls, Proc. Cambridge Phil. Soc. 32, 477 (1936)
R.B.Griffith, Phys. Rev. 136A, 437 (1964)

24 M. Kac, Proc. Amer. Math. Soc. Symposium on Mathematical Statistics and Probability, Berkeley, 1951; page 189

25 Besides Refs. 20 , see
R. Palais,"Seminar on the Atiyah-Singer Index Theorem," Ann. of Math. Studies 57, Princeton, 1965

also

M.F.Atiyah, R. Bott, and I.M.Singer, Topology Seminar, Harvard, 1962
M.F.Atiyah, Lectures on K-Theory, Harvard, 1965
F. Hirzebruch, Topological Methods in Algebraic Geometry, Springer-Verlag, Berlin, 1966

26 R. Douglas and H. Widom, Indiana Univ. Maths. Journ. 20, 385 (1970)

27 J.T.Lewis and P.N. Sisson, Commun. Math. Phys. 44, 279 (1975)

28 L. Onsager, Phys. Rev. 65, 117 (1944)

29 C.N.Yang, Phys. Rev. 85, 808 (1952)

The 180° Rule at Triple Points: A Consideration Based on a Modified Butterfly Model

Günter Dukek

Department for Theoretical Physics II, University of Ulm
D-7900 Ulm, Fed. Rep. of Germany

1. Introduction

1.1 The disposition of phase boundaries at triple and multiple points in phase diagrams of single and multicomponent systems is governed by a class of rules called collectively "The 180^0 Rule" in the literature of physics. These rules state that in a phase diagram with thermodynamically proper independent variables no phase occupies more than 180^0 of angle at a triple point. For simple one component systems (such as pure substances) described thermodynamically by two independent intensive variables, say temperature T and pressure p, the 180^0 rule is usually stated in the following form[1] : The metastable extension of each two-phase equilibrium curve beyond the triple point must lie between the other two stable curves (see Fig. 1).

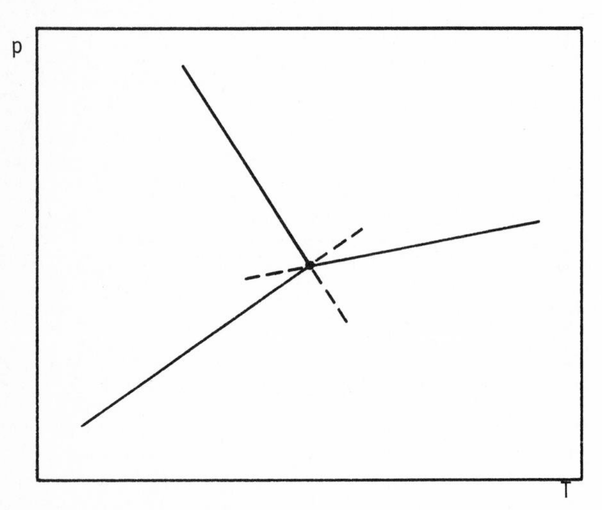

Fig.1 In this (T,p)-diagram each phase occupies less then 180^0, and the extension of each two-phase equilibrium curve passes into the stable region of the third phase

A rigorous proof of the 180^0 rule which makes no reference to metastable states has been presented recently by WHEELER [1].

Most of the known one-component systems described by two independent intensive variables possess the interesting property that the 180^0 rule applies not only in the (T,p)-plane, as shown in Fig. 1, but also in the (μ,T) and (μ,p) planes with μ being the chemical potential. As three independent intensive variables enter into the description of a system, the locus where 3 phases are in mutual equilibrium is

1 This version of the rule will be referred to later in this note.

not merely a point but rather a curve in the space of these variables. If a plane section is taken through this triple curve and the two-phase equilibrium surfaces which meet there, the result will in gereral be a diagram of precisely the same nature as that shown in Fig. 1, i.e. the 180^0 rule will normally apply in such a plane section. Under certain circumstances, however, violations of this rule can occur, for instance, if the triple curve exhibits a terminal point in the space of intensive variables. As four and more intensive variables enter into the description the situation becomes even more complex.

1.2 In view of these facts it should be interesting to have available a fairly elementary model system which allows one to test the validity of the 180^0 rule under various conditions. Such a model will be presented in this note, and attention will be focused upon the circumstances under which the 180^0 rule may cease to be valid. The model is patterned after THOM's "Butterfly"[2], which is associated with a potential of the form

$$V = X^6 + \xi_4 X^4 + \xi_3 X^3 + \xi_2 X^2 + \xi_1 X \ .$$

In our considerations we use a chemical potential $\mu_{T,u,w,\xi}(x)$ of almost the same mathematical structure for the description of the model system. Due to an inherent instability the model system possesses the ability to break up into two, respectively three different phases. This may happen as the system is taken along a thermodynamic path in the space of intensive variables T,u,w,ξ reaching a point of the "catastrophe set" associated with the model. The fact that the system can break up into three different phases allows us to analyze the validity of the 180^0 rule. Before stating this model we present in a somewhat lengthy preparatory consideration some of the basic thermodynamic concepts, and we introduce some of the definitions and conventions given by THOM in the context with a treatment of gradient dynamical systems [2].

2. Thermodynamic Potentials

2.1 It may happen that, for an appropriate choice of external conditions, a system in its homogeneous phase does not satisfy the conditions of thermodynamic stability and breaks up into different parts called phases. The resulting heterogeneous system again satisfies the conditions of stability, and the coexistent phases show no unusual properties. In order to describe such phenomena in a convenient manner we introduce a particular class of thermodynamic potentials used by GIBBS. These potentials - chemical potentials - fit into a general scheme given by THOM.

2.2 According to GIBBS all thermodynamic information concerning a system is summarized in a relation

$$E = E(S,V,N,\dots,X) \ , \tag{1}$$

where E is the internal energy, and $S,V,N,\dots,X$ are $n+1$ extensive variables. The internal energy, expressed in terms of these variables, is a well known thermodynamic potential. To each extensive variable X there belongs a conjugate variable

$$\xi = \frac{\partial E(S,V,N,\dots,X)}{\partial X} \tag{2}$$

called intensive variable. The couples of conjugate variables (X,ξ), relevant for any particular system, are usually chosen from the following set:

(S,T)	entropy and temperature
$(V,-p)$	volume and negative pressure
(N,μ)	particle number and chemical potential,...

A basic assumption is that the internal energy (1) is a first-order homogeneous function of its arguments

$$E = \frac{\partial E}{\partial S} S + \frac{\partial E}{\partial V} V + \frac{\partial E}{\partial N} N + \ldots + \frac{\partial E}{\partial X} X \ . \tag{3}$$

It is convenient to introduce the quantities $e = E/N$, $s = S/N$, $v = V/N$, ..., $x = X/N$. Then (1) appears as a thermodynamic potential with only n independent variables instead of n + 1

$$e = e(s,v,\ldots,x) \tag{4}$$

and

$$\xi = \frac{\partial e(s,v,\ldots,x)}{\partial x} \ . \tag{5}$$

It may be convenient to illustrate GIBBS' view of a thermodynamic system underlying the above considerations in the simple case where $E = E(S,N)$ (see Fig.2).

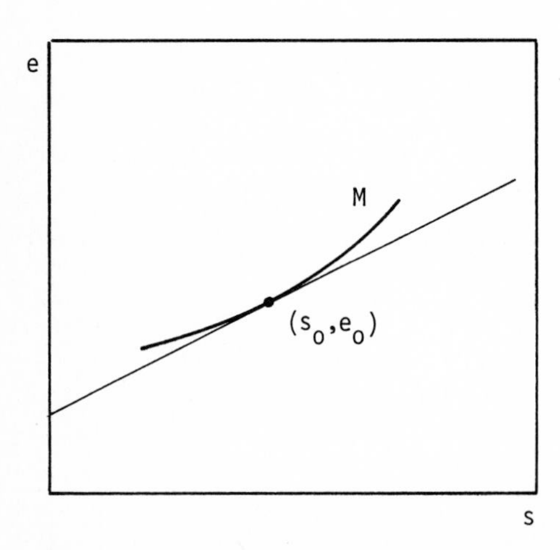

Fig.2 Graphical representation of the manifold M of points defined by the internal energy $E = E(S,N) = Ne(s)$ in a plane with coordinates $e = E/N$, $s = S/N$. A state of the system is represented by a point, e.g. (s_0,e_0). The manifold M of all such points constitutes a graphical representation of the system. At a point (s_0,e_0) the thermodynamic temperature T_0 is defined by the slope of the tangent. This graphical representation conforms to the view of a system as described in 2.2

2.3 An alternative view of thermodynamic systems, due to GIBBS, is presented in the following consideration. In consequence of this view there arise less familiar types of thermodynamic potentials. According to the idea of GIBBS we rewrite (3) as

$$E(S,V,N,\ldots,X) - TS + pV - \mu N - \ldots - \xi X = 0 \ . \tag{6}$$

The internal energy shall characterize the given system. Now it is assumed that one can vary n of the intensive variables independently, e.g. $T,p,\ldots,\xi$. Then one can solve the fundamental relation for the remaining intensive variable μ and put the resulting equation into the form

$$\mu_{T,p,\ldots,\xi}(s,v,\ldots,x) = e(s,v,\ldots,x) - Ts + pv - \ldots - \xi x \ . \tag{7}$$

GIBBS observed that (7) constitutes a complete description of the original system, defined by $E(S,V,N,\ldots,X)$, provided that the equation is used in connection with an appropriate extremum principle. The extremum principle requires that, for given values of the n independent intensive variables $T,p,\ldots,\xi$, the chemical potential shall be a minimum. The chemical potential $\mu_{T,p,\ldots,\xi}(s,v,\ldots,x)$, used in connection with this minimum principle, is a thermodynamic potential of the system. We illustrate GIBBS' observation with aid of Fig. 3 using the same system $E = E(S,N)$ as in Fig. 2.

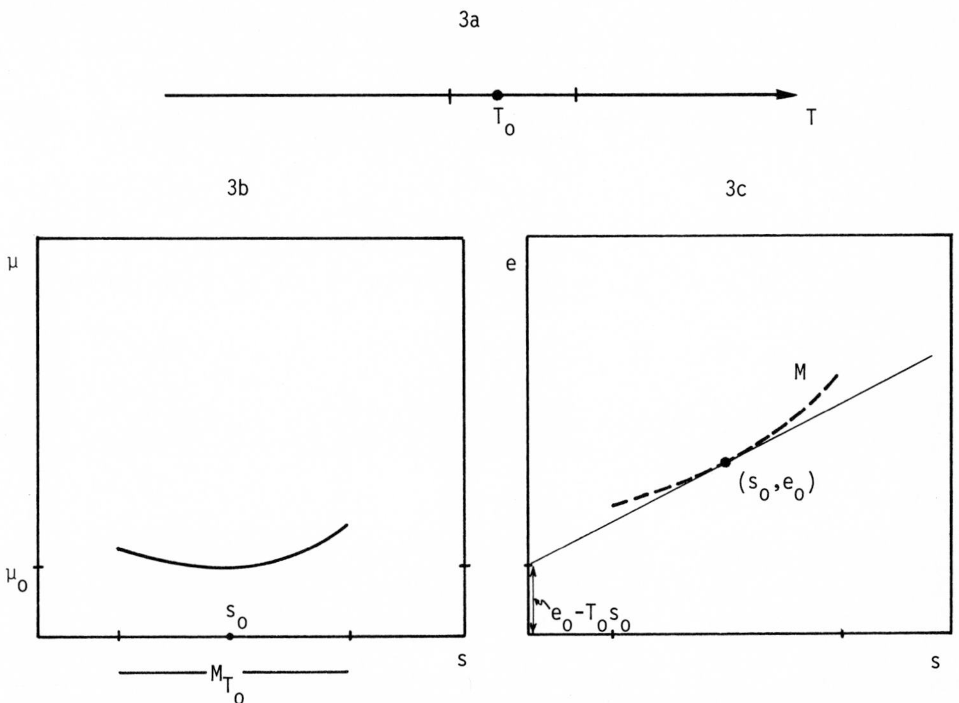

Fig.3 Graphical representation of the previously defined system as described by the chemical potential $\mu_T(s) = e(s)-Ts$ in connection with the corresponding minimum principle. At a given temperature T_o the system has available a manifold M_{T_o} of points on the s-axis. The condition that μ shall be minimal determines the proper choice of s in M_{T_o}. Denoting with μ_o and s_o the minimal value of μ and the corresponding value of s the internal energy is $e_o = \mu_o + T_o s_o$. The two values s_o,e_o belong to a point in the (s,e) plane which is a point of M. By varying the temperature T within certain limits the original manifold M of Fig. 2 is obtained.
From 3 b,c one can easily infer the equivalence of both views provided that the considerations are restricted to the minima of μ

2.4 So far we have presented two alternative views of a thermodynamic system. Taking the first view one can use the internal energy $E(S,V,N,\ldots,X)$ or, equivalently, $S(E,V,N,\ldots,X)$ or $e(s,v,\ldots,x)$ for the description of a system. This view is normally taken in statistical mechanics when one generates the entropy of a system with aid of the microcanonical ensemble. Taking the latter view one can use for the description of the same system the fundamental relation (6) or, equivalently, the chemical potential (7) in connection with the above minimum principle. GIBBS used such chemical potentials, and they fit into a general scheme given in THOM's work.

2.5 We introduce now some of THOM's ideas and definitions. For sake of simplicity they are presented in a fashion which is adapted to our previous considerations.

Let $V_\xi(x)$ be a potential where ξ labels a set of ℓ external variables (e.g. "intensive" variables) and x a set of m internal variables (e.g.s,v,...).For a given point ξ in the space of external variables the system described by $V_\xi(x)$ has available a manifold M_ξ of points in the space of internal variables. The potential $V_\xi(x)$ is a real-valued function on this manifold $V_\xi: M_\xi \to R$. For a given point ξ the internal equilibrium states of the system are defined by the minima of the potential. In general there exist several of these minima, and it is necessary to specify by which of the theoretically possible minima the behaviour of the system is dominated. THOM makes the following assumption, called MAXWELL's convention: For a point ξ where several minima of the potential are in competition the dominant equilibrium state of the system corresponds to the minimum of the potential at x_i where

$$V_\xi(x_i) < V_\xi \; (x_j), \; j \neq i.$$

MAXWELL's convention gives rise to the definition of catastrophe points in the space of external variables and induces a distinction between conflict points and bifurcation points. According to MAXWELL's convention a point ξ is a catastrophe point in two cases only:

1. If the absolute minimum of the potential is obtained for at least two distinct points x_1, x_2.

2. If the absolute minimum of the potential, obtained at a unique point x_0,ceases to be stable (i.e. is no longer quadratic).
   The point ξ is said to be a conflict point in the first and a bifurcation point in the latter case.

2.6 These definitions are important for such thermodynamic systems which can break up into different parts, i.e. different phases. It may happen that, for an appropriate choice of external conditions, a substance, e.g. water, in its homogeneous phase does not satisfy the conditions of thermodynamic stability and breaks up into two different parts- vapour and liquid. The resulting heterogeneous system again satisfies the conditions of stability, and the coexistent phases vapour and liquid show no unusual properties. These two phases can either be viewed as autonomous systems or as subsystems of the substance water. It is well known that VAN DER WAALS adopted the latter view, and he justified this by his famous continuity argument termed the continuity of vapour and liquid states. Provided that this view is taken, one can ask what happens to such a state of the system where vapour and liquid are in mutual equilibrium as the external conditions are varied. Under these circumstances some kind of competition between the phases will take place. The question which of the phases will be dominant depends on the new external conditions, i.e. in praxi on the new values of the temperature T and the pressure p. An interesting way of studying in a model the competition between the phases is to consider the chemical potential

$$\mu_{T,p} \; (s,v) = e \; (s,v) - Ts + pv \tag{8}$$

or the reduced chemical potential

$$\mu_{T,p} \; (v) = f \; (T,v) + pv \tag{9}$$

with e(s,v) respectively f (T,v) given by the corresponding expressions for the VAN DER WAALS gas [3]. The function f (T,v) introduced in (9) is the free energy per particle f = e - Ts, a LEGENDRE transform [2] of the internal energy per par-

---

[2] If e = e(s,v,...,x) is a thermodynamic potential, then the LEGENDRE transforms of e like $e^{[s]} \equiv f = e - Ts$, $e^{[s,v]} \equiv g = e - Ts + pv$, ... etc. are also thermodynamic potentials in the variables T,v,...,x respectively T,p,...,x,...etc. [3].

ticle e(s,v). With regard to a graphical representation the latter function (9) appears to be the convenient one as there enter less independent variables. The competition between the phases vapour and liquid as described by (9) is illustrated in Fig. 4, and the physical significance of catastrophe points is explained thereby.

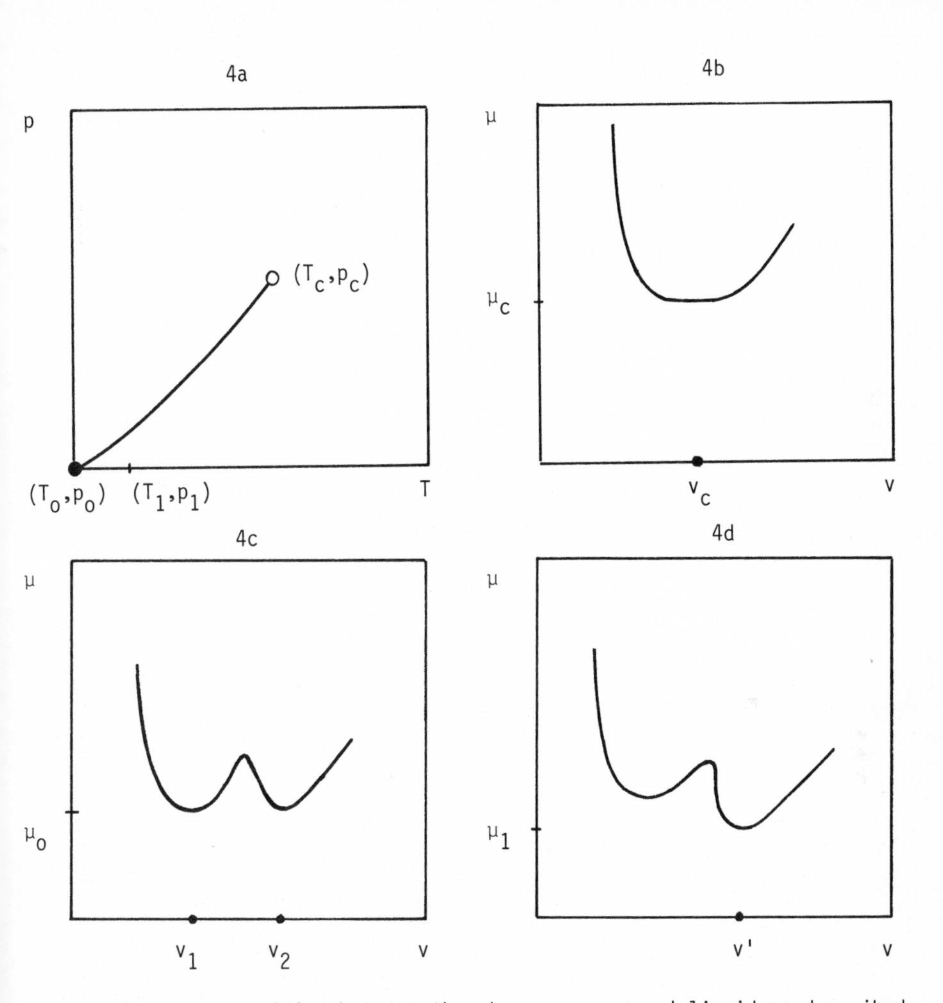

Figure 4. The competition between the phases vapour and liquid as described by the reduced chemical potential (9) of VAN DER WAALS' model.
Fig. 4a is the (T,p)-diagram of the model restricted to the set of points $T > T_o > o$ and $p > p_o > o$ where (T_o,p_o) marks a conflict point. The line of conflict points ends at the bifurcation point (T_c,p_c). This line is usually interpreted as the two-phase equilibrium curve (coexistence curve) vapour-liquid with (T_c,p_c) as critical point.
The Figs. 4 b,c,d illustrate the competition between the phases vapour and liquid as a competition between the minima of the reduced chemical potential. 3 representative situations (T_c,p_c), (T_o,p_o) and (T_1,p_1) are shown. For (T_c,p_c) there is a critical phase characterized by $v = v_c$. For (T_o,p_o) the points v_1,v_2 characterize liquid and vapour in a situation where both phases coexist, i.e. are in mutual equilibrium. For (T_1,p_1) the system takes the state with absolute minimal value of μ corresponding to the point v = v' (vapour)

2.7 As is well known, VAN DER WAALS' model exhibits a particular thermodynamic instability at $T = T_c$, $v = v_c$ where the stability condition [3]

$$\frac{\partial^2 f(T,v)}{\partial v^2} > 0 \tag{10}$$

is violated. For, T_c, v_c we have $\partial^2 f/\partial v^2 = o$, $\partial^3 f/\partial v^3 = o$ and $\partial^4 f/\partial v^4 > o$. The thermodynamic instability induces a structural instability. In fact, for $T = T_c$, $p = p_c$ the point $v = v_c$ (see Fig. 4b) is a singular point [3] of the chemical potential $\mu_{T_c,p_c}(v)$, more precisely a non-structurally stable point [4] of the function. A slight variation of T and p near T_c and p_c will lead to a perturbation of $\mu_{T_c,p_c}(v)$ such that there arises a well defined number of topologically different types of functions $\mu_{T,p}(v)$, e.g. functions with only one minimum or functions with two minima [5]. The occurrence of conflict points near a bifurcation point can easily be understood from this point of view [6].

3. The 180° Rule

3.1 VAN DER WAALS' model of a real gas demonstrates the important fact that a thermodynamic system can break up into different phases if there exists an appropriate thermodynamic instability. The type of instability underlying VAN DER WAALS' model is of a fairly simple nature and leads to the occurrence of only two different phases. More difficult types of thermodynamic instabilities (respectively structural instabilities) are necessary if one aims to give models of three-phase systems. A somewhat more complicated type of instability underlies the following model of a thermodynamic system defined by the reduced chemical potential

$$\begin{aligned}\mu_{T,u,w,\xi}(x) &= h(T,u,w,x) - \xi x \\ &= \alpha x^6 + ux^4 + wx^3 + \beta(T-T_o)x^2 + \mu_o(T) - \xi x \ .\end{aligned} \tag{11}$$

This potential is a modified "Butterfly" with α,β positive constants besides T_o, a characteristic temperature of the system. The quantity ξ is the intensive variable conjugate to x, T the temperature, and u,w are two intensive variables being not specified further. The function μ_o (T) gives the values of μ at $x = o$ as depending on T, and h is a LEGENDRE transform of e. For $u = o$, $w = o$ and $T = T_o$ the potential h exhibits a particular thermodynamic instability at $x = o$, where all derivatives of h with respect to x up to the order five vanish. The point $x = o$ is a singular point of codimension 4 of the chemical potential $\mu_{T_o,0,0,0}(x)$, i.e. a non-structurally stable point of this function. Due to this property of the model one may expect that the system, defined by (11), can break up into more than two different phases. All information about the external conditions under which this may happen is summarized in the set of catastrophe points associated with the model in the space of intensive variables T,u,w,ξ. The mathematical properties of the set of catastrophe points associated with THOM's "Butterfly" are well established. In [2,4], for instance, different plane sections of this set are presented. As the essential features of the "Butterfly" are not altered by the ansatz (11) [4] we may partially refer to these results given in the literature.

[3] The singular point $v = v_c$ is said to be of codimension two [2,4] .

[4] It should be noted here that DEVONSHIRE [7] arrived at a similar ansatz (for h) in the context with the theory of first-order-phase transitions in ferroelectrics, i.e. from a markedly different point of view. In this theory DEVONSHIRE studied the possibility of describing the discontinuous or first-order transition from the non polar phase I to the polar phase II of a crystal in the absence of an external field with aid of a LANDAU type theory. In DEVONSHIRE's theory the quantities u,w are not variables but merely characteristic constants of the ferroelectric crystal under consideration. For reasons of crystal symmetry w is identically zero in the theory.

3.2 With regard to a discussion of the 180^0 rule it is natural to study plane sections for which u and w are kept constant. Attention will be focused on the breakdowns of the 180^0 rule as the condition

$$D(u,w) = 2u^3 + 27w^2 < 0 \tag{12}$$

is violated, i.e., stands with the equality sign.
First we consider the particular case where $w = \text{const} = o$, $u \doteqdot o$ (see Figs. 5a,b,c), then we consider the case where $u = \text{const} < o$ and $D(u,w) \doteqdot 0$ (see Figs. 5d,e,f)

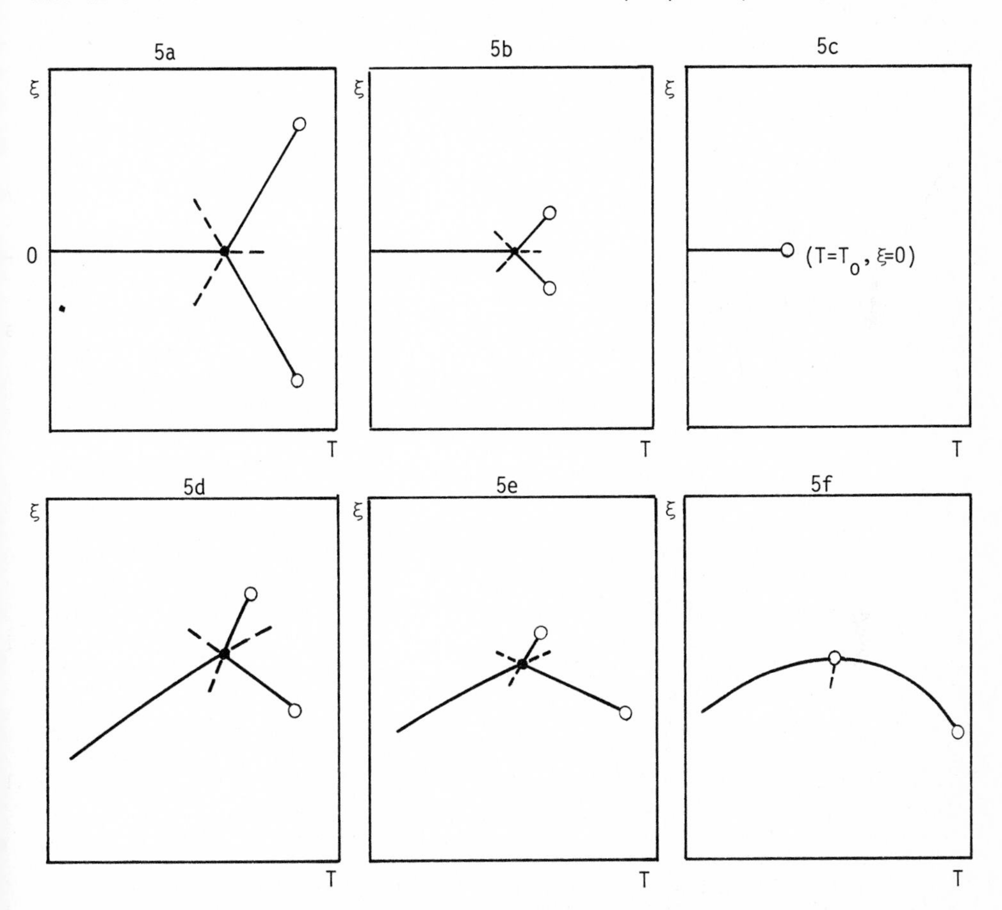

Fig. 5. Plane sections of the set of catastrophe points associated with (11) in diagrams with T and ξ as coordinate axes for $w = o$, $u < o$(a,b); $w = u = o$ (c); $u < o$, $D(u,w) < 0$ (d,e) and $u < o$, $D(u,w) = 0$ (f). Bifurcation points are marked in all diagrams by open circles. Contrary to MAXWELL's convention metastable extensions are allowed and indicated by broken lines. With ξ equal to the external electric field E (being the conjugate to the polarization P) the diagrams 5a,b,c are physically relevant in the theory of ferroelectrics. Due to the crystal symmetry (i.e. $w \equiv o$) of ferroelectrics the diagrams 5d,e,f are not relevant for these substances.

These different sections of the catastrophe set associated with the model, defined by (11), demonstrate that the system can break up into two respectively three different phases. The normal case met, for instance, at a conflict point on the straight line $\xi = o$ of Fig. 5a, is that the system breaks up into two different phases, characterized by two different values of the quantity x, say x_1 and x_2. The same is true at a conflict point lying on one of the other two lines of Fig. 5a. On such a line, however, the exceptional case can occur that the phases become gradually more similar and become identical at the terminal point. At such a point, being a bifurcation point, the phases are critical in the same sense as discussed before in the case of VAN DER WAALS' model. Another exceptional case arises due to the fact that the three lines meet at a common point, a "triple" point in the (T,ξ)-diagram. At such a "triple" point (i.e. conflict point where the absolute minimum of the chemical potential is obtained for three distinct points) the system can break up into three different phases characterized by different values x_1, x_2, x_3 of the quantity x. The way in which the metastable extensions pass beyond the "triple" point clearly indicates that the 180o rule is satisfied for the conditions met in Fig. 5a. The situation changes, however, gradually as u tends to zero, and for u equal to zero the 180^0 rule ceases to be valid (see Figs. 5b,c). The final situation of Fig. 5c resembles in some respects to the situation of Fig. 4a with the important difference, however, that three phases (instead of two) have become identical, i.e. are critical phases with $x = x_c = 0$. This value corresponds to the singular point of codimension 4 of the chemical potential $\mu_{T_0,0,0,0}$ (x).

Another way how the 180^0 rule may cease to be valid is shown in the Figs. 5d,e,f. Now u and w are different from zero, and in the last diagram the exceptional situation arises that two of the three phases being in mutual equilibrium have become identical, i.e. are critical.

These considerations show that the 180^0 rule is valid in the (T,ξ)-diagrams for fixed values of u and w provided that the condition (12) is met. Apparently, this condition expresses the fact that none of the three phases being in mutual equilibrium at the "triple" point is a critical phase. As the condition (12) is violated some kind of higher-order "triple" point arises as the limiting case where two, respectively three phases become identical, i.e. critical. Under these cirumstances the 180^0 rule in its simple version as presented here ceases to be valid. The fact that the 180^0 rule ceases to be valid as the phases become critical can, of course, be demonstrated analytically under quite general hypotheses. Nevertheless it is interesting to make this observation directly within the framework of a fairly elementary model.

Finally it should be noted that there are more sophisticated versions of the 180^0 rule taking into account also the limiting cases where the phases become critical [1,8]. These limiting cases, in particular the so-called tricritical points at which the densities of all three phases become identical, have excited considerable attention in the literature, e.g. [9,10]. The above model may serve as a guide for these complicated problems of 3-phase equilibria.

Acknowledgment

Professor G. Falk kindly read the manuscript and his comments are gratefully acknowledged.

References

1. Wheeler, J.C.: Geometric constraints at triple points, Journal of Chemical Physics, 61, 4474 (1974)

   Wheeler, J.C.: The 180^0 rule at a class of higher-order triple points, Physical Review A, 12, 267 (1975)

2. Thom, R.: Stabilité Structurelle et Morphogénèse, Benjamin, New York, 1972

3. Falk, G.: Theoretische Physik II, Thermodynamik, Springer Verlag, Heidelberger Taschenbücher Band 27

4. Thom, R.: Topological models in biology, Topology, Vol. 8, pp 313-335 (1969)

Güttinger, W.: Catastrophe geometry in physics and biology. In: Lecture Notes in Biomathematics 4 Springer, Berlin and New York, 1974, pp 2-30

5. Fowler, D.H.: The Riemann-Hugoniot catastrophe and van der Waals' equation. In: Towards a Theoretical Biology, Vol 4, ed. by C.H. Waddington, Edingburgh University Press 1972

6. Thom, R.: Phase transitions as catastrophes. In: Statistical Mechanics, ed. by S.A. Rice, K.F. Freed and J.C. Light, Chicago 1972

7. Devonshire, A.F.: Theory of ferroelectrics. Philosophical Magazine, Suppl. 3, 85 (1954)

8. Leggett, J.: Implications of the ^{3}He phase diagram below 3 mK, Progress Theoretical Physics, Vol 51, 1275 (1974)

9. Griffiths, R.B.: Thermodynamics near the two-fluid critical mixing point in ^{3}He - ^{4}He, Physical Review Letters, Vol. 24, 715 (1970)

10. Schulman, L.S.: Tricritical points and type-three phase transitions, Physical Review B, Vol. 7, 1960 (1972)

Phase Diagrams and Catastrophes

K. Keller, G. Dangelmayr, and H. Eikemeier

Institute for Information Sciences, University of Tübingen
D-7400 Tübingen, Fed. Rep. of Germany

Abstract

Maxwell sets of catastrophe polynomials are put into correspondence with thermodynamic phase diagrams near higher-order critical points. An abstract lattice model is set up which, using scaling principles, provides a theoretical basis for making catastrophe-theoretic techniques capable of reproducing empirical critical exponents.

Introduction

The objective of this paper is to compare the Maxwell sets of various catastrophe polynomials with thermodynamical phase diagrams for liquids, ferroelectrics, ferromagnetics, superfluid helium, multicomponent mixtures, etc., and to abstract from this some ideas concerning the feasibility of making catastrophe theory into a scheme capable of quantitative predictions about phase transition phenomena. To this end, in Section 1 we review some basic definitions. In Section 2 we relate some of the above concepts to n-th order critical points by comparing the Maxwell sets (or coexistence sets) of catastrophe theory with physically known phase diagrams. The simplest possible dependence of the formal order parameter and the formal control variables on physical quantities will be assumed. The results of this comparison are abstracted in Section 3 in a set of heuristic principles. In Section 4 a general lattice model is introduced which leads to a nonanalytic relation between the formal and the physical quantities. This enables us to show how catastrophe-theoretic models are capable of reproducing reasonable numerical results for critical exponents.

1. Basic concepts

The concept of 'order parameter' enables one to distinguish different phases and describe a variety of physical systems on the same formal basis. An n-th order phase transition is defined by the coalescence of n phases. Let x be the order parameter of an r-component thermodynamic system with at most n different phases. If p phases are observed, the number f of independent variables of the system is given by Gibb's phase rule, $f = r + 2 - p$. Since there exists at least one phase, the maximum number of variables is

$$f^* = r + 1. \tag{1.1}$$

Let X be the space of order parameter, U the space of formal control variables, and suppose $G: X \times U \to R$ to be a smooth family of functions--bounded below--whose minima with respect to the order parameter x describe the phases of the system. According to Eq. (1.1) we have $\dim U = f^*$. The coalescence of p phasesimplies p-1 supplementary conditions, i.e., the number of free variables in U is reduced to $f_p = r + 3 - 2p$. The requirement for an n-th order critical phase transition point to appear at an isolated point u_0 in U implies that $f_n = 0$ and therefore,

$$\dim U = 2n - 2. \tag{1.2}$$

We take this singular point u_o as the origin u=0 of U. The coalescence of n phases (i.e., minima of G) implies that G(x,0) is a singularity but not a Morse function. Therefore, we assume G(x,u) to be a universal unfolding (minimal dimension) of the singularity G(x,0), i.e.,

$$G(x,u) = G(x,0) + Q(x,u). \tag{1.3}$$

The minimal dimension of this unfolding is dim U. It is well known (Thom [3], resp. Arnold [4]) that the stable unfoldings can (for codimensions ≤ 4, resp. ≤ 16) be classified in terms of a few normal forms $V(x',u')$, where $x' = x'(x,u)$ and $u' = u'(u)$. We relabel x',u' as x,u and thus consider normal forms V(x,u). Since many physical systems are insensitive to small perturbations, one may expect them to be to some extent also modeled by structurally stable catastrophe schemes. Therefore, we tentatively suppose that the qualitative (and perhaps quantitative) description of an n-th order critical point is governed by a normal form V(x,u) with codimension = dim U. In order to identify this normal form with one of the thermodynamic potentials, it is necessary to specify the dependence of the order parameters and the variables in parameter space on physical quantities. Without any further specific assumption about this dependence, the only natural restriction is its invertibility near the critical point. The consequences of this assumption will be discussed in Sections 3 and 4.

The formal, catastrophe-theoretic equation of state is defined by $\Sigma(V) = \{(x,u) \mid d_x V(x,u) = 0\}$. Let $\Delta(V) = \{(x,u) \in \Sigma(V) \mid d_x^2 V(x,u) = 0\}$ and let π be the projection $X \times U \rightarrow U$. The bifurcation diagram $\pi(\Delta(V))$ divides U into regions such that V has a constant number of minima in any region. The usual convention to single out a stable state in thermodynamics is the 'Maxwell convention' [3], viz., to select the absolute minimum of V(x,u) for a given u. All points of Σ satisfying this convention form the restricted formal equation of state Ω. If, for a given $u \in U$, there are p different values $x_1, \ldots, x_p$ with $V(x_1,u) = \ldots = V(x_p,u)$ absolutely minimal, the coexistence of p phases is guaranteed. This set $\{x_1, \ldots, x_p, u\}$ will be denoted by ω' and its projection onto U by ω. ω is the coexistence set, describing phase transitions of first order. Sometimes we will refer to ω or ω' as Maxwell set. The set where the minima of V coalesce $(2 \leq p \leq n)$ will be denoted by Γ'. Its projection $\Gamma = \pi(\Gamma')$ describes higher order phase transitions.

2. Phase diagrams

A. The cusp catastrophe

For second order critical points it follows from (1.2) that dim U = 2. In this case, the isolated critical point is the origin of the unfolding parameters, i.e. $V(x,u) = x^4/4 + u_2x^2/2 + u_1x$, [3, 5 - 8]. We have $\Sigma = \{(x,u_1,u_2) \mid x^3 + u_2x + u_1 = 0\} = \{(\theta, -\theta^3 - \eta\theta, \eta) \mid \theta,\eta \in \mathbb{R}\}$ and $\Delta = \{(\theta, 2\theta^3, -3\theta^2) \mid \theta \in \mathbb{R}\}$. The coexistence set is the negative u_2-axis, and $\omega' = \{(\theta, 0, -\theta^2) \mid \theta \in \mathbb{R} \setminus \{0\}\}$. In Fig. 2.1 we sketch Σ, Δ and ω' and in Fig. 2.2 the projection $\pi(\Delta)$ and ω.

To compare Figs. 2.1 and 2.2 with known thermodynamics, we have to specify the dependence of control variables and order parameter on measurable quantities. One quantity important for all problems is the reduced temperature $t = (T - T_c)/T_c$. Examples for other reduced measurable quantities are the pressure $p = (P - P_c)/P_c$ or the magnetic field h. A measurable quantity such as density or magnetization which undergoes a jump during a first order phase transition represents a physical order parameter m.

(1) van der Waals model:
To describe the critical point of a liquid-gas transition we have to specify $u_1(p,t)$, $u_2(p,t)$ and $x(v)$, where $v = (V - V_c)/V_c$ is the reduced volume. In particular, $x(v) = -v/(1+v)$, $u_1(p,t) = (8t - 2p)/3$ and $u_2(p,t) = (8t + p)/3$ identifies Σ with the van der Waals equation

$$v^3/(1+v)^3 + (8t+p)\, v/3(1+v) + (2p-8t)/3 = 0. \qquad (2.1)$$

With $(P_c, V_c, T_c) = (a/27b^2, 3b, 8a/27Nkb)$ Eq. (2.1) reproduces the familiar standard form $(P + a/V^2)(V - b) = NkT$. The coexistence line $\omega(p,T)$ is [6]:

$$P = 4TP_c/T_c - 3P_c. \qquad (2.2)$$

In Fig. 2.3 we compare Eq. (2.2) with the coexistence line for the van der Waals model calculated numerically by C. Kittel [9]. It is seen that the Maxwell convention gives an approximation which is good only near (P_c, T_c).

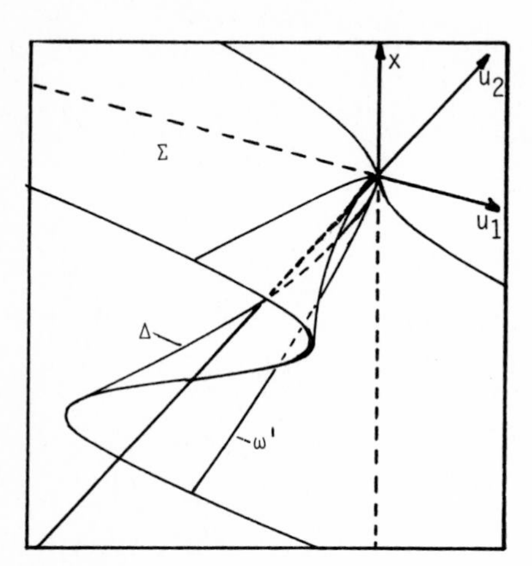

Fig. 2.1 Σ, Δ, ω' for the cusp catastrophe

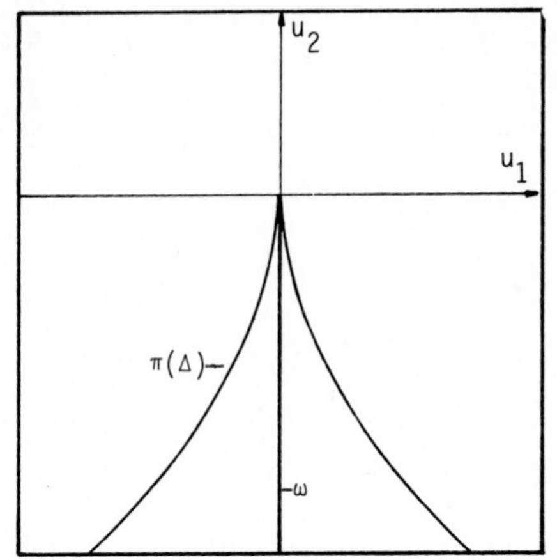

Fig. 2.2 The projections of bifurcation set Δ and Maxwell set ω' in U-space

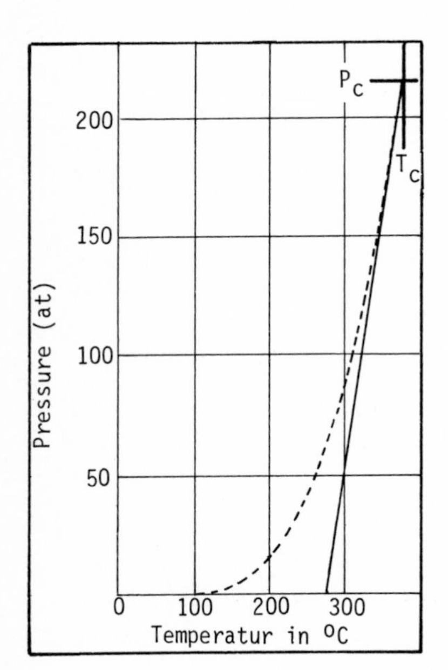

Fig. 2.3 Coexistence lines of van der Waals model dashed line due to C. Kittel [9], solid line due to Eq. (2.2)

Other choices for $x(v)$, $u_1(p,t)$, $u_2(p,t)$ lead to different equations of state, different coexistence lines, etc. (see, e.g. [10]). Hence, many physical models of the liquid-gas critical point can be fit into the cusp catastrophe.

(2) Laser radiation:
Additional physical requirements impose restrictions on the parameter space U. Put $u_1 \equiv 0$, $u_2 = (x-g_0)/g_1$ and $x = |E|$, where $|E|^2$ is proportional to the number of emitted photons, x is the pump rate and $g_0(\omega)$, $g_1(\omega)$ are the lowest order coefficients in the gain $g(\omega) = g_0(\omega) - g_1(\omega)|E|^2$. Then the cusp catastrophe reproduces a simple model for the onset of laser emission [11]. Even a more complicated model for laser radiation based on quantum mechanical considerations is governed by the cusp with $u_1 \equiv 0$ [1]. However, it is possible to find a physical variable enabling one to observe the full unfolding of the cusp [8]: Applying an external field α shifts the energy states of laser atoms and thereby changes the distance ε between energy states (Zeeman effect, Stark effect). For the relation of catastrophe parameters and physical quantities, one finds $x = \langle a \rangle + \alpha$ (a: annihilation operator), $u_1 = \varepsilon \hbar \omega \alpha / c$, $u_2 = (N \lambda^2 \tanh(\beta \varepsilon / 2) - \varepsilon \hbar \omega) c$ with experimentally variable pump rate λ and field α.

(3) Binary mixtures:
More complicated are binary fluid mixtures and alloys. If we choose the order parameter to identify one phase with one component, Eq. (1.1) gives $f^* = 3$, i.e. $W = \{(t, p, \mu_1(\mu_2))\}$. In this space, binary mixtures or alloys will exhibit an infinite line of second order phase transitions ([2,12,13]). To observe an isolated critical point and apply the catastrophe scheme, one has to introduce an artificial restriction in W, e.g. P = 1at. Only this restriction enables one to use the cusp catastrophe as a model for binary mixtures.

B. The butterfly catastrophe

According to Eq. (1.2) a tricritical point implies that dim U = 4. Therefore, the butterfly

$$V(x,u) = x^6/6 + u_4 x^4/4 + u_3 x^3/3 + u_2 x^2/2 + u_1 x \qquad (2.3)$$

should supply a model for tricritical points. In this case, Fig. 2.4 shows, for different values of u_4, a series of three sections [5] of the ω,Γ corresponding to Eq. (2.3) in (u_1,u_2,u_3)-space.

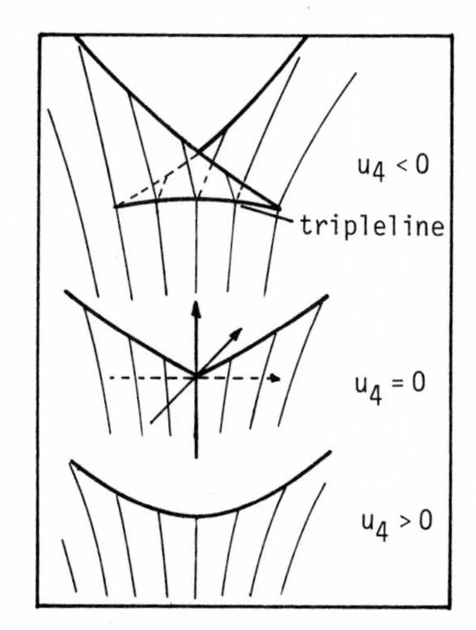

Fig. 2.4 Sketch of the coexistence sets for the butterfly [5]

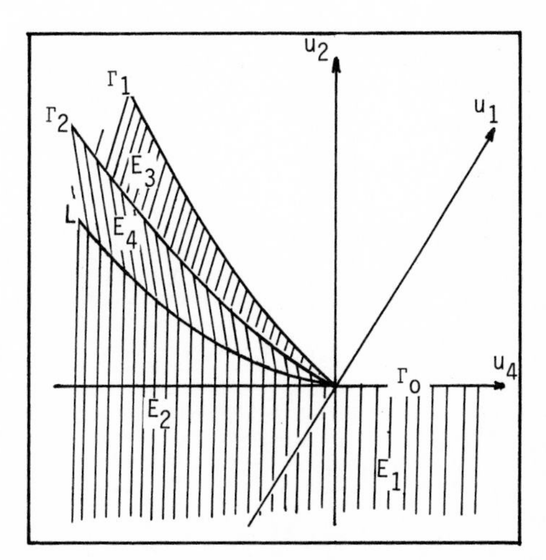

Fig. 2.5 Sketch of the coexistence set for the butterfly with $u_3 = 0$ [6]

If $a_3 \equiv 0$, Eq. (2.3) at once gives the coexistence sets ω and Γ by $\omega = \bigcup_{i=1}^{4} E_i$, $\Gamma = \Gamma_0 \cup \Gamma_1 \cup \Gamma_2$ [6] where

$$E_1 := \{(u_1,u_2,u_4) \mid u_1 = 0,\ u_2 < 0,\ -\infty \leq u_4 \leq \infty\}$$

$$E_2 := \{(u_1,u_2,u_4) \mid u_1 = 0,\ 0 \leq u_2 \leq \tfrac{3}{16}u_4^2,\ u_4 \leq 0\}$$

$$E_3 := \{(u_1,u_2,u_4) \mid u_2 = \tfrac{35}{32}\sqrt{-10/3u_4}\,u_1 + \tfrac{3}{16}u_4^2,\ 0 \leq u_1 < \tfrac{6}{25}u_4^2\sqrt{-3u_4/10},\ u_4 \leq 0\}$$

$$E_4 := \{(u_1,u_2,u_4) \mid u_2 = -\tfrac{35}{32}\sqrt{-10/3u_4}\,u_1 + \tfrac{3}{16}u_4^2,\ -\tfrac{6}{25}u_4^2\sqrt{-3u_4/10} < u_1 \leq 0,\ u_4 \leq 0\}$$

$$\Gamma_{1,2} := \{(u_1,u_2,u_4) \mid u_1 = \pm\tfrac{6}{25}u_4^2\sqrt{-3u_4/10},\ u_2 = \tfrac{9}{20}u_4^2,\ u_4 \leq 0\}$$

$$\Gamma_0 := \{(u_1,u_2,u_4) \mid u_1 = u_2 = 0,\ u_4 \geq 0\}$$

These sets are shown in Fig. 2.5. A first order transition between three phases is given by the intersection of E_2, E_3, E_4 forming the triple line $L = \{(u_1,u_2,u_4) \mid u_1 = 0,\ u_2 = 3u_4^2/16,\ u_4 \leq 0\}$.

(1) Ternary mixtures:
For ternary mixtures (e.g. CH_4, H_2O, CO_2), the physically measurable quantities are the total molar density per unit volume ρ, the reduced temperature t, the reduced pressure p and two independent chemical potentials, e.g. μ_1, μ_2. By virtue of Eqs. (1.1) and (1.2) such mixtures possess an isolated tricritical point. If $x(\rho)$ and $u_i(p,t,\mu_1,\mu_2)$ $(i = 1,2,3,4)$ are analytic functions, Eq. (2.3) leads to a Landau-type model and the essential features of Fig. 2.4 are reproduced [14], because none of the u_i vanish identically.

(2) Ferroelectrics:
a) If $x = x(P) = P$, $u_1 = u_1(E) = -E$, $u_2 = u_2(t,w)$, $u_3 \equiv 0$, and $u_4 = u_4(t,w)$, Eq. (2.3) at once gives the Fatuzzo-Merz model for ferroelectrics around a tricritical point [16]. Here P is the polarization, E the electric field and the variable w is pressure or concentration.

b) An example in which w is the concentration is given by the ferroelectric $NH_4Cl_{1-w}Br_w$ studied by Jahn and Neumann [15]. The experimental phase diagram is represented by a curved two-dimensional cut, for which $u_2 = 0$ for $u_4 > 0$ and $u_2 \sim |u_4|^n$ with $n > 2$ for $u_4 < 0$. Therefore, since for the triple line $u_2 = 3u_4^2/16$, the curved cut intersects L with increasing $|u_4|$.

(3) Ferromagnetics:
Replacing P, E and w by magnetization M, magnetic field H, and crystal field energy Δ, Eq. (2.3) with $x(M)$, $u_1(H)$, $u_2(t,\Delta)$, $u_3 \equiv 0$, $u_4(t,\Delta)$ describes a model of tricritical points in ferromagnets [6,17]. In Eq. (2.3), let us again specify $x(M) = M$, $u_1(H) = -H$, and let $u_2(t,\Delta)$, $u_4(t,\Delta)$ be analytic in their arguments. Then one obtains a mean field theory of ferromagnetic tricritical points.

(4) Superfluid helium:
A somewhat unusual example is shown in Fig. 2.7 (due to [2]). We conjecture [6] that the tricritical point of He_4 is just beyond experimentally reachable pressure. Then, in the phase diagram of Fig. 2.7, the λ-line is equivalent to Γ_0, and the liquid-vapor coexistence curve is equivalent to a curve in E_4, which increases faster than Γ_2. In this case the order parameter x depends on m_s, the mass of superfluid He (He II), on the mass of the normal fluid He (He I) and on m_g, the mass of the gaseous component of He. With $x = m_s/(m_n + m_g) - m_g/(m_n + m_g)$ we see that $x > 0$ for He II, $x = 0$ for He I and $x < 0$ for gaseous He. The coördinates of the tricritical point are $P_c \simeq 0$ and $T_c \simeq T_\lambda(0)$, whereby $T_\lambda(P)$ denotes the λ-line.

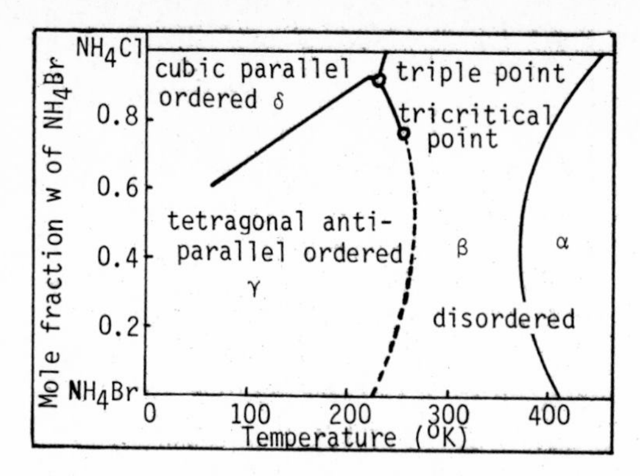

Fig. 2.6 Phase diagram of $NH_4Cl_{1-w}Br_w$ [15] with dashed line corresponding to second order phase transitions, and solid lines corresponding to first order phase transitions

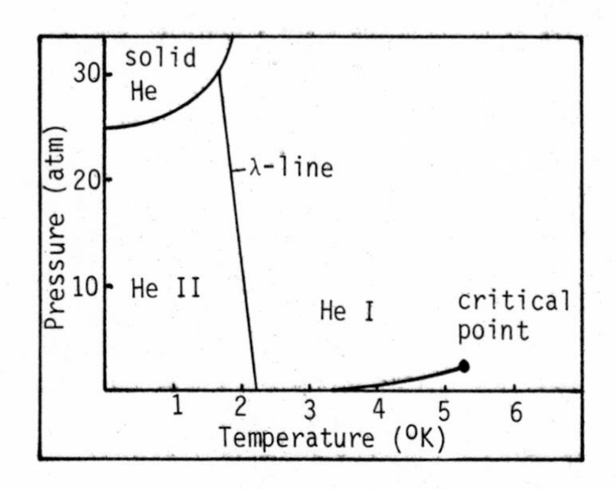

Fig. 2.7 Phase diagram of He_4 [2]

In the above physical examples (2b) and (4) only two parameters have been considered as variable. Therefore, one can only find two independent functions of control variables.

(5) Quaternary mixtures:
For quaternary mixtures (e.g., water, phenol, pyridine, n-hexane) we have $f^* = 5$. Therefore, these mixtures show a whole line of tricritical points, some having been discovered experimentally [18]. As in the case of binary mixtures, we have to impose a further appropriate restriction on U which enables us to describe quaternary mixtures with the butterfly model. Without this restriction, we cannot apply the butterfly model, because in that case the tricritical points are not isolated. The line of tricritical points does not end and, consequently, there is no tetracritical point in a quaternary mixture [14,18]. Thus, the subsequent star catastrophe will not give a possible model for mixtures with four components [5].

C. The star catastrophe

A model for tetracritical points is given by the star catastrophe defined by $V(x,u) := x^8/8 + \sum_{k=1}^{6} u_k x^k/k$. Four phases coalesce at the tetracritical point. In particular, a tetracritical point is expected to occur in 5-component fluid mixtures [5,18]. Then the control variable are functions of t, p, and of four chemical potentials μ_i. The corresponding phase diagrams have been discussed by Dangelmayr and Keller [19]. The results are shown in Figs. 2.8 and 2.9. Note that taking $u_3 \neq 0$, $u_5 \neq 0$ merely gives rise to a distortion of the phase diagrams (if u_3, u_5 are not too large).

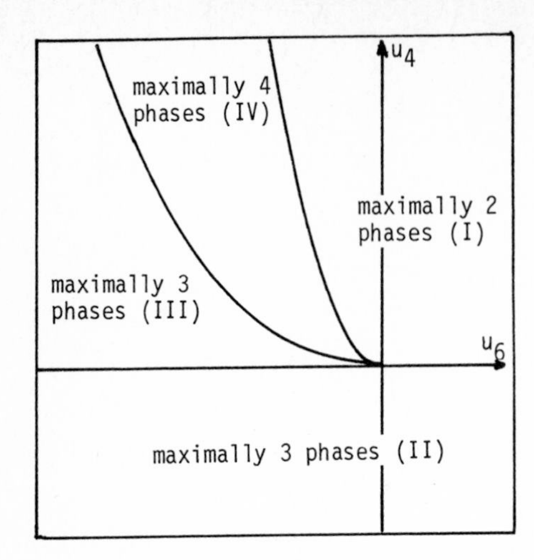

Fig. 2.8 The qualitatively different regions I - IV in u_4-u_6-plane ($u_3 \equiv 0$, $u_5 \equiv 0$) [5]

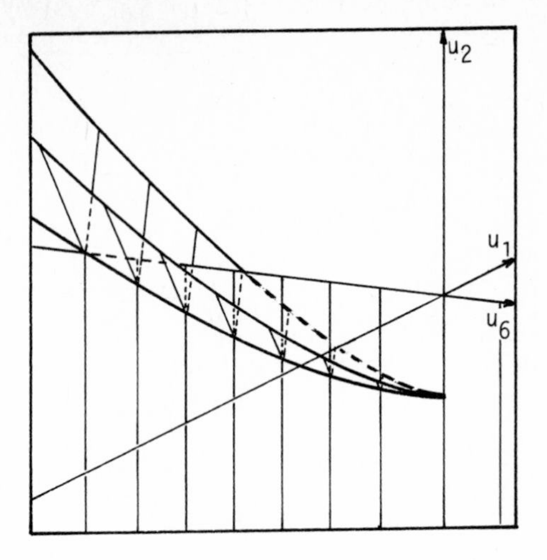

Fig. 2.9 The projection of the coexistence set in u_1-u_2-u_6-space ($u_4 > 0$, $u_3 \equiv 0$, $u_5 \equiv 0$) [5]

D. Two order parameters

Two coupled order parameters are needed to describe some thermodynamic systems. Examples for such systems are [20]: ferromagnetic-ferroelectric systems, ferromagnetic-antiferromagnetic systems, ferroelectric-piezoelectric systems, crystalline-superfluid systems, and the orientation-position ordering in liquid crystals. The basic concept is not restricted to one order parameter and can naturally be extended. Since very little is known about a catastrophe-theoretic description of these systems, we confine ourselves to the 'reduced double cusp' as an apparently typical example. The catastrophe polynomial of the reduced double cusp is given by

$$V(x,y,u) := x^4/4 + y^4/4 - 6x^2y^2 + u_1(x^3 - 3xy^2) + u_2(x^2 + y^2) - u_3x - u_4y. \qquad (2.4)$$

V is a non-universal unfolding of the singularity $V(x,y,0)$, which belongs to the unimodal family X_9 [28,29].

(1) Model system with two scalar order parameters:
The transformation $x \to \alpha(x-x_0(T))$, $y \to \beta y$ and $u_i = u_i(T)$ turns (2.4) into

$$V = Ax^4/2 + By^4/2 + Cx^2 + Dy^2 - \lambda x^2y^2 + F(x_0) \qquad (2.5)$$

where $A = \alpha^4/2$, $B = \beta^4/2$, $C = 3\alpha^4x_0^2/2 + u_2\alpha^2 + 3u_1\alpha x_0$, $D = u_2\beta^2 + 6\alpha^2\beta^2x_0 - 3\alpha\beta^2x_0$ and $\lambda = 6\alpha^2\beta^2$. In order to eliminate the odd terms in Eq. (2.4), three additional conditions $\alpha x_0 + u_1 = 0$, $12\alpha^2\beta x_0xy - 3\alpha\beta xy + u_4 = 0$, $\alpha^4x_0 + 2u_2\alpha^2x_0 + \alpha u_3 + 3u_1\alpha^3x_0^2 = 0$ must be satisfied. These conditions fix a curve in (u_1,u_2,u_3,u_4)-space and therefore, Eq. (2.5) is a very restricted unfolding. The term $F(x_0)$ contains only $x_0(T)$. If one neglects $F(x_0)$, then Eq. (2.5) describes the model of Imry, Scalapino, and Gunther analyzed in [20].

(2) Binary mixtures:
It may also be possible to describe binary mixtures [12] by the reduced double cusp by assuming a dependence of the order parameters x,y on the density of each component. Because $f^* = 3$, the reduced double cusp is not fully unfolded for binary mixtures, but in contradistinction to example (A3), we do not have to impose artificial restrictions in parameter space. Nevertheless, the natural restriction $f^* = 3 < \dim U$ may have the serious consequence that the center of the unfolding is not physically accessible in special systems. However, Kohnstamm's phase diagrams [12] indicate that parabolic umbilics occur in multicomponent mixtures [21] and the former are subsets of the double cusp.

3. Consequences

The comparison between the Maxwell sets of various catastrophe polynomials and physical phase diagrams indicates that:
(a) catastrophe theory reproduces the phase transition phenomena near an isolated n-th order critical point quite well;
(b) there is a maximum number of parameters independently variable experimentally given by the codimension k of the normal form $V(x,u)$. Not all independent variables may be experimentally accessible, a fact imposing restrictions on U;
(c) since catstrophe theory is a local theory, the relation between measurable quantities and control parameters should be only locally invertible. Let $M = \{m\}$ be the space of physical order parameters and W the space of physical control variables. Then each ansatz $x = x(m)$, $u_i = u_i(w_1, \ldots, w_k)$ $(1 \leqslant i \leqslant k)$ should be locally invertible near $x(m) = 0$, $u(w) = 0$.

From this one may conclude that the concept of structural stability hints at a description of an isolated n-th order critical point by exactly one catastrophe polynomial depending on n. Our conjecture is, therefore, that every model for an n-th order phase transition should be contained in this special polynomial. The term 'contained' means that it should be possible to find a local, invertible transformation between $M \times W$ and $X \times U$ making both phase diagrams locally identical. The assumption that the physical parameter m is mapped diffeomorphically into the formal order parameter x and the physical variables $(w_1, \ldots, w_k)$ are mapped diffeomorphically into the formal control variables $(u_1, \ldots, u_k)$ ensures a one-to-one correspondence between catastrophe theory and Landau theory. However, local invertibility ought to be enough. If the measurable quantity m does not double its value if the system's size is doubled, then m is not a true extensive variable [2]. With m being more closely akin to an extensive variable, a function $\phi(m)$ should exist, which is extensive and which is locally taken as our formal parameter x. Catastrophe theory dictates the power of x in $V(x,0)$ at the n-th order critical point.

Near higher order phase transition points the dependence of the relevant physical quantities on the reduced temperature is given--in first approximation--by power laws. The exponents of these power laws are the critical exponents [13,17,22]. The critical exponents, computed from the Landau theory are in rough agreement with experimental values. A diffeomorphic dependence $x(m)$, $u(w)$ does not, therefore, alter the Landau exponents [8], and neither does a change of the Maxwell convention [8,10]. Hence, we expect a good model of critical phenomena (based, e.g., on a renormalization group analysis [13]) to not be restricted to a diffeomorphic dependence $x(m)$, $u(w)$. The following simple lattice model establishes such a non-diffeomorphic dependence.

4. A general lattice model

Suppose a thermodynamic system may be divided into N subsystems (e.g., a crystalline ferroelectric into the crystal's elementary cells). Let us denote by $\psi(z)$ the measurable order parameter of the subsystem. Each subsystem will be identified with exactly one point in an abstract D-dimensional lattice [23] and for each lattice point z_i the variable $\psi(z)$ has the value $\psi(z_i)$. In what follows, the bracket $\langle\cdot\rangle$ denotes the statistical average over all lattice points. Then, the macroscopic measurable order parameter is given by

$$m = \langle\psi\rangle. \tag{4.1}$$

If $H(\mu,\psi)$ is the microscopic Hamiltonian and $\sigma(m)$ the natural logarithm of the structure function $g(m)$, which measures the degeneracy of the macroscopic state m, the partition function Z is given by

$$Z = \int \mathcal{D}\,\psi(z)\; e^{-\beta\int d^D z H(\mu,\psi)} = \int dm\, g(\mu,m)\; e^{-\beta E(\mu,m)} = \int dm\; e^{-\beta \mathcal{F}(\mu,m)}$$

where E is the internal and $\mathcal{F}$ the free energy. The probability distribution $P(\mu,m) = Z^{-1} \exp(-\beta\mathcal{F}(\mu,m))$ has sharp peaks at the minima of $\mathcal{F}(\mu,m)$ because the total free energy $\mathcal{F}$ is of the order of the system volume and large from the microscopic point of view. The absolute minimum of $\mathcal{F}$ therefore has the largest probability. Assuming the Maxwell convention, it is then natural to establish some relation between $\mathcal{F}$ and a catastrophe polynomial. The free energy per unit volume V_0 is given by $F(\mu,m) = \mathcal{F}(\mu,m)/V_0$ and F has the same minimum properties as $\mathcal{F}$.

Let us now assume that the spatial extension of the subsystems is of the order of the correlation length ξ [13,22,23]. Furthermore, the set of parameters $\{\mu\}$ shall depend on physical control variables such as reduced temperature t, i.e., we suppose that higher order phase transitions exist. Variation of t gives a curve $\mu(t)$ in parameter space $\{\mu\}$. To each point of this curve there belongs one special lattice, with the lattice points equivalent to subsystems of extension $\xi(t)$. The point $\mu(0)$ gives rise to a phase transition of n-th order since $t = (T - T_c)/T_c$. Let us confine ourselves to $n = 2$. Then according to the hypotheses stated in Section 3, this phase transition will necessarily be described by an unfolding of $x^4/4$ and the unfolding starts to develop as soon as $t \neq 0$. Therefore, we put $F(0,m) = x^4/4$. To infer the dependence of x on m, we apply a dimensional analytic argument [13]. This dimensional analysis will lead to scaling laws, and for this reason our lattice model may be termed a 'scaling model.' Now, let us transform the unit of length from 1 to s. Then the volume transforms as

$$V_s = s^{-D} V_1. \tag{4.2}$$

The total free energy $\mathcal{F}$ is independent of any special choice of the unit of length. Consequently, the free energy per unit volume transforms according to

$$F_s = s^D F_1 \tag{4.3}$$

The correlation function G(k) is given by [22]

$$G(k) = \langle|\psi_k|^2\rangle = \frac{1}{V}\int d^D z\, d^D y \langle\psi(z)\,\psi(y)\rangle\, e^{-ik(z-y)} \sim \langle\psi\rangle^2 \times \text{volume}, \tag{4.4}$$

where ψ_k is the Fourier transform of $\psi(z)$. The scale transformation implies that

$$G(sk) \sim (s^{2y})(s^{-D}) = s^{-D+2y} \tag{4.5}$$

for (4.4) in the limit $t \to 0$. Experimentally, it is known that $G(k) \sim k^{-2+\eta}$ for $t \to 0$. The above scaling implies a scaling in momentum space: $k \to sk$. From this we obtain

$$G(k) \longrightarrow G(sk) \sim s^{-2+\eta} G(k). \tag{4.6}$$

Comparing (4.5) and (4.6), one concludes that

$$y = (D - 2 + \eta)/2. \tag{4.7}$$

Eq. (4.1) implies that m has the form

$$m \sim s^y. \tag{4.8}$$

For $t \simeq 0$, $F(\mu,m) \simeq x^4/4$ and with (4.3) we find that x^4 scales proportional to s^D. Therefore,

$$x \sim s^{D/4}. \tag{4.9}$$

Comparison of (4.8) and (4.9) gives the result

$$x \sim m^\theta, \qquad \theta = D/4y. \tag{4.10}$$

The construction of the lattice points fixes θ. To see this, we compare two different lattices by varying t, choosing s such that $s \cdot \xi(t) = 1$ for all t. Since $\xi(t) \sim |t|^{-\nu}$ [13,17,22], s is proportional to $|t|^{\nu}$. Inserting $s \sim |t|^{\nu}$ into Eq. (4.3) gives $F \sim |t|^{\nu D}$.

The specific heat is $c_\nu = -T\,\partial^2 F/\partial T^2 \sim |t|^{-\alpha'}$ (if $t < 0$). With Eq. (4.3) it follows that

$$\nu D - 2 = -\alpha'. \tag{4.11}$$

According to [25] we have $\nu = 0.638 \pm 0.002$ and $\eta = 0.041 \pm 0.003$ and the mean experimental value of α' is $\alpha' \simeq 0.03$ [23]. (Our abstract lattice neither registers a difference between fluids and magnets nor between α $(t > 0)$ and α' $(t < 0)$). Assuming the above values for ν, η and α' we find $D = 3.09$, $y = 0.56$, and $\theta = 1.36$.

On the coexistence line of the cusp (due to Maxwell's convention), we have $|u_2(t)| = x^2$ and by virtue of Eq. (4.3), $x^2 \sim |t|^{\nu D/2}$. The above values for D and ν imply that $|u_2(t)| \sim |t|^{\lambda}$ with $\lambda = 0.98$.

The abstract lattice shows that the behavior near an isolated critical point is governed by the cusp with a nonstandard relation

$$\begin{aligned} x(m) &\sim m^{\theta} \\ u_2(t) &\sim t^{\lambda} \end{aligned} \tag{4.12}$$

Since θ,λ are nonintegers, the dependence (4.12) is nonanalytic. Such a dependence was already proposed by Benguigui and Schulman [24] and independently by Dangelmayr [5] and Keller [6]. Calculating critical exponents from the cusp with (4.12) inserted, yields the results given in the second column of Table I [23]. Inserting numerical values for λ,θ gives the numerical critical exponents of the subsequent columns. These exponents agree with the exponents of the physical model mentioned below.

λ		0.98	1	1	1	$\frac{15}{16} \simeq 0.94$	1.1	
θ		1.36	1	4	$\frac{3}{2}$	$\frac{3}{2}$	1.43	
α',α	$2(1-\lambda)$	0.04	0	0	0	$\frac{1}{8} \simeq 0.12$	-0.2	0.03 ± 0.14
β	$\lambda/2\theta$	0.36	0.5	$\frac{1}{8} \simeq 0.12$	$\frac{1}{3} \simeq 0.33$	$\frac{5}{16} \simeq 0.31$	0.38	0.34 ± 0.04
δ	$4\theta - 1$	4.44	3	15	5	5	4.72	4.25 ± 0.13
γ',γ	$\lambda(2\theta-1)/\theta$	1.24	1	$\frac{7}{4} = 1.75$	$\frac{4}{3} \simeq 1.33$	$\frac{5}{4} = 1.25$	1.42	1.20 ± 0.12
	CT	Scaling model	Landau theory	2 dim. Ising model	3 dim. Heisenberg model($n=\infty$)	3 dim. Ising model	3 dim. Heisenberg model($s=\pm\frac{1}{2}$)	Experiment

Table I Numerical values of critical exponents for the cusp catastrophe, in dependence on λ and θ

A discussion of the tricritical point in ferroelectrics shows that one can determine critical exponents by using the butterfly catastrophe and a nonanalytic relation between formal and measurable variables, defined by [6] $x(P) \sim P^{\theta}$, $u_2(t,0) \sim t^{\lambda}$, and $u_4(t,0) \sim t^{\lambda/2}$. Then for the tricritical exponents, one obtains $\alpha = 0$, $\alpha' = 2 - 3\lambda/2$, $\beta = \lambda/4\theta$, $\delta = 6\theta - 1$, $\gamma = \lambda(3\theta - 1)/2\theta = \gamma'$. Taking $\theta = 1.14$ and $\lambda = 0.96$ yields good agreement with experimental results, while $\theta = 1$, $\lambda = 1$ produces the results of the Landau theory.

λ		0.96	1	
θ		1.14	1	
α	0	0	0	$\simeq 0$
α'	$2 - 3\lambda/2$	0.56	0.50	0.56 ± 0.06
β	$\lambda/4\theta$	0.21	0.25	0.21 ± 0.05
δ	$6\theta - 1$	5.84	5	?
$\gamma = \gamma'$	$\lambda(3\theta - 1)/2\theta$	1.02	1	1.01 ± 0.03
	Butterfly model		Landau theory	Experiment

Table II Tricritical exponents for the butterfly catastrophe in dependence on λ, θ

5. Conclusions

The general lattice model sketched in Section 4 confirms our hypothesis that all familiar models for critical points ($n = 2$) are locally present in the cusp. Locally means that the choice $x(m) \sim m^\theta$ and $u_2(t) \sim t^\lambda$ holds only locally. Different values for λ, θ give the exponents for quite different models. Another confirmation of our hypothesis is the possibility of fitting λ, θ for the butterfly model of tricritical points.

In our lattice-scaling model a strange event is involved, viz., the anomalous lattice dimension $D = 3.09$ for the critical point. The abstract lattice, however, is a mathematical model and it need not necessarily have the same dimension as the real physical system under consideration. Geometrical objects with noninteger dimension can be described by fractals [26,27] whose macroscopic forms are self-similar to their microscopic parts. Each part of a fractal exhibits the same structure as the whole. This reflects the physical fact that critical behavior is governed by diverging correlation lengths. At the critical point, therefore, thermodynamic systems look the same, whatever the scale of observation [13,22]. Near critical points one finds an intrinsic length of unit, the correlation length ξ and all details finer than ξ become irrelevant. In the theory of fractals to the extent it exists, equations such as (4.2) and (4.4) remain valid for noninteger dimension D. The Hausdorff-Besicovitch dimension D raises the question what renormalization group concepts, catastrophes, and geometry have in common in the context of phase transition phenomena.

The authors thank Prof. W. Güttinger, E. Obermayer and V. Schmidt-Ramsin for many helpful discussions.

References

1. Haken, H.: Synergetics, Springer 1978
Haken, H.: Cooperative phenomena in systems far from thermal equilibrium and in nonphysical systems, Rev. Mod. Physics 47 (1975), 67

2. Falk, G.: Theoretische Physik II: Thermodynamik, Springer 1968

3. Thom, R.: Structural Stability and Morphogenesis, Benjamin 1975

4. Arnold, V.: Critical Points of Smooth Functions, Proc. of the International Congress of Mathematicians, Vol. 1, Vancouver 1974

5. Dangelmayr, G.: M.S. thesis, Tübingen 1975

6. Keller, K.: M.S. thesis, Tübingen 1975

7. Vendrik, M.C.M.: A Classification of Phase Diagrams by Means of Catastrophe Theory, thesis, Nijmegen 1977

8. Poston, T., J. Stewart: Catastrophe Theory and its Applications, Pitman 1978

9. Kittel, C.: Physik der Wärme, Oldenbourg 1973

10. Fowler, D.H.: The Riemann-Hugeniot catastrophe and van der Waals equation, in: Towards a Theoretical Biology, Vol. 4 (ed. H. Waddington), Edinburgh 1972

11. Haken, H.: Laserlicht - Ein neues Beispiel für eine Phasenumwandlung? in: Festkörperprobleme X, Pergamon, Vieweg 1970

12. Kohnstamm, P.: Thermodynamik der Gemische, Handbuch der Physik X, Springer 1926

13. Ma, S.K.: Modern Theory of Critical Phenomena, Benjamin 1976

14. Griffiths, R.B.: Thermodynamic models for tricritical points in ternary and quaternary fluid mixtures, Journal of Chemical Physics 60, 195 (1974)

15. Jahn, I.R., E. Neumann: Phase diagram of the solid state solution system $NH_4Cl_{1-x}Br_x$ as determined by optical methods, Solid State Communications 12, 721 (1973)

16. Fatuzzo, E., W.J. Merz: Ferroelectricity, in: Selected Topics in Solid State Physics, Vol. 7,(ed. E.P. Wolfarth), North-Holland 1967

17. Stanley, H.E.: Phase Transitions and Critical Phenomena, Clarendon Press 1971

18. Griffiths, R.B., B. Widom: Multicomponent-fluid tricritical points, Phys. Rev. A 8, 2173 (1973)

19. Dangelmayr, G., K. Keller: Thermodynamic Example for the Star Catastrophe (in preparation)

20. Imry, Y., D.J. Scalapino, L. Gunther: Phase Transitions in Systems with Coupled Order Parameters, preprints, University of California, Santa Barbara 1973 and Tufts University, Medford 1973

21. Obermayer, E.: private communication, 1978

22. Kadanoff, L.P.: Critical behavior, universality and scaling, in: Proceedings of the Int. School of Phys. 'Enrico Fermi,' Varenna 1970 (ed. M.S. Green), Academic Press 1971

23. Keller, K.: Ph.D. thesis, Tübingen 1979

24. Benguigui, L., L.S. Schulman: Topological classification of phase transitions, Phys. Letters 45 A, 315 (1973)

25. Stanley, H.E., A. Hankey, M.H. Lee: Scaling, Transformation Methods and Universality, Academic Press 1971

26. Mandelbrot, B.B.: Fractals. Form, Chance and Dimension, Freeman 1977

27. Berry, M.V.: Catastrophe and fractal regimes in random waves, and Distribution of modes in fractal resonators, this volume

28. Zeeman, E.C.: The Umbilic Bracelet and the Double-Cusp Catastrophe, in: Structural Stability, the Theory of Catastrophes and Application in the Sciences, (ed. P. Hilton), Lecture Notes in Math. 525, Springer 1976

29. Callahan, J.: Special Bifurcations of the Double Cusp, preprint, University of Warwick 1978

Information Measures and Thermodynamic Criteria of Motion

F. Schlögl

Institut für Theoretische Physik, Technische Hochschule Aachen
D-5100 Aachen, Fed. Rep. of Germany

Abstract

The well known connection between entropy and Shannon's definition of information finds a generalization in the connection between Kullback's information gain and thermodynamic quantities which are essential for entropy production, for the stability criterion of Glansdorff and Prigogine, or for the probability and dynamics of fluctuations in a steady state. Another generalization leads to a set of ordered correlation measures. In particular the measure of second order leads to specific heat and generalizations which show a characteristic critical behaviour in non-equilibrium phase transitions.

1. Introduction

Thermodynamics of non-equilibrium processes has got a relatively close form in the linear region. This region of "linear thermodynamics" is restricted on processes which occur so close to an equilibrium state that the macroscopic dynamic equations can be linearized with respect to the quantities describing the deviations from the equilibrium. The latter are the thermodynamic "forces" and "fluxes", introduced by ONSAGER, who has developed linear thermodynamics in its essentials. The situation is totally different in "non-linear thermodynamics" which is concerned with processes which are yet describable by macroscopic thermodynamic means which, however, don't fulfill the linearization condition of the linear region. In non-linear thermodynamics a lot of interesting and often surprising phenomena have been observed and studied. In this respect it became especially in the last two decades a field of intensive study with many results which go into very different branches of application. Typical phenomena are the occurence of instabilities, multi-stability, cycles in time, and selfreproducing structures, which by the school of GLANSDORFF and PRIGOGINE have been generally called "dissipative structures". Whereas the study of features of special systems or special classes of systems has been very successful and has brought a lot of new insights, we know very few about new general laws in non-linear thermodynamics which could be comparable in its generality with the principles in thermostatics and linear thermodynamics.

Of course the question is open of whether such general laws can be found. It seems, however, reasonable to seek in the framework of statistical physics for general features of dynamics of probability distributions in GIBBS' phase space, which are in a large extend independent of individual features of the special physical system. As we can understand very general laws of macroscopic thermostatics and linear thermodynamics in this way, there can be some hope for success also in non-linear thermodynamics.

Fundamental quantities in thermodynamics, as entropy and entropy production are related to measures over probability distributions which have been worked out in information theory and which allow a valuating comparison of different probability distributions. Important are inequalities restricting the change of such measures in time as a consequence of dynamical equations for the probability distribution in GIBBS' phase space. It should be stressed that they are mathematical consequences. No arguments are necessary for their derivation which are based on an interpretation of these quantities as measures of information. Therefore they are not dependent on an "informational interpretation" of thermodynamics which developed, after the connection between an information measure and thermodynamics was detected, but which is not accepted by everybody. Nevertheless, the main ideas of such an interpretation shall also be mentioned in the following.

2. Thermal Equilibria

The well known connection between entropy S in equilibrium thermodynamics and the probability density $w(\xi)$ in GIBBS' phase space

$$(Q,..Q_f;P,..P_f) = \xi \tag{2.1}$$

was given by BOLTZMAN:

$$S = -k \int d\xi \; w \ln w \tag{2.2}$$

It was called by v. LAUE "one of the deepest ideas of whole physics" [1]. BOLTZMANN's result is closely related to the SHANNON-information [2]

$$I(p) = \Sigma_\alpha p_\alpha \ln p_\alpha \tag{2.3}$$

of a probability distribution

$$p = (p_1, p_2,..p_\alpha...) \tag{2.4}$$

over states α of a system which form a complete disjunction of events. The measure I(p) is introduced as the mean number of "yes-no" decisions which are necessary to give as communication the knowledge contained in p about the question which state α is to be expected. We call it the mean bit-number of the distribution p. It is a matter of mathematical technic to extend such a measure to a continuous set of microstates ξ leading to a measure over probability densities $w(\xi)$:

$$I(w) = \int d\xi \; w \ln w. \tag{2.5}$$

Up to the BOLTZMANN factor k which in the following may be put equal to unity by an adequate choice of units, entropy S is equal to the negative of I(w). So it can be interpreted as the lack of information or the lack of knowledge about the microstate of the system. This is a more precise definition than the explanation of entropy as a "measure of disorder" because only in very special cases a scheme of order can be given.

The equilibrium distribution belonging to given macroscopic variables M which are interpreted in a statistical theory as mean values of phase space functions $M(\xi)$ is found by the prescription of maximizing S to fixed values M what gives

$$w(\xi) = \exp\left[\Phi - \lambda_\nu M^\nu(\xi)\right]. \tag{2.6}$$

This expression, in which - as in the following - the summation has to be carried out over equal greec super- and subscripts designating the macroscopic variables M^ν, shall be called a "generalized canonical distribution". Special cases are the canonical, the grand canonical, the pressure, or the magnetic field ensemble. The construction of the generalized canonical distribution is interpreted by JAYNES [3] as a general method of an "unbiased guess" of a probability distribution to given mean values M^ν of random variables. This method is based on the requirement of minimizing information measure I(w) under given conditions to avoid any unjustified amount of information as a biased and unwarranted prejudice.

Second law, in the form that entropy in an isolated system never decreases, corresponds to the statement that information I(w) after the last observation never can increase. Against such an interpretation often arguments are directed that entropy as a fundamental quantity in physics must not be dependent on the knowledge of the observer. It should, however, be realized that all such arguments are directed also against the conception of the probability distribution w being dependent on this knowledge. Otherwise one has to deny the connection "2-1" between entropy and probability. Once more it shall be emphasized that the following considerations are totally independent of these interpretations.

Eq. (2.1) leads (with k=1) for the generalized canonical distribution (2.6) to the relation

$$S = -\Phi + \lambda_\nu M^\nu. \tag{2.7}$$

As a consequence of the normalization condition for w we find

$$\frac{\partial \Phi}{\partial \lambda_\nu} = M^\nu . \qquad (2.8)$$

This yields in the form

$$\frac{\partial S}{\partial M^\nu} = \lambda_\nu \qquad (2.9)$$

the statement of GIBBS' fundamental equation for dS, which fixes the structure of thermostatics to a great extent.

A relative information measure over two probability distributions w, w^o in the same space ξ of microstates has been introduced by KULLBACK [4]:

$$K(w,w^o) = \int d\xi \; w \ln \frac{w}{w^o} \qquad (2.10)$$

which is called "KULLBACK information" or "information gain". It leads to a measure for the information which is necessary to change w^o into w. It can be interpreted as the mean value of the bit-number which is necessary for the corresponding communication. The mean value is formed with the "corrected" probability distribution w. For instance, such a change from w^o into w is caused by a new observation. This measure always is positive and vanishing only if w and w^o are totally identical.

If we take for w^o a generalized canonical distribution belonging to mean values $M^{\nu o}$ and corresponding parameters λ_ν^o, we get

$$K(w,w^o) = S^o - S - \lambda_\nu^o \,(M^{\nu o} - M^\nu) \geqq 0. \qquad (2.11)$$

This is a well known inequality in macroscopic thermodynamics expressing the stability of the equilibrium w^o. If w also is of the form (2.6)' with values M^ν deviating from $M^{\nu o}$ by small values ∂M^ν, we get

$$K(w,w^o) = - \frac{\partial \lambda_\nu}{\partial M^\mu} \, \delta M^\mu \delta M^\nu \geqq 0. \qquad (2.12)$$

In particular we get for the generalized susceptibilities:

$$\frac{\partial \Lambda_\nu}{\partial M^\nu} \leqq 0. \qquad (2.13)$$

This inequality determines the sign of magnetic susceptibility, specific heat, and so on. It also leads to the principle of LE CHATELIER and BRAUN [5].

3. MORI Distribution

Now we shall discuss the question of which probability distribution $w(\xi)$ in GIBBS' phase space belongs to a non-equilibrium state which in macroscopic thermodynamics is described by a set of local densities $m^\nu(\underline{r},t)$ at time t and at point $\underline{r}$ in ordinary three dimensional space. These densities of quantities of type M^ν shall be interpreted in a statistical theory as mean values of phase space functions $m^\nu(\underline{r};\xi)$. The time dependence of the mean values is introduced by a time dependent w. This introduction is called the SCHRÖDINGER picture.

If we would apply JAYNES' construction using only the instantaneous mean values at time t, we should obtain a distribution

$$w^{\ell}(\xi,t) = \exp\left[\Phi(t)-\int d^3\underline{r}\lambda_{\nu}(r,t)\; m^{\nu}\;(\underline{r},\xi)\right] \tag{3.1}$$

which is called the "local equilibrium distribution". The summation over ν and the integration over $\underline{r}$ will be abbreviated in the following by a scalar product symbol. So we write

$$w^{\ell}(\xi,t) = \exp\left[\,\Phi(t)-\lambda(t)\cdot m(\xi)\,\right]. \tag{3.2}$$

This distribution is constructed without taking into account any knowledge about preceding states. Therefore it cannot describe correlations in time correctly. In particular it fails totally to describe transport phenomena.

For a very important and large class of systems we can find a better description. It is the class of systems which show a well separated behaviour on a microscopic short time scale and a macroscopic long time scale what is signified by a separation of the frequency spectrum of all occuring functions. In a time τ which is small on the long time scale, but large on the short time scale such a system practically can be regarded as isolated so that a LIOUVILLE operator L exists ruling the changes of the system in phase space. On the other side in the distribution

$$w(t) = e^{iL\tau}\; w^{\ell}(t-\tau) \tag{3.3}$$

all correlations and transport flows between neighbouring volume cells in $\underline{r}$-space will be fully developed after the time τ, whereas they were suppressed in w^{ℓ}. The so constructed MORI-distribution w indeed has been proving a success in describing transport phenomena. This distribution was given by MORI [6] and later on different arguments by MC LENNAN [7] and ZUBAREV [8].

If w, w^{o} are transformed by a pure LIOUVILLE motion, i.e. by a reversible change of macroscopic states, then $I(w)$ and $K(w,w^{o})$ remain unchanged. They are "reversibility invariants". Therefore we can replace these quantities at time t by $I(w^{\ell})$ and $K(w^{\ell},w^{\ell o})$ at time $t-\tau$. On the macroscopic time scale we can replace $t-\tau$ by t. This means that we can replace the MORI distribution in $I(w)$ and $K(w,w^{o})$ by the local equilibrium distributions. In particular, GIBBS' fundamental equation is valid for the local quantities. The assumption of that this is valid in macroscopic physics is called the approximation by "local equilibria". We see that it is justified for the class of systems with the mentioned separation of long and short time behaviour. They are often called "hydrodynamic systems" what is in some respect misleading because turbulence phenomena must be excluded.

The MORI states are not so much an approximation, as rather the result of an unbiased guess which takes into account not only the instantaneous mean values but also the connection to immediately preceding values.

4. Positivity of Entropy Production

The information gain (2.10) is a special case of a generalized measure which was first introduced by CSISZAR [9] into information theory

$$J(w,w^{o}) = \int d\xi\; w^{o} G\,\frac{w}{w^{o}}\quad , \tag{4.1}$$

where $G(u)$ is a convex function. The convexity means in particular that

$$G(u)-G(\hat{u}) \geqq G'(\hat{u})\cdot(u-\hat{u}). \tag{4.2}$$

A linear transformation of $w(\xi)$ into another probability distribution $\hat{w}(\xi)$

$$\hat{w}(\xi) = \int d\xi' \ R(\xi,\xi')w(\xi') \tag{4.3}$$

is originated by a "stochastic" kernel R which is non-negative and satisfies

$$\int d\xi R(\xi,\xi') = 1. \tag{4.4}$$

If w^o is transformed into $\hat{w}^o$ by the same R, we get

$$J(w,w^o)-J(\hat{w},\hat{w}^o) = \int d\xi d\xi' \ R(\xi,\xi')w^o(\xi') \left[G\frac{w(\xi')}{w^o(\xi')} - G\frac{\hat{w}(\xi)}{\hat{w}^o(\xi)} \right]. \tag{4.5}$$

Inequality (4.2) then shows that this expression never is negative.

Of particular importance are linear equations of motion for $w(\xi)$:

$$\dot{w}(\xi) = \int d\xi' \ \Lambda(\xi,\xi')w(\xi') \tag{4.6}$$

where the kernel Λ is independent of w. Such equations are called "master equations". If ξ are points of GIBBS' phase space, we shall call them master equations on GIBBS' level. We see that $J(w,w^o)$ never increases

$$J(w,w^o) \leqq 0 \tag{4.7}$$

if w,w^o are solutions of the same master equation.

We now are interested in the change with time of $w(\xi)$ of a system in a bath. Let the phase space variables of the bath be X. If we combine the system and the bath to a new system, the probability distribution can be factorized

$$W(\xi,X) = w(X|\xi)w(\xi), \tag{4.8}$$

where $x(X|\xi)$ is the conditional probability distribution of the bath being in the state X if the original system is in the state ξ. We can characterize the thermal contact by the statement [10] that on the macroscopic long time scale $w(X|\xi)$ is independent of the distribution $w(\xi)$. This involves that $w(\xi)$ fulfills a master equation if $W(\xi,X)$ satisfies a master equation in (ξ,X)-space. The latter, however, is the case when the combined system is so large that it can be described as isolated.

In particular $K(w,w^o)$ never decreases on the long time scale when w is the thermal nonequilibrium state of a system in a bath and w^o is the equilibrium with the bath. The bath can be any large system in thermal contact with the original system. The bath has to be so large that its macroscopic state remains practically unchanged. The contact can include exchange of energy and matter. For a system in a bath the quantitiy

$$P = -\dot{K}(w,w^o) \geqq 0 \tag{4.9}$$

is non-negative. Using MORI states we get

$$P = \int d^3\underline{r} \ \Sigma_\nu(\lambda_\nu-\lambda_\nu^o)\dot{m}^\nu = X\cdot\dot{m} \tag{4.10}$$

where $m^\nu(\underline{r})$ are the mean values of the phase space functions $m^\nu(\underline{r};\xi)$.

We get [11] not only the positivity of entropy production P but also the bilinearity in "forces"

$$X_\nu = \lambda_\nu - \lambda_\nu^o \qquad (4.11)$$

and "rate fluxes" $\dot{m}^\nu$. When we introduce vector flows $\underline{s}^\nu$ by

$$\dot{m}^\nu = -\underline{\nabla}\,\underline{s}^\nu + w^\nu, \qquad (4.12)$$

we can write P in the form

$$P = \int d\underline{r}\, \Sigma_\nu\, (\underline{s}^\nu \nabla X_\nu + w^\nu X_\nu). \qquad (4.13)$$

Only the positivity of the global entropy production P is shown in this argumentation. No conclusion about the sign of the local integrand has been made and, therefore, no difference to the interpretation of PFAFFELHUBER [12] exists.

5. Stability of a Steady State

Now we are interested in thermodynamic states of a system which can be far away from an equilibrium. They may be describable by macroscopic densities $\tilde{m}^\nu(\underline{r})$ and corresponding MORI distributions $\tilde{w}(\xi)$ in phase space. $\tilde{m}^\nu(\underline{r})$ and $\tilde{w}(\xi)$ may belong to a steady state. That means, these quantities are constant in time. As always

$$K(w+\delta w, w) \geqq 0, \qquad (5.1)$$

the quantity

$$K(w,\tilde{w}) = \Phi - \tilde{\Phi} - (\lambda - \tilde{\lambda})\cdot m \qquad (5.2)$$

is a convex function of the variables m. Using the entropy

$$S = -\Phi + \lambda \cdot m \qquad (5.3)$$

and defining

$$\Delta S = S - \tilde{S}, \quad \Delta m = m - \tilde{m} \qquad (5.4)$$

we get

$$K(w,\tilde{w}) = -\Delta S + \left(\frac{\partial \tilde{S}}{\partial m}\right)\Delta m = -\Delta_{NL} S \qquad (5.5)$$

what is the non-linear term of a TAYLOR expansion of $-\Delta S$ with respect to Δm. As it is a convex function in m-space, we can apply LIAPUNOV's theory getting the following result: The system never will leave a region in m-space which is given by

$$\Delta_{NL} S \geqq \text{const} \qquad (5.6)$$

and surrounds the steady state m, if always and everywhere in its interior is fulfilled

$$\Delta_{NL} \dot{S} \geqq 0. \qquad (5.7)$$

This is a sufficient criterion on LIAPUNOV stability for a finite region surrounding the steady state. Expressed by the macroscopic variables one gets

$$\Delta_{NL} \dot{S} = (\lambda - \tilde{\lambda}) \cdot \dot{m} = \dot{m} \cdot \Delta X. \qquad (5.8)$$

The restriction on infinitesimal deviations δm from the steady state, instead of finite deviations Δm, leads to a criterion for the so called "local stability":

$$\delta^2 \dot{S} \geqq 0. \qquad (5.9)$$

It means that the second variation of S is non-negative. This is the stability criterion of GLANSDORFF and PRIGOGINE [13] which was given in macroscopic theory in a more intuitive way. The bilinear form (5.8) in fluxes and the variation of forces is the well known X-variation of P introduced by the same authors.

If the steady state, in particular, lies in the region of linear thermodynamics and if moreover ONSAGER-symmetry is fulfilled, then (5.9) gives the principle of minimum entropy production. This principle says that in such a steady state the value of entropy production P assumes a minimum to fixed surroundings. It was given by GLANSDORFF and PRIGOGINE already 1945. Often attention has not been paid to the restrictive conditions for its validity.

6. Bit-number Cumulants

To any random quantity b the cumulants $\langle b^n \rangle_c$ of order in n are defined by the power expansion

$$\ln \langle e^{i\alpha b} \rangle = \sum_{n=0}^{\infty} \frac{(i\alpha)^n}{n!} \langle b^n \rangle_c \quad . \tag{6.1}$$

We can extend this definition to the bit-number

$$b(\xi) = \ln w(\xi) \tag{6.2}$$

which belongs to the knowledge about the question of whether ξ will occur or not when $w(\xi)$ is known [15]. It may be mentioned, by the way, that then the expression of (6.1), the "generating function of the cumulants", is, up to a factor, a RÉNYI information [14].

As we can see directly by the generating function, we can make two statements which are basic for introducing the bit number cumulants.

The first one is concerned with two physical systems with phase space variables ξ_I and ξ_{II} respectively which may be combined to a new system with phase space (ξ_I, ξ_{II}). If the two systems are uncorrelated, then we get for the combined system

$$\langle (b_I + b_{II})^n \rangle_c = \langle b_I^n \rangle_c + \langle b_{II}^n \rangle_c \quad . \tag{6.3}$$

That means, the cumulants become additive. So we can generally define an ordered set of measures for the correlations between the subsystems by the difference of the left and right hand side in (6.3).

The second statement is that the bit-number cumulants don't change if $w(\xi)$ is changed by a LIOUVILLE motion. That means, that they can only change during an irreversible process. Therefore let them be called "reversibility invariants".

As a rule, the cumulants of the lowest orders are the most important. The order zero always vanishes. The first order cumulant is the mean value of b, what is SHANNON information. New is the second order cumulant. It is the bit-number variance

$$\langle b^2 \rangle_c = \langle b^2 \rangle - \langle b \rangle^2 \tag{6.4}$$

which may be called Q.

We shall discuss this quantity for the generalized canonical distribution (2.6) of a thermodynamic equilibrium [16]:

$$Q = -\lambda_\mu \lambda_\nu \frac{\partial M^\mu}{\partial \lambda_\nu} = -\lambda_\nu \frac{\partial S}{\partial \lambda_\nu} \tag{6.5}$$

In thermostatics always one M, it may be called M^o, is internal energy. Then λ_o^{-1} is temperature T. If we write for the other ν

$$\lambda_\nu = -\zeta_\nu / T, \tag{6.6}$$

we obtain

$$-T^{-1} \left(\frac{\partial S}{\partial T}\right)_\zeta = \lambda_\nu \frac{\partial S}{\partial \lambda_\nu} \tag{6.7}$$

where the subscript ζ designates that the derivative with respect to T has to be formed with fixed ζ_ν. So we obtain

$$Q = T^{-1} \left(\frac{\partial S}{\partial T}\right)_\zeta \quad . \tag{6.8}$$

If all occuring M^ν are - like internal energy, volume or magnetization - of such a kind that they can be changed without a change of the material compostion of the system, then Q is the heat capacity for fixed ζ. If, however, also M^ν - like particle numbers of certain species - occur in the generalized canonical distribution the changes of which always are connected with a change of the material composition, then Q is a new quantity. For instance, for a grand canonical ensemble of a chemical substance with mole number n and chemical potential one obtains

$$\frac{1}{n} Q = c_v - \frac{T}{v^2} \left(\frac{\partial v}{\partial p}\right)_T \left[s - v \left(\frac{\partial p}{\partial T}\right)_v\right]^{-2} \quad . \tag{6.9}$$

p is pressure, v, s, c_v are volume, entropy and specific heat to fixed volume per mole [17].

That specific heat changes in a non-analytic way by passing the critical point of a phase transition, can be understood as caused by the building up of critical correlations between parts of the system. Specific heat is a bit-number variance Q and therefore susceptible to such correlations. As a rule some susceptibilities, i.e. derivatives of M^μ with respect to λ_ν, show a more dramatic critical behaviour than specific heat, to which they contribute corresponding to (6.5).

The diagram of steady non-equilibrium states of some systems shows close analogies to equilibrium phase transitions. These systems show so called "non-equilibrium phase transitions". Special chemical models show such transitions [18] and can be described by thermodynamic means. The order parameter here is mole number n of a certain chemical component. The generalized canonical distribution in which this number is a random quantity is the grand canonical ensemble. The Q of this distribution belongs also to the corresponding MORI distribution. The same is not true for the susceptibilities. If we seek for quantities which show a critical behaviour in the non-equilibrium phase transition, we don't have simple analogies to the equilibrium susceptibilities. We therefore are led to Q of (6.9). A detailed study shows [17] that not only Q but moreover Q/n shows a non-analytic critical behaviour in the considered non-equilibrium phase transitions. This is an indication for critical correlations also in these transitions.

7. Fluctuations in a Steady State

If $\tilde{w}(\xi)$ is the probability distribution belonging to a macroscopic stable steady state, we can ask for the probability of finding a certain fluctuation in the interior of the stability region around the steady state. As can be shown in different ways [19,20], this probability is up to a factor given by

$$W \sim \exp\left[-K(w,\tilde{w})\right] \quad , \tag{7.1}$$

where $w(\xi)$ is the distribution belonging to the fluctuation, i.e. to the macroscopic state which deviates from the steady state $\tilde{w}$.

We again assume the hydrodynamic character of the system for which MORI distributions are appropriate. In K we can replace these distributions by local distributions and can write

$$w(\xi) = \tilde{w}(\xi) \quad \exp\left[\Xi-\zeta\cdot\nu(\xi)\right] \tag{7.2}$$

where

$$v^{\nu}(\xi,\underline{r}) = m^{\nu}(\xi,\underline{r}) - m^{\nu}(\underline{r}) \tag{7.3}$$

$$\zeta_{\nu}(\underline{r}) = \lambda_{\nu}(\underline{r})-\tilde{\lambda}_{\nu}(\underline{r}) \tag{7.4}$$

$$\frac{\delta\Xi}{\delta\zeta_{\nu}}(\underline{r}) = v^{\nu}(\underline{r}) \tag{7.5}$$

This leads to

$$K(w,\tilde{w}) = \Xi -\zeta\cdot v \tag{7.6}$$

$$\frac{\delta K}{\delta v^{\nu}(\underline{r})} = -\zeta_{\nu}(\underline{r}) \quad . \tag{7.7}$$

Now we can form correlation functions for the distribution W(v) in v-space, which is connected with K by (7.1) and get in particular

$$< v^{\mu}(\underline{r})\zeta_{\nu}(\underline{r}') > = \int dv \; v^{\mu}(\underline{r}) \frac{\delta W}{\delta v^{\nu}(\underline{r}')} \tag{7.8}$$

$$= -\delta^{\mu}_{\nu} \; \delta(\underline{r}-\underline{r}') \tag{7.9}$$

In the last step a partial integration was applied.

If we now look for macroscopic dynamics of small deviations from the steady state, we can assume linear equations between v and ζ in the form

$$\dot{v}^{\mu}(\underline{r},t) = \int_{-\infty}^{t} dt' \int d^3\underline{r}'\Lambda^{\mu\nu}(\underline{r},\underline{r}',t-t')\cdot\zeta_{\nu}(\underline{r}',t') \tag{7.10}$$

Written in a shorter way:

$$\dot{v}(t) = \int_{-\infty}^{t} dt' \; \Lambda(t-t')\cdot\zeta(t') \tag{7.11}$$

$\Lambda(t)$ is defined only for positive t and therefore it is put equal to zero for negative t. We especially assume forces of the form

$$\zeta(t') = \zeta(t_0)\, h(t_0 - t'), \tag{7.12}$$

where h(t) is the HEAVISIDE step function, which is equal to unity for positive t and vanishing for negative t. Then we get, in particular,

$$\dot{v}(t) = \int_{-\infty}^{t_0} dt'\, \Lambda(t-t')\cdot\zeta(t_0). \tag{7.13}$$

In statistics we have to replace v(t) in this equation by its conditional mean value under the condition that all $\zeta(t_0)$ at time t_0 are given. Then we can form

$$< v(t)v(t_0) > = \int_{-\infty}^{t_0} dt'\, \Lambda(t-t')\, <\zeta(t_0)v(t_0)>. \tag{7.14}$$

Eq. (7.9) then leads for $t_0<t$ to:

$$< v(t)v(t_0) > = -\int_{-\infty}^{t_0} dt'\, \Lambda(t-t'). \tag{7.15}$$

Differentiation with respect to t_0 yields the result

$$\Lambda(t-t_0) = -<\dot{v}(t)\dot{v}(t_0)> \quad . \tag{7.16}$$

That means explicitly

$$\Lambda^{\mu\nu}(\underline{r},\underline{r}',t) = -\ <\dot{v}^{\mu}(\underline{r},t)v^{\nu}(\underline{r},o)>\ \ h(t). \tag{7.17}$$

This is a fluctuation-dissipation theorem in a stable steady state which can be far away from an equilibrium. The autocorrelation function of v has to be formed in this steady state. Here of course the full MORI distribution has to be used (or at least another adequate distribution which is able to describe the correlations). It cannot be replaced by the local equilibrium distribution because the correlation function is not a reversibility invariant, whereas K is one. The local equilibrium distribution would suppress certain correlations. This fluctuation dissipation theorem often was postulated as a generalization of the well known equilibrium theorem, but it was not proved in a consequent way in the general form. Also counter examples against it were given. These, however, were not systems with the separation of long and short time behaviour. We have to except processes of turbulence, in radiation fields, processes near a critical point of a phase transition of any kind, or in low temperature physics. The time scale separation is essential for the validity of the theorem.

References

[1] v. Laue, M: Geschichte der Physik
Universitäts-Verlag Bonn 1946.

[2] Shannon, C., Weaver, W.: The Mathematical Theory of Communication.
Urbana: University of Illinois Press 1949.

[3] Jaynes, E.T.: Phys. Rev., 106, 620 (1957).

[4] Kullback, S.: Annals of Math. Statistics 22, 79 (1951).

[5] Schlögl, F.: Z. Physik 198, 559 (1967).

[6] Mori, H.: Phys. Rev. 115, 298 (1959).

[7] McLennan, J.A.: Adv. Chem. Physics,
Interscience, New York 1963.

[8] Zubarev, D.N.: Soviet Phys. Doklady 10, 526 (1965).

[9] Csiszar, I.: Stud. Sci. Math. Hung. 2, 299 (1967).

[10] Schlögl, F.: Z. Physik 193, 163 (1966).

[11] Schlögl, F.: Z. Physik 191, 81 (1966).

[12] Pfaffelhuber, E.: J. Statistical Physics 16, 69 (1977).

[13] Glansdorff, P., Prigogine, I.: Physica 46, 344 (1970).

[14] Rênyi, A.: Wahrscheinlichkeitsrechnung.
VEB Deutscher Verlag der Wissenschaften, Berlin 1966.

[15] Schlögl, F.: Z. Physik B20, 177 (1975).

[16] Schlögl, F.: Z. Physik 267, 77 (1974).

[17] Schlögl, F.: Z. Physik B22, 301 (1975).

[18] Schlögl, F.: Z. Physik 253, 147 (1972).

[19] Schlögl, F.: Z. Physik 244, 199 (1971).

[20] Schlögl, F.: Z. Physik 249, 1 (1971).

Phase Transitions as Catastrophes: A Perspective

W.T. Grandy, Jr.*

Department of Physics and Astronomy, University of Wyoming
Laramie, WY 82071, USA

and

Institute for Information Sciences, University of Tübingen
D-7400 Tübingen, Fed. Rep. of Germany

Despite its evident elegance and very beautiful geometrization of a broad spectrum of natural phenomena, catastrophe theory is often charged as being capable of bringing only qualitative insights to various physical processes. Although this charge can no longer be sustained with respect to optics and scattering theory, say, it must be admitted that the application to critical phenomena has been less rewarding. It has been conventional to choose as catastrophe manifold the number-density surface $n(P,T)$, say, with pressure (P) and temperature (T) as control variables. This is a manifestly macroscopic description and, as is well known, leads to the classical mean-field-theory results for critical exponents. Indeed, it now seems that this is all that can ever arise when the problem is formulated in this way, at least without introducing a number of *ad hoc* extensions.

Yet one has the impression that catastrophe theory should do much better than this, for it has the right "physical feel" to it. The effects of structural instability and morphogenesis are exhibited quite explicitly by phase transitions, but any useful topological description must focus on the ultimate physical origins of the phenomena. It is suggested here that the theory simply has not been applied at the proper level, in terms of the fundamentally important variables. In order to illustrate the essential points the discussion is centered on a simple model of a first-order phase transition, as exemplified by the liquid-gas phases of a system of simple molecules.

The mechanical properties of a classical system are conveniently described in a multidimensional phase space determined by the degrees of freedom of the system. When isolated at a fixed total energy E the system can be represented by a point on the energy surface bounding a total phase volume $\Omega(E)$. Evolution of the system, as described by Hamilton's equations, yields a trajectory on this surface which is unique if the initial conditions are completely specified. The derivative of $\Omega(E)$ is called the structure function, $g(E)$, and is essentially the measure of the energy surface. The important physical properties of the system can be obtained from a knowledge of the structure function. Rather than calculate $g(E)$, it is often easier to calculate its Laplace transform

$$Z(\beta) = \int_0^\infty g(E)\, e^{-\beta E}\, dE \quad , \tag{1}$$

where it is presumed that the total system energy has a finite lower bound. Usually $Z(\beta)$ is called the generating function, although for our purposes it is more useful

---

* Richard Merton Visiting Professor of the Deutsche Forschungsgemeinschaft, Universität Tübingen, 1978-79.

to refer to it as the partition function. While $g(E)$ must be positive, it is sometimes claimed to be monotonic increasing, but this does not follow from anything else.

The parameter β has no immediate physical significance for a system with few degrees of freedom, and for which initial conditions are readily specified. It attains considerable meaning, however, for large systems in which it is fundamentally impossible to specify all the microscopic initial conditions. In that case β is related to a macroscopic initial condition on the total system energy and is, in fact, just a Lagrange multiplier (inverse temperature). One is then able to interpret the integrand of Eq. (1) as an unnormalized probability distribution.

These remarks are offered so as to emphasize the physical structure of the partition function ordinarily derived in statistical mechanics by direct and more rigorous probability methods. For a system of N particles contained in a volume V at temperature $T = (k\beta)^{-1}$, where k is Boltzmann's constant, the object of major interest is the canonical partition function

$$Z_N(\beta,V) \equiv \sum_i e^{-\beta E_i(N,V)} \quad , \tag{2}$$

where the sum is over all possible arrangements of the microscopic states of the system. Some insight into this expression is gained in the form of two modifications, the first of which stems from the observation that in a large system most arrangements i are enormously degenerate, so that E_i will have the same value for many arrangements i. Thus it is convenient to rewrite the sum in terms of distinct global energy levels of the system, and for that purpose we introduce numbers $g_j \equiv g(E_j)$ describing the degeneracy of each global state E_j. Then,

$$\begin{aligned} Z_N(\beta,V) &= \sum_j g_j \, e^{-\beta E_j(N,V)} \\ &= \int_0^\infty g_N(E,V) \, e^{-\beta E} \, dE \ , \end{aligned} \tag{3}$$

where the second line constitutes the second modification, to which we now turn.

In a many-particle system one can only recognize changes in energy that are huge compared to the energy intervals between neighboring states. As a consequence, it is not unrealistic to presume that the energy levels vary continuously, and this is noted explicitly in Eq. (3) by introduction of a formal spectral density, or density-of-states function. More mathematical rigor might be gained by first considering a Stieltjes integral, but intuition suggests that the relevant measure is differentiable at all but the lowest energies. And even that potential shortcoming is easily remedied, as will be noted below.

The importance of Eq. (3) stems not only from the fact that it has the same mathematical form as found in Eq. (1), but its physical structure and meaning are the same. In particular, all of the interesting physics is contained in the function $g_N(E,V)$, while the exponential is a simple monotonic decreasing function. Indeed, the Laplace transform has almost no interesting properties: for fixed N and V it is a monotonic decreasing function of β, logarithmically convex, holomorphic in the entire right half-plane of convergence, and for all finite N has no real zeros. These comments are most strikingly illustrated by the case of noninteracting charged particles in the presence of an external uniform magnetic field. Although Z_N is a smooth monotonic function, g_N exhibits a very complicated oscillatory behavior (de Haas-van Alphen effect). The very rich structure of g_N has been folded into the smooth function Z_N, in terms of which it is very difficult to observe the interesting physics directly.

Equation (3) yields the normalized probability distribution over the system energy states:

$$P_N(E) = \frac{g_N(E,V)\, e^{-\beta E}}{Z_N(\beta,V)} \quad , \tag{4}$$

with which one can calculate all kinds of expectation values yielding the thermodynamic properties of the system. Many of these can also be obtained, of course, by calculating logarithmic derivatives of $Z_N(\beta,V)$. As an example, for a Boltzmann gas of noninteracting particles the density-of-states function has the form

$$g_N(E,V) = A\,\frac{V^N\, E^N}{\Gamma(N+1)} \quad , \tag{5}$$

which is also the correct quantum mechanical expression when A contains Planck's constant. (Actually, N should be replaced by 3N/2 -1, but it is only the form that is of interest here.) From Fig. 1 it is observed that the product $E^N \exp(-\beta E)$ produces, for very large N, an extremely sharp probability distribution peaked at the average energy N/β. This point can also be ascertained from Eq. (4) in terms of the condition for a maximum,

$$g_N'(E) = \beta g_N(E) \quad , \tag{6}$$

which has a unique solution in this case.

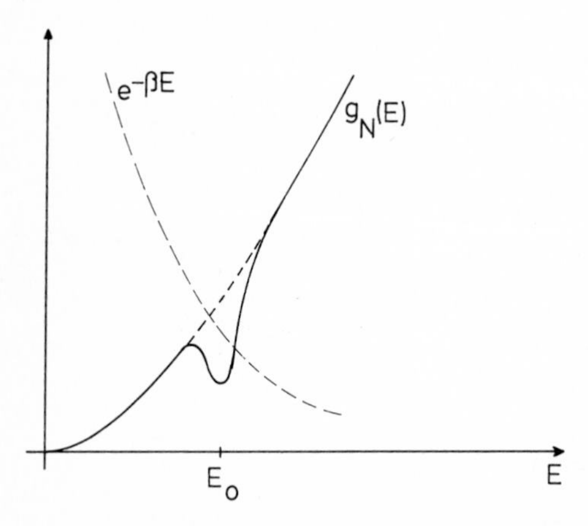

Fig. 1 Qualitative behavior of the density-of-states function for a fluid. The dotted curve represents a weakly-interacting system, the solid curve characterizes a dynamical fold, and the exponential curve determines the probability distribution.

Although it is rather difficult to calculate g_N exactly when particle interactions are included, approximate calculations provide some qualitative information. One gains the impression that g_N retains the global form (5), with the addition of some small-amplitude distortions. A significant modification arises when the two-body interaction contains an attractive portion, for then it is possible for the energy spectrum to extend to negative values. Stability requires the spectrum to be bounded below, and we shall represent this lower bound by (-a). By definition g_N must be positive, so that now one must write $g_N = g_N(E+a, V)$. The total effect of these considerations is to multiply the right-hand side of Eq. (3) by a factor $\exp(\beta a)$.

The preceding discussion serves to emphasize that in the context of statistical mechanics any dynamical instability in the system can only arise and be manifested macroscopically in $g_N(E,V)$. If this function contains a fold of the type depicted qualitatively in Fig. 1, then the system can exhibit a phase transition. For most values of β this portion of the spectrum will contribute essentially nothing to the integral of Eq. (3), nor to expectation values. But at a certain value of β (temperature) Eq. (6) will have three roots and $P_N(E)$ in Eq. (4) becomes bimodal, describing a phase transition with associated large fluctuations (standard deviation).

Reflection upon the physical causes of this kind of instability leads one to conclude that this is precisely the qualitative way in which g_N should behave. If, as is commonly believed, the two-body interaction contains a hard core and a repulsive tail, there will exist for an appropriate density $n = N/V$ an energy E_0 for which the competition between kinetic and potential energy becomes severe; i.e. catastrophic. One intuitively expects that the degeneracy, or number of microscopic arrangements, is considerably reduced at this point. It is a balancing act which the system, as well as the experimentalist, finds difficult to maintain!

A simple, but useful model illustrating these remarks has been studied in some detail. Consider the straightforward modification of Eq. (5) given by

$$g_N(E,V) = A\,\frac{V^N\,E^N}{\Gamma(N+1)}\,\{1 - b\,\exp[-\,\lambda(E - E_0)^2]\}^N \quad , \tag{7}$$

where $0 \leq b < 1$, and λ is a parameter of $O(1)$. This introduces the qualitative effect shown by the solid line in Fig. 1 and specifically reproduces the behavior discussed above. With some effort one can evaluate the partition function, but the resulting expression is illuminating in only one respect: no matter how large but finite a value N is given, $Z_N(\beta,V)$ is still an uninteresting, monotonic decreasing function of β. If, however, one examines the limit $N\to\infty$, the integral is readily evaluated owing to the δ-function character of part of the integrand. The limiting form is just

$$Z_N(\beta,V) = A\frac{V^N\,e^{\beta a}}{\beta^{N+1}}\;\{\,1 - b\,\exp[-\lambda(E_0 - N/\beta)^2]\}^N \quad , \tag{8}$$

including the lower bound to the spectrum. When $\beta \simeq N/E_0$, one sees that $Z_N(\beta,V)$ possesses a zero at $b = 1$. This possibility arises, of course, because the Laplace transform itself tends to become ill-defined in the limit and the associated theorems break down.

It has long been known that in the thermodynamic limit $\log Z_N(\beta,V)$ exists, is convex, and remains continuous, but that in this limit one or more zeros of Z_N may approach the real axis. It is then the zeros of Z_N which may introduce singular behavior into the logarithmic derivatives, and so only in the limit $N\to\infty$ does Z_N become interesting. Equation (8) provides an explicit representation of this behavior. But, even in the limit, Z_N must be known quite well in order to locate its zeros, and this seems an intractable problem (except for the two-dimensional Ising model). Rather, one might now be able to circumvent this procedure by studying directly the structural properties of $g_N(E,V)$, which are not necessarily strongly dependent on the limit.

A somewhat remarkable consequence of the model provided by Eq. (8) is revealed by writing the lower bound on the energy more explicitly as (-anN), and making what we shall call the van der Waals approximation near the critical point:

$$b \exp [-\lambda(E_0 - N/\beta)^2] \simeq cn = c/v \quad , \tag{9}$$

where $v \equiv V/N$ is the specific volume. Then Eq. (8) becomes

$$Z_{vdW} = \frac{A}{\beta^{N+1}} v^N (1-c/v)^N e^{\beta aN/v} \quad , \tag{10}$$

from which we obtain the van der Waals equation of state:

$$\begin{aligned} P_{vdW} &= \frac{1}{\beta N} \frac{\partial}{\partial v} \log Z_{vdW} \\ &= \frac{kT}{v - c} - \frac{a}{v^2} \end{aligned} \tag{11}$$

This does not include the Maxwell construction, but if that device is employed one finds a phase transition at $v = 3c$. Note that the transition does <u>not</u> arise from the zero in Eq. (10) at $v = c$, but primarily from the lower bound on the energy spectrum (which itself is revealing). The nonanalytic behavior of the exponential factor in Eq. (10) near $v = 0$ also illustrates why perturbation techniques cannot reach the phase transition. The zero, rather, probably corresponds to the well-known transition in the hard-sphere fluid, thought to be a liquid-solid transition. If one takes $c = \pi r_0^3/3$, where r_0 is the hard-sphere radius, then $c/v \simeq 0.94$ comes close to predicting the known critical point. Because of the <u>ad hoc</u> nature of the model (7) and the crudeness of the approximation (9), however, one should not take these results too seriously. Nevertheless, they are interesting!

The insight into first-order phase transitions provided by focusing on the structure of the density-of-states function is not yet deep enough to provide a complete theory. For fixed V, however, one notices that $g_N(E,V)$ defines a <u>family</u> of functions as N varies, suggesting that it is the density-of-states <u>surface</u> that should be considered. In fact, it appears more useful to maintain N fixed and employ the volume V as the control variable. This can be accomplished formally by using the pressure ensemble, rather than the grand canonical ensemble, wherein the expectation value of the volume is specified instead of the total particle number, and the related Lagrange multiplier is essentially the pressure, P. The appropriate partition function is now the double Laplace transform

$$W_N(\beta,s) = \int_0^\infty dV\, e^{-sV} \int_0^\infty dE\, e^{-\beta E}\, g_N(E,V) \quad , \tag{12}$$

and one readily makes the identification $s = \beta P$. Equation (12) is easily derived by maximizing the entropy subject to constraints in the form $\langle E\rangle$, $\langle V\rangle$. The probability distribution is now determined by the intersection of the surfaces defined by $g_N(E,V)$, $\exp(-\beta E)$, and $\exp(-sV)$. Clearly the appropriate catastrophe surface is $g_N(E,V)$, as illustrated schematically in Fig. 2. This is oversimplified, of course, since one is interested in extraordinarily large N.

We mention two further generalizations. First, in order to extend these ideas to second-order phase transitions it is necessary to recognize that g_N has a fundamentally different structure for essentially-localized particles, as in spin systems. In these cases the topology of the density-of-states surface can be quite

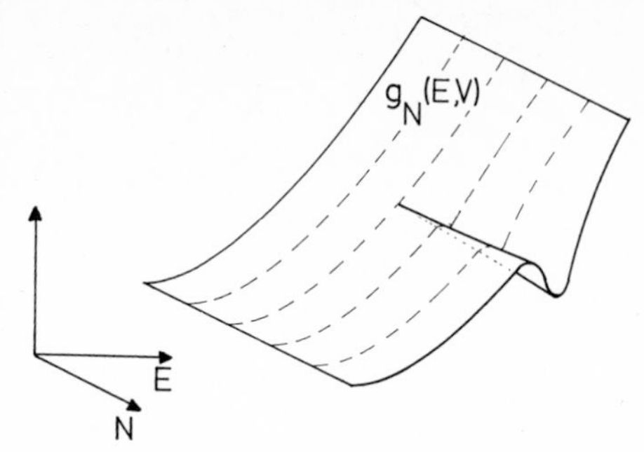

Fig. 2 The density-of states surface leading to a first-order phase transition.

different from that described above, and the universality of phase transitions may not be as striking at this level as it is macroscopically.

A second generalization may be needed when considering critical phenomena on the completely quantum level, as with superfluids. Rather than deal with the spectral density, one may wish to proceed more carefully in terms of the high-probability manifold (HPM) of states in the underlying Hilbert space. The HPM is very sharply defined by the principle of maximum entropy in the limit $N\to\infty$, as is its dimension, W. As the catastrophe unfolds the HPM bifurcates and W is considerably reduced in the bifurcating region of the manifold. These events surely induce a geometrical description completely equivalent to that above, but when quantum phenomena are manifest it may be prudent to keep the Hilbert-space picture in mind.

These brief remarks suggest that the application of catastrophe theory to phase transitions is a subject still very much alive and viable. It is the topology of $g_N(E,V)$ which is of primary interest, and which is directly related to the dynamical physical processes underlying the phenomena. The dynamical unfolding of this surface is intimately related to the interplay between kinetic and potential energy on the microscopic level, which is in turn manifested in the geometrical and topological properties of the surface. It is now necessary, of course, to develop a dynamical theory of $g_N(E,V)$, but in this task one at least has direct contact with the relevant physical processes. Although this is a difficult undertaking, there is an enormous gain over sole preoccupation with the partition function, in that it is only necessary for the theory to produce macroscopically-manifested geometrical and topological properties, rather than possibly unattainable analytic details. These, in turn, will predict related properties of the partition function which should become very sharp in the limit $N\to\infty$, and one therefore obtains the desired thermodynamic behavior. Some progress with these various problems is being made.

Part VI

Solitons

Solitons in Physics: Basic Concepts

R.K. Bullough and R.K. Dodd

Department of Mathematics, University of Manchester,
Institute of Science and Technology, P.O. Box 88
Manchester M60 1QD, U.K.

and

School of Mathematics, Trinity College, Dublin 2, Ireland

1. Introduction

Solitons are mathematical objects which arise as solutions of certain non-linear wave equations. This class is a rapidly growing one and the reasons for current interest are threefold. First, methods have been found for finding these solutions analytically and therefore exactly, and ideas on their "Completeness" have emerged. Second, solitons have mathematical properties which make them particle-like and in those contexts of non-linear physics where they arise they are the natural elementary excitations. Third, those contexts of non-linear physics are proving to embrace the whole of modern physics itself.

In this lecture we want to do two things: to introduce the soliton solutions of a number of physically important non-linear wave equations in a very simple way; and to illustrate by some relatively simple examples the range of application of solitons to modern non-linear physics. Abstract soliton theory requires considerable mathematical technique. In a second lecture (II) we shall show how the underlying mathematics of the subject has developed since the first strictly experimental studies of JOHN SCOTT RUSSELL [1] in the ten year period from 1834. RUSSELL studied the propagation of water waves in channels of finite depth. He arrived empirically at the formula $c^2 = g(h+k)$ connecting the speed c of propagation of a wave, the depth h of the water, and the maximum height k of the disturbance above the free water surface: g is the acceleration due to gravity. RUSSELL's wave was an example of a 'solitary' wave, and he originated the name. It has the nonlinear property that its speed c depends on its amplitude.

2. Solitons as Solitary Waves

A working definition of a soliton is that of a solitary wave with a particular collision property. Therefore we certainly first need the concept of the solitary wave. This is typically a bell-shaped disturbance which not only does not break up but has an unchanging shape: the disturbance $u(x,t)$ depends on the argument $x-ct$ where c is a constant. We consider plane waves - there is <u>one</u> space variable x as well as the time t.

The distortionless disturbance $u(x-ct)$ (sometimes called a wave of permanent profile) moves up the x-axis with speed c. It is a solution of the linear dispersionless wave equation

$$u_t + c\,u_x = 0 \,. \tag{1}$$

In contrast the linear equation

$$u_t - c\,u_{xxx} = 0 \tag{2}$$

has harmonic wave solutions $u^k = \cos(\omega t - kx)$ with dispersion $\omega = ck^3$ and speed ck^2. The large k modes run away from the small k modes and an initial bell shaped disturbance disperses (breaks up). The harmonic waves are the only distortionless solutions of (2).

The simplest non-linear wave equation is perhaps

$$u_t + 6u\, u_x = 0 \tag{3}$$

(the number 6 is conventional: notice that by scaling x, t and u itself the 6 can be scaled to any value). This is the equation of a 'simple wave' with an implicit solution

$$u = u(x-6ut) \,, \tag{4}$$

that is the solution depends upon u itself (see II). This solution shocks. Points where u is large and positive have large effective speed $6u$ along x and u develops a rising front where the large values of u pile up.

It is possible to balance the non-linearity in (3) against the dispersion in (2) to get a solitary wave $u(x-ct)$. We take the Korteweg-de Vries equation (KdV) [2]

$$u_t + 6uu_x + u_{xxx} = 0 \tag{5}$$

which has the solitary wave solution

$$u(x,t) = 2\xi^2 \operatorname{sech}^2 [\xi(x-4\xi^2 t)] \tag{6}$$

in which $c = 4\xi^2$ and ξ is a constant. This is a single bell shaped pulse which propagates along x without changing shape. It was the first solitary wave recorded for it is in fact essentially the solitary wave observed by RUSSELL [1]. It is a one parameter solution and its speed c is proportional to its amplitude $2\xi^2$ (the KdV describes the propagation of gravity waves in shallow water: it is restricted to one-way propagation, in this case the positive x-axis: Russell's formula $c^2 = g(h+k)$ can be obtained only approximately from (6) [3]). That the speed should depend on the amplitude is a common but by no means necessary characteristic of a soliton. In fact (6) happens to be both a solitary wave and a soliton.

Evidently at $x = -\infty$ we can 'superpose' N distinct solutions (6) characterised by parameters $\xi_1, \dots, \xi_N$ by spacing these out an infinite distance apart without overlap (there is no superposition principle for solutions of non-linear wave equations although we are about to find one for solitons). Conveniently we order the solutions $1, 2, \dots, N$ from the left with $\xi_1 > \xi_2 > \dots > \xi_N$. Then (since solutions can be placed a finite distance apart with only exponential damage) the larger pulses overtake the smaller ones and the set undergoes a complicated collision.

For an arbitrary non-linear wave equation it would be difficult to say more. Non-linear scattering takes place between pulses and in perturbation theory, for example, terms would grow dramatically in number. The behaviour of the KdV is very simple however (though perturbation theory is not the ideal vehicle to show this - although see [4]): at $x = +\infty$ there emerges the N pulses (6) in the reverse order $\xi_N, \xi_{N-1}, \dots, \xi_1$ [5]. The only change is that the argument $\xi_i\,(x - 4\xi_i{}^2 t - x_i) \to \xi_i(x - 4\xi_i{}^2 t - x_i - \delta_i)$ (x_i is the position of the i-th pulse at the time $t = 0$). The small "phase shift" δ_i is made up of strictly pairwise shifts: $\delta_i = \sum_{j\neq i} \delta_{ij}$

and there are no 3-body shifts. Total phase shift is conserved: $\sum_i \delta_i = \sum_i \sum_{j \neq i} \delta_{ij} = 0$. This is one of an infinite number of conserved quantities associated with the simpler solitons (see II).

We can interpret this result of collisions in two ways: either solitons pass through each other with small pairwise interactions inducing the shifts δ_{ij}; or they bump like particles transferring energy, momentum and therefore amplitude, without losing their essential form. Either point of view is tenable: computer solutions of a soft collision between solitons show that a trough always exists between the two disturbances and amplitude leaks mysteriously between one side of the trough and the other. In a hard collision the pulses overlap and the trough disappears however (for some rather specialised illustrations of the troughs in soft collisions see [6,7]).

Not all solitary waves are solitons. For example the non-linear diffusion equation, the Burgers equation [8],

$$u_t + 6uu_x - b\,u_{xx} = 0, \quad b > 0 \tag{7}$$

has a permanent profile solution

$$u(x,t) = u_1 + (u_2 - u_1)[1 + \exp\{3b^{-1}(u_2 - u_1)(x - ct)\}]^{-1} \tag{8}$$

with $c = \frac{1}{2}(u_1 + u_2)$. It is a constant step-like shock, with asymptotes u_1 and u_2 as $x \to \mp\infty$, moving along x. Two such shocks with parameters (u_3, u_2) and (u_2, u_1) and $u_3 > u_2 > u_1$ collide: they fuse and emerge as the single confluent shock (u_3, u_1) as $t \to \infty$ [4].

Steplike solutions like (8) are called "kinks." Their derivatives are bell-shaped, and are solitary waves. These solitary waves may be solitons and in this case the kinks also have the soliton collision property. In this case we may also loosely refer to the kinks as solitons.

The supreme example of the kink of this kind is the kink solution of the sine-Gordon equation (s-G)

$$u_{xx} - u_{tt} = \sin u. \tag{9}$$

The kink is

$$u(x,t) = 4\tan^{-1} \exp\{(x - vt)(1 - v^2)^{-\frac{1}{2}}\} \; ; \quad 1 > v \geq 0. \tag{10}$$

The derivative of the kink is

$$- u_t = 2v[1 - v^2]^{-\frac{1}{2}} \operatorname{sech}\{(x - vt)(1 - v^2)^{-\frac{1}{2}}\} \tag{11}$$

and the speed is v. The kink is a "2π-kink": $u \to 0$ as $x \to -\infty$, $u \to 2\pi$ as $x \to +\infty$, and u_x, u_{xx}, etc. $\to 0$ as $|x| \to \infty$. Two such kinks satisfy $u \to 4\pi$ as $x \to +\infty$. They collide and pass through each other if $v_1 > v_2$ and v_1 is the parameter of the kink on the left at $t=0$.

Notice that $0, 2\pi, 4\pi, \ldots$ are zeros of $\sin u$. If we Lorentz transform (9) to the rest frame that equation is the ordinary differential equation $u_{xx} = \sin u$ with a first integral $\frac{1}{2}u_x^2 + \cos u = \text{constant}$. If the constant is one, the zeros of u_x are zeros of $\sin u$. These are points in the 2-dimensional phase space (u, u_x) characterised by "zero 'velocity' u_x means zero 'force' u_{xx}". They are therefore equilibrium points.

It is easy to see they are unstable equilibrium points since, e.g. $\frac{1}{2}u_x^2 + \cos u \approx \frac{1}{2}u_x^2 + 1 - \frac{1}{2}u^2$ for small u and trajectories are rectangular hyperbolas near $u = 0$. The points π, 3π, etc are stable: trajectories near π are circles about this value. The trajectory $\frac{1}{2}u_x^2 + \cos u = 1$ describes a motion between $u = 0$ and $u = 2\pi$ determined from $u_x = 2 \sin \frac{1}{2} u$ or

$$\int^{\frac{1}{2}u} \frac{dv}{\sin v} = x + \text{constant} \tag{12}$$

so $u = 4\tan^{-1} e^{x+\delta}$. This is therefore the trajectory of a kink.

It is worthwhile understanding the unstable equilibrium points. We take the related "ϕ-four" equation

$$\phi_{xx} - \phi_{tt} = - \phi + \phi^3 \tag{13}$$

which in the rest frame is $\phi_{xx} = - \phi + \phi^3$ with kink trajectories $\frac{1}{2}\phi_x^2 + \frac{1}{2}\phi^2 - \frac{1}{4}\phi^4 = \frac{1}{4}$ between $\phi = -1$ and $\phi = +1$. The kinks are

$$\phi = \pm \tanh\left\{ \frac{1}{\sqrt{2}} \frac{x - vt}{(1 - v^2)^{\frac{1}{2}}} \right\} \tag{14}$$

after Lorentz transformation to the moving frame. The two signs allow us to define a kink (+) and an antikink (-): ϕ goes from -1 to $+1$ in the kink and from $+1$ to -1 in the antikink as x goes from $-\infty$ to $+\infty$. The points $\phi_x = 0$, $\phi = \pm 1$ are unstable equilibrium points at the maxima ± 1 of the 'potential' $\frac{1}{2}\phi^2 - \frac{1}{4}\phi^4$. However the ϕ-four Hamiltonian which yields (13) is

$$H = \int_{-\infty}^{\infty} [\tfrac{1}{2}\phi_t^2 + \tfrac{1}{2}\phi_x^2 + \tfrac{1}{4} - \tfrac{1}{2}\phi^2 + \tfrac{1}{4}\phi^4]dx. \tag{15}$$

The true potential energy is of opposite sign and the points $\phi = \pm 1$ are stable equilibrium points energetically. Thus the kink (14) takes the system from a stable equilibrium point to a stable equilibrium point of the same energy. These points are distinct degenerate vacuum states in particle physics, for example [9]. Notice that the kink solution of $\phi_{xx} = -\phi+\phi^3$ has the character of an instanton solution of the nonlinear oscillator $\phi_{tt} = +\phi-\phi^3$ [10]. Instantons are more typically solutions of Yang-Mills theories in 4-dimensional Euclidean space [9]. In contrast solitons are typically solutions in one space and one time dimension of nonlinear evolution equations $u_t = K[u]$ where $K[u]$ is some nonlinear functional of u. The connections between the solutions of these two different types of system are not fully understood (but see [11]).

Although the kink (10) of the s-G has the same property as the kink of the ϕ-four, there is an important difference between the two cases. The ϕ-four has only the two singular points $\phi = \pm 1$ and a kink must be followed by an antikink. The s-G has singular points at 0, $\pm 2\pi$, $\pm 4\pi$, etc. and kinks or antikinks can follow kinks or antikinks. This is obviously necessary for soliton solutions. The s-G does have soliton solutions whilst ϕ-four does not. In so far as the ϕ-four has been much used in theoretical physics (Landau-Ginsberg expansions, the laser phase transition analogy, bifurcation theory [12], displacive phase transitions in 1-D crystals [7] and 'central peak phenomena' [13], and particle physics [10,14] have all been concerned with ϕ-four type potentials) it may seem surprising that the ϕ-four does not have soliton solutions. Approximate soliton-like kinks like (14) seem however [7] to be important physical entities.

Another useful property of (14) is that it has a mass (= 2/3). This is a consequence of the Lorentz invariant form of (13) rather than any soliton or solitary wave property. The solitary wave character keeps this mass finite however and the kink as a particle model has a finite self-energy. The kink (10) has mass 8. Notice how 'close' to the ϕ^4 is the s-G (the negative sign corresponds to displacing the origin by π):

$$\sqrt{6}\,(u_{xx} - u_{tt}) = -\sin\sqrt{6}\,u \approx -\sqrt{6}\,u\,[1 - u^2].$$

3. Multisoliton Solutions

We give here the analytical forms of the so called multisoliton solutions of the s-G equation (9). These take the forms [15]:

2π-kink
$$u(x,t) = 4\tan^{-1}\exp\Theta_1\ ; \tag{16}$$

4π-kink
$$u(x,t) = 4\tan^{-1}\frac{\sinh\frac{1}{2}(\Theta_1 + \Theta_2)}{(a_{12})^{\frac{1}{2}}\cosh\frac{1}{2}(\Theta_1 + \Theta_2)}$$

$$a_{12} = (a_1 - a_2)^2(a_1 + a_2)^{-2}\ ; \tag{17}$$

0π-kink
(or "breather")
$$u(x,t) = 4\tan^{-1} r\sin\Theta_I\,\mathrm{sech}\,\Theta_R$$

$$r = a_R\,a_I^{-1}$$

$$\Theta_R = \tfrac{1}{2}\,a_R[(1 + \frac{1}{a_R^2 + a_I^2})x + (1 - \frac{1}{a_R^2 + a_I^2})t + x_R]$$

$$\Theta_I = \tfrac{1}{2}\,a_I[(1 - \frac{1}{a_R^2 + a_I^2})x + (1 + \frac{1}{a_R^2 + a_I^2})t + x_I] \tag{18}$$

$2N\pi$-kink
$$\cos u(x,t) = 1 - 2\left(\frac{\partial^2}{\partial x^2} - \frac{\partial^2}{\partial t^2}\right)\ \ln f(x,t)$$
$$f(x,t) = \det||M||. \tag{19}$$

The $N\times N$ matrix $||M||$ has elements

$$M_{ij} = 2(a_i + a_j)^{-1}\cosh\{\tfrac{1}{2}(\Theta_i + \Theta_j)\},$$
$$\Theta_i = \pm\,\gamma_i(x - v_i t) + x_i$$
$$a_i^2 = (1 - v_i)(1 + v_i)^{-1}$$
$$\gamma_i^2 = (1 - v_i^2)^{-1}\ , \tag{20}$$

in (19). In (16) Θ_1 is given by Θ_i as in (20). In (17) Θ_1, Θ_2 are similar: a_1 and a_2 are the two (real) parameters of the 2-parameter solution (17) related to velocities v_1, v_2 as in (20). Asymptotically (17) is [15]

$$u(x,t) \sim 4\tan^{-1}\exp[\Theta_1 + \eta_1^{\pm}] + 4\tan^{-1}[\Theta_2 + \eta_2^{\pm}]$$

as $x \to \pm\infty$:

$$\eta_1^{\pm} = -\eta_2^{\pm} = \pm \tfrac{1}{2} \ln a_{12} . \tag{21}$$

The 0π or "breather" solution (18) is of a type not discussed before. It is a 2-parameter solution which does not break up. Instead it has an internal oscillation which progresses and is modulated by a moving external envelope. It is obtained from (19) by choosing a_1 and a_2 complex with $a_1 = a_2^* = a_R + i\, a_I$ and $x_1 = x_2^* = x_R + i\, x_I$. Its energy is $16\gamma_R\{r^2(1 + r^2)^{-1}\}^{\frac{1}{2}} = 16\gamma_R \sin\mu$ (for $a_1 = a\, e^{i\mu}$): $\gamma_R = (1 - v_R^2)^{-\frac{1}{2}}$ and $v_R = (1 - a^2)(1 + a^2)^{-1}$. In terms of μ and v_R, $\Theta_R = \gamma_R \cos\mu.(x - v_R t) - \frac{1}{2} a_R\, x_R$ and $\Theta_I = \gamma_R \sin\mu.(t - v_x) - \frac{1}{2} a_I\, x_I$. Clearly in the rest frame the breather oscillates with internal frequency $\sin\mu$. This frequency becomes discrete when the breather solution is quantised (see II). The 0π acts like a soliton: it can collide with 2π-kinks, 2π-antikinks and other breathers and passes through them with only a phase shift. It can be thought of as a bound kink-antikink pair in which ϕ moves from 0 to $2\pi-\varepsilon$ and back to 0 as x moves from $-\infty$ to $+\infty$. It therefore has a total kink angle of zero.

The N-kink solution (19) contains all the other solutions as special cases. It was derived in the form equivalent to (19) [16,17] by a difficult analogy drawn between the KdV and s-G equations. HIROTA [18] had given a multisoliton solution of the KdV. We now know how to obtain this solution by the so-called inverse scattering method [19,20]. We develop the theory of the ISM in II. The ISM solves the initial value problem $u(x,0) = f_1(x)$, $u_t(x,0) = f_2(x)$; $u \to 0 \pmod{2\pi}$, u_x, u_{xx}, etc. $\to 0$ as $|x| \to \infty$; for the s-G. It shows that the solution consists of two parts: the multisoliton part like (19) and the 'background'. The background diffuses away so that asymptotically only the multisoliton part is significant. In this sense all initial conditions ultimately produce only solitons: these solitons are 2π-kinks, 2π-antikinks or breathers. All kinks and antikinks must travel at different speeds (they either attract or repel each other); but any number of breathers can travel at the same speed as any one kink or antikink. The kinks have a 'topological charge' 2π and mutually repel: antikinks have charge -2π and kink-antikink pairs attract; breathers have no charge. There are no other solutions of the s-G apart from the background and the solution is "complete" in this sense.[1]

4. Applications of Solitons to Physics

A less than definitive list of applications of solitons in physics already includes

(i) Gravity waves in deep and shallow water.
(ii) The theory of plasmas and the interaction of radiation with plasmas.
(iii) Superconductivity: the theory of Josephson junctions.
(iv) Fermi liquid theory; spin waves in the A- and B-phases of liquid ^{3}He below 2.6 mK.
(v) Ferromagnetics: Bloch wall motion.
(vi) Resonant and non-resonant non-linear optics and laser physics.
(vii) Non-linear crystal physics: theory of dislocations, anharmonic crystals; recurrence phenomena in thermal transport and non-ergodic behaviour (Fermi-Pasta-Ulam problem [21]); displacive and other phase transitions and central peak phenomena; linear conductors (like TTF-TCNQ).
(viii) Theory of fundamental particles.
(ix) Astrophysics: solitons in the solar corona have been suggested (the Great Red Spot in Jupiter has been called a soliton also!).

---

[1] The remarks of this paragraph assume that the eigenspectrum of the scattering problem is nondegenerate (no repeated eigenvalues).

Examples which embrace almost all of this list are treated in [7]. A large number of illustrative diagrams is given there.

The key equations are:

The KdV
$$u_t + 6uu_x + u_{xxx} = 0 \tag{22}$$

The modified KdV
$$u_t + 6u^2u_x + u_{xxx} = 0 \tag{23}$$

The non-linear Schrödinger eqn. (NLS)
$$i\,u_t + 2u|u|^2 + u_{xx} = 0 \tag{24}$$

The s-G
$$u_{xt} - \sin u = 0 \tag{25}$$

Other important equations are

The Hirota eqn.
$$i\,u_t + 3i\alpha|u|^2\,u_x + \rho\,u_{xx} + i\,\sigma u_{xxx} + \delta|u|^2u = 0 \quad (\alpha\rho = \sigma\delta) \tag{26}$$

The reduced Maxwell-Bloch (RMB) eqns.
$$\begin{aligned} u_x &= -\mu s \\ v_x &= Ew + \mu u \\ w_x &= -Ev \\ E_t &= v \qquad (\mu = \text{const.}) \end{aligned} \tag{27}$$

The Boussinesq eqn.
$$u_{tt} - (12uu_x + u_{xxx})_x = 0 \tag{28}$$

The 3-wave interaction (decay type)
$$\begin{aligned} u_{1,x} + c_1\,u_{1,t} &= iq\,u_2\,u_3^* \\ u_{2,x} + c_2\,u_{2,t} &= iq\,u_1\,u_3 \\ u_{3,x} + c_3\,u_{3,t} &= iq\,u_1^*\,u_2 \end{aligned} \tag{29}$$

The Toda lattice
$$u_{n,tt} = e^{-(u_n - u_{n+1})} - e^{-(u_{n-1} - u_n)} \tag{30}$$

The Kadomtsev-Petviashvili eqn.
$$\tfrac{3}{4}\beta^2\,u_{yy} + \{\alpha u_t + \lambda u_x + \tfrac{1}{4}(u_{xxx} + 6uu_x)\}_x = 0. \tag{31}$$

All these equations have multisoliton solutions. The majority are in the form of systems of non-linear evolution equations (NEEs) $u_t = K[u]$. These are characterised by requiring only the initial data $u(x,0)$ to determine the motion. This is why the s-G is placed in characteristics form in (25) (by transforming to new variables $x \pm t$). In II we work from the NEE form.[2]

---

[2] Eqn. (28) takes the form $u_t = p$, $u_x = q$, $q_x = r$, $r_x = y$, $p_t = q_x + y_x + 12(ur + q^2)$. Other examples are similar.

We quote a few physical applications: the KdV to shallow water waves, lattice recurrences, plasma ion acoustic waves; the modified KdV to Alfven waves in a cold collisionless plasma; the NLS to 1-dimensional self-focussing, self-phase modulation, Langmuir turbulence in plasmas [22] (Zakharov's caverns [23]), laser-plasma interactions (optical filament formation); the s-G to spin waves, Josephson junctions, non-linear optics (self induced transparency [24,25,26], lattice theory and particle physics [26]; the RMB to self-induced transparency [25,26]; the Boussinesq to hydrodynamics and plasmas; the 3-wave to stimulated Raman back scattering in plasmas; the Toda lattice as a soluble lattice with hard core and harmonic limits; the Kadomtsev-Petviashvili equation as a two dimensional non-stationary problem in a weakly dispersive medium [28,29,30]. We refer to [7] for more comprehensive references on the other topics.

5. Some Particular Applications of Solitons in Physics

We shall develop four particular examples in more detail. These are the application of the multisoliton solutions of the NLS to optical filament formation in laser irradiated neutral dielectrics and the application of the s-G to spin waves, optical pulses and Josephson junctions. The three applications of the s-G can be taken in one jump!

Consider the neutral dielectric: in a (scalar) field $E(\underset{\sim}{x},t)$ the dipole $P(\underset{\sim}{x},t)$ is

$$P(\underset{\sim}{x},t) = \alpha\, E(\underset{\sim}{x},t) + \alpha_{NL}\, \{E(x,t)\}^3 + \cdots\cdots \tag{32}$$

and α and α_{NL} are the constant linear and first non-linear susceptibilities. Maxwell's equation is linear and is

$$\nabla^2 E(\underset{\sim}{x},t) - c^{-2}\partial^2 E(\underset{\sim}{x},t)/\partial t^2 = 4\pi n c^{-2}\, \partial^2 P(\underset{\sim}{x},t)/\partial t^2 \tag{33}$$

where n is the atomic number density. We look for complex envelope solutions $\varepsilon(x,y,z,t)$ modulating carrier waves $e^{i(\omega t-kz)}$:

$$E(x,t) = \varepsilon(x,y,z,t)e^{i(\omega t-kz)} + \text{c.c} \quad . \tag{34}$$

We impose the linear dispersion relation $\omega^2 = c^2k^2 - 4\pi n\alpha\omega^2$. We equate coefficients of all terms in $e^{i(\omega t-kz)}$ (in principle one should go on and obtain a coupled sequence of equations for the complex envelopes of the harmonics). We find

$$\varepsilon_{xx} + \varepsilon_{yy} + \omega^2 c^{-2}\, 12\pi n\alpha_{NL}|\varepsilon|^2\varepsilon + 2\, i\, c^{-2}[\omega\varepsilon_t(1 + 4\pi n\alpha) - c^2k\varepsilon_z] = 0 \tag{35}$$

We look for steady state solutions (ε does not depend on t). By suitable scaling

$$\varepsilon_{xx} + \varepsilon_{yy} + 2\, |\varepsilon|^2\, \varepsilon - i\, \varepsilon_z = 0 \tag{36}$$

which is the 2-dimensional NLS equation in (x,y) and a 'time' (z).

In one space dimension (x) the NLS (36) has the 1-soliton solution

$$\varepsilon(x,z) = \frac{2\eta\, \exp\{4i(\xi^2 - \eta^2)z - 2\, i\, \xi x + i\delta\}}{\cosh[2\eta(x - x_0) - 8\eta\xi z]} \quad . \tag{37}$$

It contains a carrier which corrects the linearised carrier in (35). Solitons (37) are typical of the so-called envelope solitons. This particular one can be seen in experiments on deep water [31]. As an electric field it carries intensity $4\eta^2\, \text{sech}^2[2\eta(x - x_0) - 8\eta\xi z]$ This is the intensity across a wave-guide like channel induced in the medium by the intense field making an angle $+ \tan^{-1} 4\xi$ to the z-axis. An arbitrary laser profile $\varepsilon(x,0)$ breaks up into a number of such soliton channels as z increases.

The NLS is special amongst the key equations (22) - (25) in that it describes a single complex field. A consequence is that the speed (given by 4ξ in (37)) does not determine the amplitude (given by 2η). However in II we shall see $\zeta = (\xi + i\eta)$ is an important eigenvalue of a related linear scattering problem: in the case when this eigenvalue is purely imaginary, as may be the case for the s-G and modified KdV for example, both speed and amplitude are determined by the number η.

To order E^3 (32) and (33) are the ϕ-four equation (13) for the real field $\phi \equiv E$. The ϕ-four can always be approximated by the NLS. Notice that the driving term of (32) and (33) is $\alpha E + \alpha_{NL} E^3 \to + \phi + \phi^3$. The linear term does not matter since it is removed by the linear dispersion relation. The sign of the ϕ^3 matters since $- \phi^3$ changes (24) to $i u_t + u_{xx} - 2u|u|^2 = 0$. This equation has more complicated multisoliton solutions than does (24) [32].

The envelope solitons (37) look rather like breathers. The ϕ-four has no breather but the associated envelope solitons of the NLS can play this role. The NLS is in one sense a stopping point: if one looks for envelope solutions $U e^{i(\omega t - kx)}$ of the NLS one finds U satisfies a NLS equation! The NLS turns up as the archetypical equation describing weakly non-linear strongly dispersive systems: the KdV describes weakly non-linear weakly dispersive systems.

The second application is more complicated. We consider first a collection of spin $\frac{1}{2}$ systems (n cc^{-1}) in a constant magnetic field B_0 along $-z$. We use the gyro-magnetic ratio $\gamma\hbar = e\hbar m_e^{-1} c^{-1}$ (g-factor = 2). The Larmor frequency is $\omega_L = \gamma B_0$. The $m_s = \pm\frac{1}{2}$ states have energies $\pm \frac{1}{2}\hbar\omega_L$. An inhomogeneous transverse R.F. field $B_x(\underset{\sim}{x},t)$ flips the spins and these flips propagate as spin waves. From the Hamiltonian density

$$\mathcal{H}(x) = \tfrac{1}{2}\hbar\omega_L \sigma_z(\underset{\sim}{x}) - \tfrac{1}{2}\gamma\hbar\, \sigma_x B_x \tag{38}$$

and commutation relations $\underset{\sim}{\sigma} \times \underset{\sim}{\sigma} = 2i\hbar \underset{\sim}{\sigma}\, \delta(x - x')$ one easily finds

$$\dot{\underset{\sim}{\sigma}} = \underset{\sim}{\omega} \times \underset{\sim}{\sigma}; \; \underset{\sim}{\omega} = (-\gamma B_x, 0, \omega_L). \tag{39}$$

We define the Bloch vector density $\underset{\sim}{r}(x,t)$ by the expectation value

$$\underset{\sim}{r}(\underset{\sim}{x},t) \equiv \langle \underset{\sim}{\sigma}(\underset{\sim}{x},t) \rangle \; .$$

Then

$$\dot{\underset{\sim}{r}} = \underset{\sim}{\omega} \times \underset{\sim}{r} \; . \tag{40}$$

The transverse magnetic dipole is $\frac{1}{2}\gamma\hbar\, r_1(\underset{\sim}{x},t) = \frac{1}{2}\gamma\hbar \langle \sigma_x(\underset{\sim}{x},t) \rangle$ and we assume for simplicity of comparison with the two other case that we can ignore oscillating fields due to σ_y and σ_z. Maxwell's equations are

$$\nabla^2 B_x - c^{-2} B_{x,tt} = 4\pi n c^{-2} \tfrac{1}{2}\gamma\hbar r_{1,tt} \; . \tag{41}$$

With the choice of $\underset{\sim}{\omega}$ in (39) equations (40) and (41) form a system of coupled nonlinear partial differential equations governing the non-linear propagation of spin waves. We call this the Bloch-Maxwell (BM) system. It is linearised by noting that $r_{1,t} = - \omega_L r_2$, $r_{2,t} = \omega_L r_1 + \gamma B_x r_3$, and $r_{1,tt} = - \omega_L{}^2 r_1 - \omega_L \gamma B_x r_3$. Then, for constant inversion, this equation forms with (41) the usual pseudo-Bose system.

A second equivalent problem arises in the propagation of 10^{-9} sec. optical pulses through media with a resonant non-degenerate transition. We take n 2-level atoms cc^{-1} each with the resonant frequency ω_s. Pulses are envelopes modulating carriers of frequency $\omega \approx \omega_s$. This justifies the 2-level atom approximation. This atom is a 2-state system with a spin representation: spin up (down) is occupation of the upper (lower) state. The equations take precisely the BM form with [7,25]

$$\underset{\sim}{\omega} = (-2p\hbar^{-1}E(\underset{\sim}{x},t),0,\omega_s) \ . \tag{42}$$

The electric field replaces the magnetic field in the Maxwell equation

$$\nabla^2 E - c^{-2} E_{tt} = 4\pi c^{-2} pnr_{1,tt} \tag{43}$$

(the electric dipole proves to be $pr_1 = p\langle\sigma_x\rangle$ with p the dipole matrix element).

A third equivalent problem arises in the Josephson junction of large area (such a junction is typically $\lesssim 1$ mm. long). The junction is a 2-state system (the two sides of the junction). If a voltage V is applied between these two sides the two states differ in energy by $2eV$ ($2e$ is the charge of the Cooper pair). The sides will couple by some tunneling parameter K (say). The junction equation will be (40) [7] with

$$\underset{\sim}{\omega} = (2K\hbar^{-1},\ 0,\ 2eV\hbar^{-1}). \tag{44}$$

Consider the plane Josephson sandwich with superconductors top and bottom separated by a thin uniform layer of oxide of effective thickness (including penetration depth) d. If V is applied across the oxide, a field $E = d^{-1}V$ exists there in the direction of the normal (called z) to the plane of the junction (the x-y plane). Consider plane E-waves propagating along x and carrying a magnetic field B along y.

The quantity $r_3 \equiv \langle\sigma_z\rangle$ is the difference in occupation number densities either side of the junction. The current density component j_z along z is thus $j_z = \beta\, r_{3,t}$ (for some constant β). This drives the (transverse) wave equation in the usual way:

$$E_{xx} - \bar{c}^{-2} E_{tt} = 4\pi\beta\bar{c}^{-2}k^{-1}r_{3,tt}. \tag{45}$$

We are in a material medium so $\bar{c} = ck^{-\frac{1}{2}}$. This arises through the displacement current $C\,V_{,t}$: $C = k/4\pi d$ is the capacitance of the junction per unit area.

There are two differences from the two BM systems considered previously: r_3 rather than r_1 drives the wave equation, and V not K will depend on t in ω (we do not argue here that K should be more complicated: our Bloch equations (40) follow FEYNMAN [33]. See [7].) We reach the BM problem with the simple switch of components $1\leftrightarrow3$, $2\leftrightarrow2$!

The BM system

$$E_{xx} - \bar{c}^{-2} E_{tt} = 4\pi\beta\bar{c}^{-2}k\ r_{3,tt} \tag{46a}$$

$$\underset{\sim}{r}_t = \underset{\sim}{\omega} \times \underset{\sim}{r}\ \ (\underset{\sim}{\omega} = (2K\hbar^{-1},\ 0,\ 2ed\ E\hbar^{-1})) \tag{46b}$$

does not have multisoliton solutions [7]. The reduced Maxwell-Bloch system which replaces (46a) by the one-way going (single characteristic) wave equation $E_x + \bar{c}^{-1}E_t = 2\pi\bar{c}^{-1}\beta k^{-1}r_{3,t}$ can be scaled to (27) [25,26] and does have multisoliton solutions. The elimination of the backward going characteristic is admissible in the short optical pulse problem because typically one is concerned with the very low densities of metal vapours ($n \sim 10^{11}\ cc^{-1}$). For the large area junction the numbers scarcely permit this approximation (although the multisoliton RMB behaviour will come through approximately).

In the Josephson junction problem one usually assumes instead that $r_3 \approx 0$ everywhere (little charge imbalance on the two sides). Then the Bloch equation (46b) becomes

$$r_{2,t} = 2ed\hbar^{-1} E\, r_1 \tag{47}$$

$$r_{1,t} = -\, 2ed\hbar^{-1} E\, r_2$$

with solution

$$r_2 = -\sin\sigma,\ r_1 = -\cos\sigma$$

$$\sigma = 2ed\hbar^{-1} \int_{-\infty}^{t} E(x,t')\, dt' \ . \tag{48}$$

Thus

$$\sigma_t = 2ed\hbar^{-1} E, \text{ or } \sigma_t = 2eV\hbar^{-1}$$

$$j_z = -\, 2\beta K\hbar^{-1} r_2\ , \text{ or } j_z = j_{zo} \sin\sigma\ . \tag{49}$$

In their forms on the right these equations are Josephson's two equations: σ is the Josephson phase.

We have still to satisfy Maxwell's equation (46a). After one integration all through by time this becomes

$$\sigma_{xx} - \bar{c}^{-2}\sigma_{tt} = \lambda_o^{-2} \sin\sigma \ . \tag{50}$$

This is the s-G scaled by the natural length

$$\lambda_o = \{k\hbar c^2/8\pi\beta K\}^{\frac{1}{2}} \ . \tag{51}$$

It is also possible to obtain the s-G from the optical problem. We look for plane wave envelope solutions modulating resonant carrier waves by setting

$$E(x,t) = \hbar p^{-1}\, \varepsilon(x,t) \cos\{\omega_s(t - c^{-1}x)\}$$

$$r_1(x,t) = Q(x,t) \cos\{\omega_s(t - c^{-1}x)\} + P(x,t) \sin\{\omega_s(t - c^{-1}x)\}. \tag{52}$$

One finds by using the fact that P, Q, ε vary on a 10^{-9} sec. time scale and $\omega_s \sim 10^{15}$ that $Q \approx 0$ and

$$P_{,t} = \varepsilon N$$

$$N_{,t} = -\, \varepsilon P \tag{53}$$

(where $N = r_3(x,t)$) whilst

$$\varepsilon_x + c^{-1}\, \varepsilon_t = \alpha\, P \tag{54}$$

($\alpha = 2\pi\, p^2 n\, \omega_s\, \hbar^{-1}\, c^{-1}$).

An "attenuator" is an initially unexcited medium. For this (53) is solved by

$$P = -\sin\sigma$$

$$N = -\cos\sigma$$

$$\sigma(x,t) = \int_{-\infty}^{t} \varepsilon(x,t')\,dt' . \tag{55}$$

From (55), (54) is

$$\sigma_{xt} + c^{-1}\sigma_{tt} = -\alpha\sin\sigma . \tag{56}$$

This is the s-G in unusual independent variables. Set

$$\sqrt{c}\,\xi = \sqrt{\alpha}\,(ct - 2x),\ \eta = \sqrt{\alpha c}\,t . \tag{57}$$

Then

$$\sigma_{\xi\xi} - \sigma_{\eta\eta} = \sin\sigma . \tag{58}$$

The crucial idea here is to exploit the resonance condition, and the approximation is very different from the assumption $r_3 \approx 0$ for the junction problem (r_3 is of course r_1 in the optical problem and this goes over to the out of phase component P of the dipole).

Both the s-G (58) for optical pulses and the s-G (50) for the large area junction have the multisoliton solutions (16) - (20). In the optical problem an arbitrary intensity envelope ($\propto \varepsilon^2$) breaks up in general into a train of sech^2 pulses. Each of these is the time derivative in the x,t coordinate system of a 2π-kink. The sech^2 pulses are solitons: they persist and the medium is transparent to them. The phenomenon is therefore called self-induced transparency (SIT) which explains this phenomenon introduced in §4. In the (x,t) system the 1-soliton is

$$\varepsilon(x,t) = p\,E_0\,\hbar^{-1}\,\text{sech}\,\tfrac{1}{2}p\,E_0\,\hbar^{-1}(t - xv^{-1}) . \tag{59}$$

It is a soliton solution of the so called SIT equations [25]. The "area" $\Theta(x,t) \equiv \sigma(x,\infty)$ is independent of x:

$$\Theta(x,t) = \int_{-\infty}^{\infty} \varepsilon(x,t')\,dt' = [4\tan^{-1} e^{\xi}]_{-\infty}^{\infty} = 2\pi . \tag{60}$$

A pulse of arbitrary area Θ entering the resonant medium undergoes a jump in Θ at the boundary of the medium to the value $2\nu\pi$ where ν is an integer. The pulse then reshapes in the medium to ν_1 2π-sech pulses (ν_1 kinks in σ), ν_2 2π-sechs of opposite sign (ν_2 antikinks in σ) and ν_3 breathers. We have $\nu_1 - \nu_2 = \nu$. Details are given in [25,7] and the references in [7].

The theory of the Josephson junction shows similar features in a slightly different way. In the rest frame the kink solution of (50) is

$$\sigma = 4\tan^{-1}\exp[(x - x_0)/\lambda_0] . \tag{61}$$

Use $\sigma_{tx} = \sigma_{xt} = 2e\hbar^{-1}V_{,x} = 2ed\hbar^{-1}c^{-1}B_{,t}$ to reach

$$\sigma_x = 2ed\,\hbar^{-1}c^{-1}B \tag{62}$$

(B is the y-component of the magnetic field). Then

$$2\pi = \sigma(\infty,t) - \sigma(-\infty,t) = 2e\hbar^{-1} c^{-1} \int B\, dx = 2e\hbar^{-1} c^{-1} \int B\, dS \quad . \tag{63}$$

Thus the kink carries one unit of flux $hc/2e$ - the single "fluxon". The boundary conditions on a large junction will be $\sigma(\infty,t) = 2\nu_1\pi$, $\sigma(-\infty,t) = 2\nu_2\pi$ in general, and the total flux is $(\nu_2 - \nu_1)\, hc/2e$. The sum of the numbers of kinks and antikinks is therefore $(\nu_2 - \nu_1)$. The breathers carry no flux.

The break up of optical pulses into their constituent solitons has been observed [34,7]. Evidence of the existence of kinks in large area Josephson junctions has been obtained by looking at changes in the voltage/current characteristics [35]. Extra current spikes have appeared with particular voltage spacings [35]. In frequency units this spacing is just the fundamental even mode frequency of the Josephson equivalent cavity. Ref. [35] argues that a kink which reaches the end of an open ended cavity is reflected as an antikink. There is a natural cavity "mode" consisting of a kink with its antikink. There is therefore one node and the mode corresponds in that respect to a harmonic cavity mode with twice the fundamental frequency. Successive kink-antikink pairs induce current spikes (or current steps) just as radiation induces current steps in small area Josephson junctions. The observations [35] are in extraordinarily good agreement with this argument. However the s-G has not been solved for open-ended or close-ended boundary conditions on finite support $-\frac{1}{2}L < x \leq \frac{1}{2}L$. All the solutions we have described in this paper I are for the real line $-\infty < x < \infty$. Certain equations, notably the KdV (22) have been solved for periodic boundary conditions $u(0,t) = u(L,t)$, $u_x(0,t) = u_x(L,t)$, etc. [36,37].

We note finally a further application of the spin wave theory of this §5. This is to spin waves in the Fermi liquid ^{3}He below 2.6 mK. This liquid has two phases, the A-phase and the B-phase. The A-phase can be thought of as two interpenetrating superfluids carrying respectively 'up' spins and 'down' spins coupled by the very weak spin dipole interaction. Spin waves therefore satisfy the s-G. The B-phase admits the symmetric up-down states (the spin is unity not zero as in a superconductor). In consequence spin waves satisfy the double s-G

$$\sigma_{xx} - \sigma_{tt} = -\left(\sin\sigma + \tfrac{1}{2}\sin\tfrac{1}{2}\sigma\right) \quad . \tag{64}$$

The evidence supports the view that this equation does not have true soliton solutions but its kink solutions have remarkable properties nevertheless. Equation (64) with positive sign on the right governs SIT with a hyperfine degeneracy quantum number $F = F' = 2$ on each of the two levels. The wobbling 4π-pulse solutions of (64) in this case have been observed [38]. The ^{3}He problem is treated in [39]. Both this and the degenerate SIT problem are studied in [7].

This completes our introduction to the soliton concept and its application to physical problems. The reader is referred to [7] and the references there for more details. In the lecture II which now follows we look at some considerably more powerful mathematical machinery. The list of references follows II.

Before we actually go on to II however it seems appropriate to offer the reader at least one illustrative diagram. The Fig. 1 which follows shows the break up of an optical pulse of area $\Theta = 8\pi$ (compare (60)) into two of the wobbling 4π-pulses. The optical medium is degenerate with the two hyperfine quantum numbers $F = F' = 2$ and the equation involved is precisely (64) with however the positive sign on the right side.

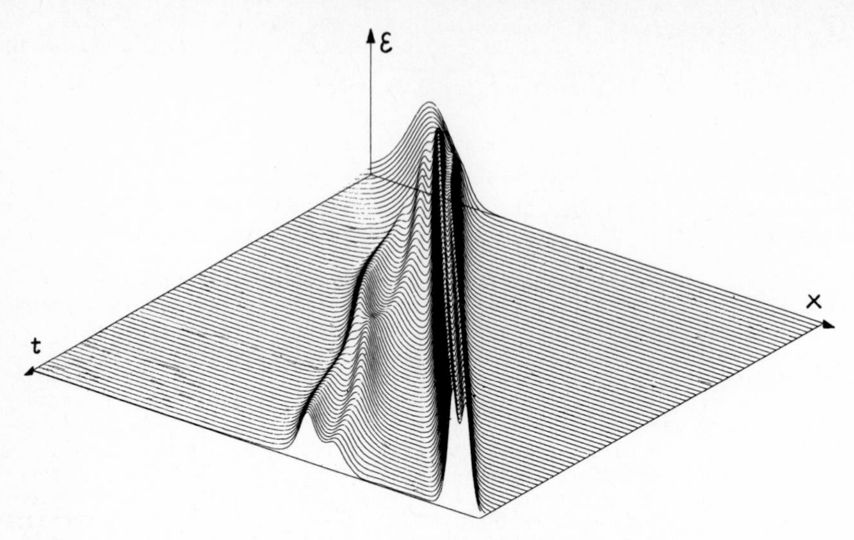

Fig.1 Break up of 8π optical pulse in a degenerate medium.

This lecture I is an improved and amplified version of that given at the International Workshop on Synergetics, Schloss Elmau, Bavaria, May 1977 [40]. In that paper it is incorrectly stated that the NLS $i\,u_t + u_{xx} - 2u|u|^2 = 0$ (with negative self interaction) does not have multisoliton solutions. For these see [32].

References appear at the end of the following article, on p. 250.

Solitons in Mathematics

R.K. Bullough and R.K. Dodd

Department of Mathematics, University of Manchester,
Institute of Science and Technology, P.O. Box 88
Manchester M60 1QD, U.K.

and

School of Mathematics, Trinity College, Dublin 2, Ireland

1. Introduction

This second article presents a brief history of the inverse scattering method for solving nonlinear evolution equations and the Hamiltonian structure associated with it. It is not a comprehensive survey of the different mathematics now concerned with soliton theory. To attempt the latter would be impossible in the present compass. In any case, soliton theory already ramifies into areas of mathematics, algebraic geometry, theory of Jacobian varieties, on the edge of the mathematical range of one of us (RKB).

The subject surely begins with the observation [1] of the single soliton solution of the Korteweg de Vries (KdV) equation, in the month of August 1834, by JOHN SCOTT RUSSELL. Since then, and until quite recently, the physics and mathematics of the subject have evolved together, each influencing the other. At the present time there are signs of schism into distinct and disparate disciplines.

RUSSELL in effect established by experiment [1] that arbitrary initial data for the KdV equation

$$u_t + 6\, uu_x + u_{xxx} = 0 \tag{65}$$

broke up into solitons. The speed of the soliton depended on its amplitude. No serious mathematical developments followed this observation until the work of ZABUSKY and KRUSKAL [41,42].

In contrast, the sine-Gordon equation

$$u_{xx} - u_{tt} = \sin u \tag{66}$$

arose in a strictly mathematical context - in differential geometry in the theory of surfaces of constant curvature [24]. It is worth remarking that LIOUVILLE gave the general solution of the nonlinear Klein-Gordon equation

$$u_{xt} = \exp m\, u \tag{67}$$

(m = parameter) in 1853 [43]. Eq. (67) is a Klein-Gordon equation $u_{xt} = F(u)$ in 'light cone' coordinates, but this is easily expressed in the Lorentz covariant form $u_{xx} - u_{tt} = F(u)$, the form of (66). LIOUVILLE's method was his own, but recently [44] we have used the general solution $u = f(x) + g(t)$ of $u_{xt} = 0$ and a Bäcklund transformation to rederive LIOUVILLE's solution in the form

$$\exp mu = \exp m\,(f-g)\,\{k \int^x \exp m\, f\, dx' + \tfrac{1}{2}\, k^{-1}\, m \int^t \exp(-mg)\, dt'\}^{-2}. \tag{68}$$

The Bäcklund transformation (BT) with free parameter k

$$u'_x = u_x + 2\,k \sin \tfrac{1}{2}(u' + u) \tag{69}$$

$$u'_t = -\,u_t + 2\,k^{-1} \sin \tfrac{1}{2}(u' - u)$$

for the sine-Gordon

$$u_{xt} = \sin u \tag{70}$$

in light cone coordinates was known to Bäcklund before 1883. We now know that a BT is characteristic of the 'integrable systems'. This phrase is used loosely here to mean those systems of nonlinear evolution equations which have soliton solutions. We define 'integrability' more precisely later in terms of 'integrable Hamiltonian systems'. We shall show elsewhere [45] how these ideas relate to the classical integrability theorem of FROBENIUS.

Integrability is concerned ultimately with integrability conditions exemplified by (69). The BT (69) relates a solution u of the sine-Gordon (s-G) (70) to another solution u' (it is strictly speaking an auto BT [44]). The integrability condition is equality of the second partial derivatives $u'_{xt} = u'_{tx}$. This means that u satisfies a partial differential equation which is (70) By adding u'_{xt} and u'_{tx} it also means that u' satisfies a partial differential equation which in this case is also (70). BT's are generalizations of contact transformations (canonical transformations) in which the derivatives of u' are expressed in terms of u, its derivatives, and u'. The structure has been extended by PIRANI [46]. The line of attack is to replace 'transformation' by 'map' in which x,t,u,u_x,u_t are independent variables. Introduce the idea of C^∞ (= infinitely differentiable) functions u equivalent to v if $u(x,t) = v(x,t)$ and all their derivatives at (x,t) are equal to order k. The k-jet of u at (x,t), $j_x^k u$, is the equivalence class thus defined. The k-jet bundle $J^k(x,t;u)$ is the set of all $j_x^k u$. This is a manifold coordinatized by, e.g., for $k = 1$, just (x,t,u,u_x,u_t). In this language, the Bäcklund map is a map $\psi : J^1(x,t,u,u_x,u_t) \times N(u') \to J^1(x,t,u',u'_x,u'_t)$ in which $N(u')$ is a C^∞ manifold coordinatized by u'. This definition can be generalized [46]. In essence J^1 is the natural manifold on which to calculate BT's like (69): we have used this convertly in finding BT's for the Klein Gordon equations [44].

Not all equations with BT's are 'integrable' in the sense of having soliton solutions: in particular LIOUVILLE's equation (67) does not because u does not vanish as $|x| \to \infty$ (it is plain that $u = 0$ is not a solution: the soliton solutions decay exponentially to zero as $|x| \to \infty$ so there can be none of these).

LUND [47] recently extended the early work on the s-G in differential geometry. We use his work here to exemplify the mathematics. Consider the problem of embedding the n-dimensional surface V_n in the $n+1$ dimensional Euclidean space E

$$E \;:\; x^i \;(i = 1,2,\dots,n+1)$$

$$V_n : y^\mu \;(\mu = 1,2,\dots,n), \qquad \text{with metric}$$

$$ds^2 = g_{\mu\nu}\, dy^\mu\, dy^\nu .$$

The embedding is isometric (preserves scalar products) if V_n can be defined through

$$x^i = x^i\,(y^1, \ldots, y^n)$$

such that $g_{\mu\nu} = \partial x^i/\partial y^\mu . \partial x^i/\partial y^\nu$. Vectors X_μ in the tangent space of V_n, and the surface normal X_{n+1} define a basis in E, $X_\mu \equiv \partial\vec{X}/\partial y^\mu$, X_{n+1}. The expression of the vectors $\partial X_\mu/\partial y^\nu, \partial X_{n+1}/\partial y^\nu$ in terms of this basis is the (linear) system of Gauss-Weingarten equations. The integrability conditions $\partial^2 X_\mu/\partial y^\nu \partial y^0 = \partial^2 X_\mu/\partial y^0 \partial y^\nu$ (for all pairs y^0, y^ν) lead to a non-linear system, the Gauss-Codazzi equations.

For a two-dimensional surface in three-dimensional Euclidean space, LUND [47] finds Gauss-Codazzi integrability conditions which ultimately take the form

$$\begin{aligned} &\frac{\partial}{\partial\xi}(\cot^2\theta\ \lambda_\eta) + \frac{\partial}{\partial\eta}(\cot^2\theta\ \lambda_\xi) = 0 \\ &\theta_{\xi\eta} - \tfrac{1}{2}\sin 2\theta + \frac{\cos\theta}{\sin^3\theta}\ \lambda_\xi\ \lambda_\xi = 0 \end{aligned} \tag{71}$$

for the Gauss-Weingarten system

$$\begin{aligned} \begin{pmatrix} v_{1,\xi} \\ v_{2,\xi} \end{pmatrix} &= \begin{pmatrix} -i\zeta + ip & q \\ -q^* & i\zeta - ip \end{pmatrix} \begin{pmatrix} v_1 \\ v_2 \end{pmatrix} \\ \begin{pmatrix} v_{1,\eta} \\ v_{2,\eta} \end{pmatrix} &= i \begin{pmatrix} r & s \\ s & -r \end{pmatrix} \begin{pmatrix} v_1 \\ v_2 \end{pmatrix} \end{aligned} \tag{72}$$

in which $p = (\cos 2\theta/2\sin^2\theta)\lambda_\xi$, $q = \theta_\xi + i\ \cot\theta\ \lambda_\xi$, $r = (1/4\zeta)\cos 2\theta - (2\sin^2\theta)^{-1}\lambda_\eta$, $s = (1/4\zeta)\sin 2\theta$. An eigenvalue has been introduced by appealing to Lie invariance (essentially invariance under interchange of ξ,η) of the pair of coupled evolution equations (71) in the two fields θ,λ. The conditions (71) are sufficient and necessary for the integration of (72). The system (72) is however a generalized ZAKHAROV-SHABAT [48] scattering solution problem as we shall see shortly. Integration of this linear system is equivalent to solving the nonlinear system (71). We thus have a natural geometrical basis to an inverse scattering problem.

However, the BT (69) already contains a similar scattering problem for the s-G. Set $\Gamma = \tan[(u + u')/4]$. Then

$$\begin{aligned} \Gamma_{,x} &= \tfrac{1}{2}\, u_x\,(1 + \Gamma^2) + k\,\Gamma \\ \Gamma_{,t} &= k^{-1}\,\Gamma\cos u - (2k)^{-1}\,(1 - \Gamma^2)\,\sin u. \end{aligned} \tag{73}$$

The Ricatti transformation $\Gamma = v_2/v_1$ now yields

$$\begin{pmatrix} v_{1,x} \\ v_{2,x} \end{pmatrix} = \begin{pmatrix} -\frac{1}{2}k & \frac{1}{2}u_x \\ -\frac{1}{2}u_x & +\frac{1}{2}k \end{pmatrix} \begin{pmatrix} v_1 \\ v_2 \end{pmatrix}$$

$$\begin{pmatrix} v_{1,t} \\ v_{2,t} \end{pmatrix} = -\frac{1}{2k} \begin{pmatrix} +\cos u & \sin u \\ \sin u & -\cos u \end{pmatrix} \begin{pmatrix} v_1 \\ v_2 \end{pmatrix} . \tag{74}$$

This is the same as (72) for $\lambda = 0$ and $u = 2\theta$ (corresponding to a surface of constant intrinsic curvature) if $i\zeta \equiv k/2$. POHLMEYER [49] has obtained a BT for the nonlinear sigma models in n fields $u_1,\ldots,u_n$ coupled only by the constraint

$$\sum_{i=1}^{n} u_i^2 = 1.$$

For $n = 4$, the O_4 invariant model, the system reduced to LUND's. For $n = 3$, the O_3 invariant case, the system is the s-G. POHLMEYER [49] has shown that the results (74) can be extended to (72) by using his BT. The BT may be much more fundamental than the idea that inverse scattering transforms like (72) or (74) are geometrical in origin. However, BT's can be derived for example, by the method of prolongation structures due to WAHLQUIST and ESTABROOK [50]. It is plain that geometrical ideas underlie this theory although the precise nature of these is not yet fully understood.

The BT can generate infinite sets of conserved quantities. We have demonstrated this elsewhere [40,44][3] and remark only that by using the BT between the s-G and

$$u'_{xt} = \{1 - k^2 u_x'^2\}^{\frac{1}{2}} \sin u'$$

which is

$$u'_x = k^{-1} \sin(u' - u) \tag{75}$$

$$u'_t = u_t + k \sin u' \tag{76}$$

one obtains the conserved densities of the s-G

$$T^2 = u_x^2,\ T^4 = u_{xx}^2 - \frac{1}{4} u_x^2,\ T^6 = u_{xxx}^2 - \frac{5}{2} u_{xx}^2 u_x^2 + \frac{1}{8} u_x^6$$

$$T^8 = u_{xxxx}^2 - \frac{7}{2} u_{xxx}^2 u_x^2 + \frac{7}{4} u_{xx}^4 + \frac{35}{8} u_{xx}^2 u_x^4 + \frac{5}{64} u_x^8,$$

$$T^{10} = \text{etc.} \tag{77}$$

---

[3] And see especially [74] below

Note that T^r has the rank r in that any term in T^r, $\Pi\, u_{nx}^{a_n}$, has rank $\sum n\, a_n = r$. The odd rank densities have been removed [40] because they are trivial. Note that, because of the assumed boundary conditions $u, u_x, u_{xx} \to 0$, etc., $d/dt \int_{-\infty}^{\infty} T\, dx = - \int_{-\infty}^{\infty} X_x\, dx = X(+\infty) - X(-\infty) = 0$ since a density T satisfies a conservation law $T_t + X_x = 0$. Thus $\int_{-\infty}^{\infty} T^n dx$ is a constant of the motion I_n (say). The existence of an infinite set of constants of the motion I_n is necessary but not sufficient for the s-G to be a completely integrable (infinite dimensional) Hamiltonian system. We return to this later.

A third equation, the nonlinear Schrödinger equation (NLS) [48]

$$i\, u_t + u_{xx} + 2u|u|^2 = 0 \tag{78}$$

has played an important role in the history of the subject. The equation of a 'simple wave' $u_t + uu_x = 0$ describes a nonlinear system without dispersion; the KdV describes a weakly dispersive, weakly nonlinear system; the NLS describes a strongly dispersive, weakly nonlinear system. The equation is of relatively recent origin [51]. Its role [48] in the development of the inverse scattering method is described below.

2. Discovery of the Inverse Spectral Transform

In §1 we mentioned three equations, the KdV, the s-G and the NLS, important to the development of the inverse scattering method. The critical step was taken by KRUSKAL and co-workers [41,42] for the KdV. It seems worthwhile sketching the argumentation adopted by these authors as we have heard it [52] since it throws up a number of points. The argument is physically based and the mathematical content has emerged later following the work of LAX [53]. We indicate the nature of this in §3 following where we also connect the theory with the inverse scattering methods associated with Eqs. (74) and (72) introduced without further comment in §1.

KRUSKAL in the period 1955-65 was apparently concerned with the periodicity exhibited by one-dimensional lattices - the FERMI-PASTA-ULAM problem [21]. First results obtained with ZABUSKY appeared in 1965 [41]. The two authors were concerned with a continuum limit of a 1D-lattice which leads to

$$y_{tt} = y_{xx}\,(\,1 + \varepsilon\, y_x). \tag{79}$$

The Riemannian invariants of the linear problem $u = \frac{1}{2}(y_x + y_t) = 0(1)$, $v = \frac{1}{2}(y_x - y_t) = 0(\varepsilon)$ lead, for $\varepsilon \ll 1$, to the one-way going wave $[\tau \equiv 4\varepsilon(t - x)]$

$$u_\tau + uu_x = 0, \tag{80}$$

the simple wave. Characteristics of this equation are [54] $dx/d\tau = u =$ constant. These lines intersect and the system shocks (the equation can be integrated by the hodograph transformation [54]). It is interesting that (80) has an infinite set of polynomial conserved densities $T^n = u^n/n$: it has some of the structure of the integrable systems like the KdV equation (the $I_n \equiv \int T^n dx$ are in involution - see below). The addition of the dispersive term $[(\Delta x)^2/12] y_{xxxx}$, a further term in the Taylor expansion used to convert the discrete lattice model to a continuum, balances dispersion against nonlinearity and leads to $u + uu_x + (\Delta x^2/12\varepsilon)u_{xxx} = 0$, the KdV. [For the record $\Delta x \sim 1/64$, $\varepsilon \sim 1/20$ and $\delta^2 \equiv \Delta x^2/12 \quad \varepsilon \sim (0.022)^2$.] Numerical solutions under periodic boundary conditions with a harmonic initial condition lead to periodic behaviour in time. It is now known that more generally, under these conditions, the KdV is periodic in space and 'almost periodic' in time [37].

The simple wave shocks and with the small dispersive term added, the resultant KdV equation also develops a (smooth) jump. KRUSKAL and co-workers looked for conservation laws across the jump and found the polynomial conserved densities u [from the equation in the form $u_\tau + (\frac{1}{2}u^2 + \delta^2 u_{xx})_x = 0$] u^2 and $2u^3 - u_x^2$. They subsequently found ten polynomial conserved densities T^r : the rank of $\Pi\, u_{nx}^{a_n}$ in this case is $r = \sum (1 + n/2)\, a_n$. T^{10} has 32 terms $(1/10)u^{10} - 36u^7u^2_{1x} - 630u^4u^4_{1x} \ldots (419904/12155)\; u^2_{8x}$ [55].

FPU also looked at a lattice with quartic rather than cubic anharmonicity. This leads to $y_{tt} = y_{xx}(1 + \varepsilon\, y_x^2)$ and thence to the modified KdV

$$v_t + v^2v_x + v_{xxx} = 0. \tag{81}$$

This also has conservation laws. MIURA found the transformation $u = v^2 + \sqrt{-6}\; v_x$ connecting these. The remarkable $\sqrt{-6}$ is actually innocuous: the MIURA transformation [56] $u = v^2 + v_x$ is a BT connecting

$$\begin{aligned} &u_t - 6\, uu_x + u_{xxx} = 0 \quad \text{and} \\ &v_t - 6\, v^2v_x + v_{xxx} = 0. \end{aligned} \tag{82}$$

The Ricatti transformation $v = \partial(\ell n\, \psi)/\partial\psi = \psi_x/\psi$ linearizes $v^2 + v_x$ and $u = \psi_{xx}/\psi$. The Galilean transformation

$$x' = x - Vt, \quad t' \to t, \quad u \to u + V \tag{83}$$

leaves $u_t - uu_x + u_{xxx} = 0$ invariant. Hence, $u \to u + \lambda = \psi_{xx}/\psi$ and the Schrödinger eigenvalue problem has emerged! Furthermore, $u = \psi_{xx}/\psi - \lambda$ into the KdV yields

$$\text{(a)} \qquad \lambda_t = 0$$

$$\text{(b) a functional } B[\psi,u] \text{ such that } B[\psi,u] = \psi_t. \tag{84}$$

GARDNER et al. [42] used these results and the property $u, u_x, u_{xx} \to 0$ as $|x| \to \infty$ to solve the KdV by the route $u(x,0)$ (initial data) $\to$ scattering data at $t = 0 \to$ (by $B[\psi,u] = \psi_t$) scattering data at time $t \to$ [via the Gel'fand-Levitan-Marchenko linear integral equation (see, e.g., [40])] $u(x,t)$, the solution of the KdV. The method is now called the 'inverse spectral transform' (IST) because (a) $\lambda_t = 0$, i.e., the spectrum of the Schrödinger operator is invariant under the KdV flow, and (b) the method generalizes the Fourier transform for linear systems [57]. The remarkable discovery by KRUSKAL and co-workers has been the source of all the developments since, with the single exception of the direct methods developed by HIROTA [58] and CAUDREY [58]. The precise connection of these with the IST is still to be established.

LAX [53] stimulated an important development and added mathematical understanding to the IST. Given the relatively prime differential operators $u \to \hat{L}_u \equiv (-\partial^2/\partial x^2+u)$, $\hat{B}_u \equiv \partial^{2n+1}/\partial x^{2n+1}$ + lower degree ($\hat{B}$ is skew symmetric, $\hat{B}^* = -\hat{B}$) such that there is a one-parameter family of unitary operators $\hat{U}$ satisfying $\hat{U}_t = \hat{B}\hat{U}$; and $\hat{L}_u$ is unitary equivalent under $\hat{U}$, i.e., $\hat{U}^{-1}(t)\hat{L}_u(t)\hat{U}(t)$ is independent of t:

(i) The eigenvalues λ_u of $\hat{L}$ are integrals of the motion, i.e. $\lambda_{u,t} = 0$. [The proof is trivial: $\hat{L}(0) = \hat{U}^{-1}(t)\hat{L}_u(t)\hat{U}(t)$, $\hat{L}_u(0)\psi(0) = \lambda_u\psi(0)$. Then $\hat{L}_u(t)\psi(t) = \lambda_u\psi(t)$ with $\psi(t) = \hat{U}(t)\psi(0)$, i.e., $\hat{U}(t)$ is the evolution operator].

(ii) $\frac{\partial}{\partial t}\hat{U}^{-1}(t)\hat{L}_u(t)\hat{U}(t) = 0 \quad \Rightarrow$

$$-\hat{U}^{-1}\hat{U}_t\hat{U}^{-1}\hat{L}_u\hat{U} + \hat{U}^{-1}\hat{L}_{u,t}\hat{U} + \hat{U}^{-1}\hat{L}_u\hat{B}\hat{U} = 0.$$

Then $\hat{U}_t = \hat{B}\hat{U} \Rightarrow -\hat{U}^{-1}\hat{B}\hat{L}_u\hat{U} + \hat{U}^{-1}\hat{L}_{u,t}\hat{U} + \hat{U}^{-1}\hat{L}_u\hat{B}\hat{U} = 0$ and

$$\hat{L}_t = [\hat{B}, \hat{L}]. \tag{85}$$

(iii) Since $\hat{L}_t = u_t$, the operator equation (85) is an evolution equation $u_t = K[u]$ in which $K[u]$ is a functional of u. By trying for the skew symmetric operator $\hat{B} = \partial^3 + b\partial + \partial b$ $(\partial \equiv \partial/\partial x)$, LAX found that with $b = -3u/4$ (our scaling is different from that used by LAX [53]) he regained from (21) the KdV equation. Further, the method generalizes to yield an infinite hierarchy of KdV equations of degrees 3 (the KdV), 5, 7, etc.

It is remarkable that LAX's operators can be found by defining the square root $\hat{R}$ of the operator $-\hat{L}_u$ by $\hat{R}^2 = -\hat{L}_u$ and $R = \partial + c_0 + c_1\partial^{-1} + c_2\partial^{-2} + \dots$ with $\partial \equiv \partial/\partial x$). Then $-\hat{L}_u = \partial^2 - u$ and $R^2 = -L_u$ is $\partial^2 + 2c_0\partial + (c_{0,x} + c_0^2 + 2c_1) + (c_{1,x} + 2c_2 + 2c_0c_1)\partial^{-1} + \cdots$, so that $c_0 = 0$, $c_1 = -\frac{1}{2}u$, $c_2 = \frac{1}{4}u_x$. Evidently, the principal part of $\hat{R}$ (that excluding powers of ∂^{-1} is ∂ and $[\hat{R}, \hat{L}] = \hat{L}_t$ is

$$u_x = u_t\,. \tag{86}$$

However, the principal part of $\hat{R}^{3/2} = (-\hat{L}_u)\hat{R}$ is

$$\partial^3 - \partial u - \tfrac{1}{2}u\,\partial + \tfrac{1}{4}u_x = \partial^3 - (\tfrac{3}{4})\,\partial u - (\tfrac{3}{4})\,u\partial \tag{87}$$

and this is LAX's operator for the KdV. Other fractional powers $\hat{R}^{5/2}$, etc., generate LAX's hierarchy.

GELFAND and DIKII [59], and other workers (NOVIKOV, DRINFELD) referenced by MANIN [60], in particular show how to solve the KdV for its multi-soliton solutions on the real line $-\infty < x < \infty$ by reducing the problem to a system of algebraic equations. Analogous but more difficult methods are used by NOVIKOV [60,61] to solve the KdV under periodic boundary conditions. MANIN [60] notices interesting new properties in the multi-soliton solutions on the real line when the operators $\hat{L}$, $\hat{B}$ are not relatively prime, i.e., are of degrees d (the order of the highest derivative) which are not relatively prime, 2 and 4 for example. He obtains remarkable 'geyseron' solutions in this case [60]!

4. Symplectic Structure

GARDNER [62] introduced the idea of a bracket (a Poisson bracket) but worked under periodic boundary conditions. For present purposes, define the functional or Frechet derivative $\delta F/\delta u$ of the functional $F[u] = \int_{\infty}^{\infty} [u]dx$ by

$$\frac{dF}{dt} = \int_{-\infty}^{\infty} \frac{\delta F}{\delta u}\frac{\partial u}{\partial t}\,dx \tag{88}$$

where t, usually the time, parametrizes u. Alternatively,

$$\lim_{\varepsilon\to 0}\frac{\partial}{\partial\varepsilon}\int_{-\infty}^{\infty} [u + \varepsilon\,\delta u]dx = \int_{-\infty}^{\infty}\frac{\delta F}{\delta u}\,\delta u\,dx. \tag{89}$$

The bracket corresponding to that found by GARDNER is the unusual one

$$\{F, G\} = \int_{-\infty}^{\infty}\frac{\delta F}{\delta u}\frac{\partial}{\partial x}\frac{\delta G}{\delta u}\,dx. \tag{90}$$

GARDNER in effect proves

$$0 = \frac{dI_n}{dt} = \int_{-\infty}^{\infty}\frac{\delta I_n}{\delta u}\frac{\partial u}{\partial t}\,dx = \{I_n, I_m\} \tag{91}$$

if $u_t = \partial/\partial x\,(\delta I_m/\delta u)$. This evolution equation is a Hamiltonian flow with Hamiltonian the mth constant of the motion I_m of the KdV equation. Since $I_m = \int_{-\infty}^{\infty} T^m\,dx$ and $T^3 = \frac{1}{2}(2u^3 - u_x{}^2)$, one finds that the KdV is a Hamiltonian flow. The constants of the motion I_m are in involution, there are precisely the right number of these, and the system constitutes a completely integrable infinite dimensional Hamiltonian system - the first example of this class.

To expand this a little more: with $\mathcal{H}_3 = \frac{1}{2}(2u^3 - u_x{}^2) \equiv T^3$ and $p_x = u$

$$\begin{aligned}\frac{\partial u}{\partial t} &= \frac{\delta H_3}{\delta p} = 6uu_x - u_{xxx}\\ \frac{-\partial u}{\partial t} &= \frac{\partial}{\partial x}\left(\frac{\delta H_3}{\delta u}\right) = -\,6uu_x + u_{xxx}\end{aligned} \tag{92}$$

can be compared with the classical prescription for a Hamiltonian flow

$$\begin{pmatrix}\dot q\\ \dot p\end{pmatrix} = \begin{pmatrix}0 & 1\\ -1 & 0\end{pmatrix}\begin{pmatrix}\partial H/\partial q\\ \partial H/\partial p\end{pmatrix} = J\ \text{grad}\ H. \tag{93}$$

In this, J is a linear map taking phase space to phase space. It is isometric (i.e., preserves inner products) and, and this is the key point, is skew. It connects the contravariant vector field $(\dot q,\dot p)$ with a covariant one (grad H). Phase space is symplectic: it carries a closed skew symmetric differential two-form ω [63]. For (93), [4] in wedge product notation [63],

$$\begin{aligned}\omega &= \omega_{ij}\, \delta x^i \wedge \delta x^j \qquad\qquad (i,j = 1,2)\\ &= \delta p \wedge \delta q \leftrightarrow \delta_1 p\, \delta_2 q - \delta_2 p\, \delta_1 q = \bar{\omega} .\end{aligned} \tag{94}$$

$\delta_1 p$ and $\delta_2 p$ ($\delta_1 q$ and $\delta_2 q$) are independent variations in $p(q)$. Notice that in the case of canonical co-ordinates $x^i \equiv (q,p)$ $2\omega_{ij}$ is just the matrix J.

The symplectic form ω is invariant (co-ordinate free) and closed, $d\omega = 0$. From Hamilton's principle if

$$\delta \int(p\, dq - H\, dt) = 0, \qquad\qquad \delta \int(\overline{p}\, d\overline{q} - \overline{H}\, dt) = 0 \tag{95}$$

describe the same trajectories in phase space,

$$\delta \int(p\, dq - H\, dt - \overline{p}\, d\overline{q} + \overline{H}\, dt + dF) = 0$$

and

$$\delta(p\, dq) - \delta(\overline{p}\, d\overline{q}) + \delta\, dF = 0$$

so that

$$d(p\, \delta q) - d(\overline{p}\, \delta\overline{q}) + d\, \delta F = 0$$

and

$$dp\ \ \delta q - d\overline{p}\, \delta\overline{q} = \delta p\ dq - \delta\overline{p}\ d\overline{q}$$

or

$$dp\ \ \delta q - \delta p\ dq = d\overline{p}\ \delta\overline{q} - \delta\overline{p}\ d\overline{q} \tag{96}$$

i.e., $\bar{\omega} = \bar{\bar{\omega}}$ under the canonical transformation $p,q \to \overline{p},\overline{q}$, $H \to \overline{H}$. In this somewhat heuristic demonstration, we have used that $\overline{H}(\overline{p},\overline{q}) = H(p,q)$, that is take the same values although their functional forms are different: hence $H\, dt = \overline{H}\, dt$ above.

In (92) the discrete variables q,p in (93) are replaced by the running set $u, p = \int_{-\infty}^{x} u(x')dx'$, for each x, and the symplectic form matrix ω_{ij} is replaced by the skew symmetric operator $\partial/\partial x$. This suggests the symplectic form defined by the skew symmetric bilinear form

$$\bar{\omega} = \int_{-\infty}^{\infty} dx \int_{-\infty}^{x} dy[\delta_1 u(x)\delta_2 u(y) - \delta_1 u(y)\delta_2 u(x)] \tag{97}$$

and the Poisson bracket (90). This bracket satisfies Jacobi's identity

$$\Big\{\{A, B\}, C\Big\} + \Big\{\{C, A\}, B\Big\} + \Big\{\{B, C\}, A\Big\} = 0, \tag{98}$$

a result which is conveniently proved via the identity

---

[4] The $x^i (i = 1,2)$ are arbitrary local co-ordinates in phase-space. The bilinear form $\bar{\omega}$ is simply associated with the symplectic form ω; but the correspondence between quadratic differential forms and bilinear forms is 1:1 (compare [68]).

$$\left\{\{A, B\}, C\right\} = \int \left[\frac{\delta^2 A}{\delta u^2} \frac{\partial}{\partial x}\left(\frac{\delta B}{\delta u}\right) - \frac{\delta^2 B}{\delta u^2} \frac{\partial}{\partial x}\left(\frac{\delta A}{\delta u}\right) \right] \frac{\partial}{\partial x}\left(\frac{\delta C}{\delta u}\right) dx. \tag{99}$$

To prove involution of the constants of the KdV motion I_n it is sufficient to know [64] that there is a skew symmetric operator $K = -\partial^3 + 2u\partial + 2\partial u$ such that the vector field $u_t = \partial_n u$ for the nth KdV flow (namely $u_t = K_n[u] = [\hat{L}, \hat{B}_n]$ with $\hat{B}_n$ LAX's nth skew symmetric operator $\hat{B}$) satisfies $\partial_{n+1} u = K\, \delta H_n/\delta u$. For then

$$\{I_n, I_m\} = \int \frac{\delta I_n}{\delta u} \frac{\partial}{\partial x} \frac{\delta I_m}{\delta u} dx = \int \frac{\delta I_n}{\delta u} \partial_m u\, dx$$

$$= \int \frac{\delta I_n}{\delta u} K \frac{\delta I_{m-1}}{\delta u} dx = - \int K \frac{\delta I_n}{\delta u} \frac{\delta I_{m-1}}{\delta u} dx = - \int \partial_{n+1} u \frac{\delta I_{m-1}}{\delta u} dx$$

$$= - \int \frac{\partial}{\partial x} \frac{\delta I_{n+1}}{\delta u} \frac{\delta I_{m-1}}{\delta u} dx = \int \frac{\delta I_{n+1}}{\delta u} \frac{\partial}{\partial x} \frac{\delta I_{m-1}}{\delta u} dx = \{I_{n+1}, I_{m-1}\}, \tag{100}$$

and continuation down to $I_1 = u$ yields zero [64].

We shall take as our definition of a completely integrable (or integrable) system with $2n$ degrees of freedom one such that (a) the system has n constants of the motion I_n, and (b) these are in involution, i.e. $\{I_n, I_m\} = 0$. Integrable systems can be given a more fundamental description and we provide this elsewhere [45]. For the KdV the infinite set of constants I_n in involution is not enough for complete integrability since precisely the right number of constants $n = \infty$ is required. However, the question of integrability was settled by ZAKHAROV and FADEEV [65] who explicitly and independently enunciated the fact that the KdV was an infinitely dimensional, completely integrable Hamiltonian system, proved the symplectic form (97) is closed under the transformation which maps $u(x,t)$ via the Schrodinger eigenvalue problem, to the scattering data $w(t)$, ζ_ℓ and c_ℓ, and integrated the equations. The scattering data $S = \{w(t), \zeta_\ell, c_\ell\}$ are constituted as follows: $w(t)$ is the reflection coefficient for Jost functions [40] $\psi \sim e^{+ikx}$, $x \to -\infty$; ζ_ℓ is the ℓth bound state eigenvalue (which proves to lie on the positive imaginary axis in the ζ-plane) and c_ℓ is the normalization of the eigenfunction. This scattering data is sufficient to reproduce the potential $u(x,t)$ via the Gel'fand-Levitan-Marchenko equation. The Hamiltonian in terms of the scattering data proves to be of action angle type:

$$\begin{aligned} H &= - \frac{8}{\pi} \int_{-\infty}^{\infty} k \log (1 - |w|^2)\, dk - \frac{32}{5} \sum_{\ell=1}^{m} \zeta_\ell^{\,5} \\ &= 8 \int_{-\infty}^{\infty} k^3 P(k)\, dk - \frac{32}{5} \sum_{\ell=1}^{m} p_\ell^{5/2}. \end{aligned} \tag{101}$$

This Hamiltonian reproduces the time evolution of the scattering data:[5]

$$P(k)_{,t} = 0, \quad p_{\ell,t} = 0; \quad Q(k)_{,t} = 8k^3, \quad q_{\ell,t} = -8\zeta_\ell^{\,3}. \tag{102}$$

These results are not peculiar to the KdV (and the TODA lattice [66]). ZAKHAROV and SHABAT almost immediately [48] integrated the NLS equation

---

[5] $Q(k)$, $P(k)$, p_ℓ, q_ℓ are new canonical coordinates defined in terms of the scattering data (see [65]).

$iu_t + u_{xx} + \chi u|u|^2 = 0$ in the LAX form $L_t = i[\hat{L}, \hat{A}]$ with

$$\hat{L} = i\begin{bmatrix} 1+p & 0 \\ 0 & 1-p \end{bmatrix} \frac{\partial}{\partial x} + \begin{bmatrix} 0 & u^* \\ u & 0 \end{bmatrix}, \quad \chi = \frac{2}{1-p^2}$$

$$\hat{A} = -p\begin{bmatrix} 1 & 0 \\ 0 & 1 \end{bmatrix} \frac{\partial^2}{\partial x^2} + \begin{bmatrix} |u|^2/(1+p) & iu_x^* \\ -iu_x & -|u|^2/(1-p) \end{bmatrix}. \tag{103}$$

The scattering problem $\hat{L}v = \zeta v$ with $\tilde{v} = [v_1, v_2]$ yields an infinite set of conserved densities. The NLS is also a Hamiltonian flow and proves to be completely integrable.

ABLOWITZ, KAUP, NEWELL and SEGUR [19,20] then provided a natural generalization: namely find $\hat{L}$, $\hat{A}$ such that

$$\hat{L}\, v = \zeta v, \quad \hat{L} = \begin{bmatrix} i\partial/\partial x & -iq \\ ir & -i\partial/\partial x \end{bmatrix}$$

$$\hat{A}\, v = v_t, \quad \hat{A} = \begin{bmatrix} A & B \\ C & -A \end{bmatrix}. \tag{104}$$

Choose A, B, C, q, r such that, under $\zeta_t = 0$, $\hat{L}_t = [\hat{A}, \hat{L}]$ is the required pair of evolution equations $q_t = K_1[q,r]$, $r_t = K_2[q,r]$ (here, $K[q,r]$ means functional of q and r). The potentials q, r form a natural canonical pair and the appropriate bracket is

$$\{F, G\} = \int_{-\infty}^{\infty} \left(\frac{\delta F}{\delta q}\frac{\delta G}{\delta r} - \frac{\delta F}{\delta p}\frac{\delta G}{\delta q} \right) dx. \tag{105}$$

An extension essentially due to CALOGERO [67] yields [68]

$$\frac{\partial}{\partial t}\hat{\sigma}\, s(x,\underset{\sim}{y},t) + h(\hat{L}^{\dagger},\underset{\sim}{y},t) \cdot \frac{\partial}{\partial y}\hat{\sigma}\, s(x,\underset{\sim}{y},t) + 2\Omega(\hat{L}^{\dagger},\underset{\sim}{y},t)\, s(x,y,t) = 0 \tag{106a}$$

where

$$s = \begin{pmatrix} r \\ q \end{pmatrix}, \quad \hat{\sigma} = \begin{bmatrix} 1 & 0 \\ 0 & -1 \end{bmatrix}$$

$$\hat{L}^{\dagger} = \frac{1}{2i}\begin{bmatrix} \partial/\partial x - 2r\int_{-\infty}^{x} dx'q & 2r\int_{-\infty}^{x} dx'r \\ -2q\int_{-\infty}^{x} dx'q & -\frac{\partial}{\partial x} + 2q\int_{-\infty}^{x} dx'r \end{bmatrix}, \tag{106b}$$

as an integrable pair of evolution equations in x,t and any number of 'time-like' variables $\underset{\sim}{y}$. The extended LAX-AKNS operator pair formulation is

$$\Delta v(x,\underset{\sim}{y},t,\zeta) = \hat{A}\, v(x,y,t,\zeta)$$

$$\Delta = \frac{\partial}{\partial t} + \underset{\sim}{h}(\zeta,\underset{\sim}{y},t) \cdot \frac{\partial}{\partial y}$$

$$\hat{L}\, v \equiv \begin{bmatrix} i\partial/\partial x & -iq \\ ir & -i\partial/\partial x \end{bmatrix} v = \zeta\, v \tag{107}$$

and $\Delta\, L = [\hat{L},\hat{A}]$ with $\Delta\zeta = 0$. We have reported a BT for this system [69] and its canonical structure is established [68]. No physical application of this extended formalism is known to us. An extension to bigger $N\times N$ matrices has brought some mathematical solutions - the boomerons - with remarkable properties. This ingenious extension is due to CALOGERO and DEGASPERIS [67] who exploit Wronskian relations for, for example, an $N\times N$ Schrodinger problem (in which u becomes an $N\times N$ matrix) to obtain linear equations of motion for the matrix reflection coefficients and to which there corresponds a nonlinear evolution equation for u. NEWELL [70] has recently extended the AKNS [19,20] scheme to $N\times N$ matrices and generalized the canonical structure to this case.

As an early example of the use of the AKNS 2×2 inverse scattering scheme consider the 'reduced MAXWELL-BLOCH' system of equations [26] (see the equivalent set (27))

$$E_t = -\, s, \qquad r_x = -\, \mu s \tag{108}$$

$$s_x = Eu + \mu r, \qquad u_x = -\, Es.$$

It is easy to check that

$$A = \frac{-i\zeta}{4\zeta^2-\mu^2}\, u, \qquad\qquad B = \frac{-i\zeta}{4\zeta^2-\mu^2}\,[s + i\,\frac{\mu r}{2\zeta}]$$

$$C = \frac{-i\zeta}{4\zeta^2-\mu^2}\,[s - i\,\frac{\mu r}{2\zeta}], \qquad\qquad r = \tfrac{1}{2}\, E = -\, q = -\, q^* \tag{109}$$

with $r_x = -\mu s$ [do not confuse the q,r potentials with the three dependent variables r,s,u in (108)], reproduces (108) in the form $\hat{L}_t = [\hat{A},\hat{L}]$ under the condition $\zeta_t = 0$. The N-soliton solution is [25,26]

$$E^2 = 4\, \frac{\partial^2}{\partial x^2}\, \ln \det |M|$$

$$M_{nm} = \frac{2}{E_m+E_n}\cosh[\tfrac{1}{2}\,(\theta_n+\theta_m)]$$

$$\theta_n = \tfrac{1}{2}\, E_n(x - 4[E_n{}^2 + 4\mu^2]^{-1}\, t + \delta_n). \tag{110}$$

found by Gibbon et. al. in 1973 [71]. Note that when $\mu=0$ the system is the s-G: equations (108) are $E_t = -s$, $s_x = Eu$, $u_x = -Es$. Put $s = -\sin\sigma$, $u = -\cos\sigma$, $E = \sigma_x$; then $\sigma_{xt} = \sin\sigma$. In this case

$$\hat{A} = \frac{i}{4\zeta}\begin{bmatrix} \cos\sigma & \sin\sigma \\ \sin\sigma & -\cos\sigma \end{bmatrix} \tag{111}$$

with q and r in the scattering problem (104) given by $q = -\,\sigma_x/2 = -\, r$. This is the form obtained from LUND's result (72) when $\lambda\equiv 0$. The $\hat{L},\hat{A}$ pair agrees with the results obtained from the BT (69) quoted in (74). This pair was given first by ABLOWITZ, KAUP, NEWELL and SEGUR [72].

Whilst the BT thus shows that the $\hat{L},\hat{A}$ pair for the s-G is a natural pair it is important to notice that this extension of the LAX pair formulation by AKNS is critically different from LAX's original formulation and from its extensions by ZAKHAROV and SHABAT [48] and those reported by MANIN [60]. For the sine-Gordon equation, a LAX pair of differential operators is a pair of 4×4 first-order (= first degree) differential operators [73]. In the 2×2 formulation in the AKNS scheme $\hat{A}$ is not a differential operator [see (111)] and it involves the eigenvalues of the scattering problem $\hat{L}\ v = \zeta\ v$. Thus, if (i) $[\hat{A},L] = L_{,t}$, (ii) $Lv = \zeta v$, (iii) $\zeta_t = 0$, then $L(\hat{A}v - v_t) = \zeta(\hat{A}v - v_t)$ and $(\hat{A}v - v_t)$ is an eigenfunction belonging to ζ, i.e., $Av - v_t = \lambda(\zeta)v$ in general.[6] It is easy to see that ζ^{-1} in $\hat{A}$ acts as an integral operator by calculating $[\hat{L},\hat{A}]$ from (111) explicitly. One finds

$$[\hat{L},\hat{A}] = \frac{-\sin\sigma}{2\zeta}\begin{bmatrix}0 & 1\\ 1 & 0\end{bmatrix}\begin{bmatrix}\partial/\partial x & \tfrac{1}{2}\sigma_x\\ \tfrac{1}{2}\sigma_x & -\partial/\partial x\end{bmatrix}$$

$$= i\,\frac{\sin\sigma}{2}\begin{bmatrix}0 & 1\\ 1 & 0\end{bmatrix} = i\begin{bmatrix}0 & \tfrac{1}{2}\sigma_{xt}\\ \tfrac{1}{2}\sigma_{xt} & 0\end{bmatrix} = \hat{L}_{,t} \qquad (112)$$

The first step uses $Lv = \zeta v$ to cancel the ζ^{-1}.

5. Theory of the s-G

The rest of this article is concerned with developing the theory of the s-G in detail.

The s-G is a Hamiltonian flow: the two potentials q,r in the AKNS generalization (104) are $q = -\sigma_x/2$, $r = \sigma_x/2$ and $q = -r = -r^*$. The pair cannot constitute an independent pair of canonical variables. However, in this case, one finds quite generally [68] that

$$H(r,q) = H(q) - \tfrac{1}{2}\int \frac{\delta H}{\delta q}(q - r_{,x})\,dx.$$

Then[7]

$$q_t = \frac{\delta H}{\delta r} = -\tfrac{1}{2}\frac{\partial}{\partial x}\frac{\delta H}{\delta q}$$

$$r_t = -\tfrac{1}{2}\frac{\delta H}{\delta q} + \tfrac{1}{2}\frac{\delta^2 H}{\delta q^2}(q - r_{,x}). \qquad (113)$$

Thus, if $q = r_{,x}$, i.e., $r = \int_{-\infty}^{x} q\,dx'$ initially, then it is true for all time. This again indicates in a natural way that the correct choice for the Poisson bracket is GARDNER's choice (90) and that for the symplectic form is (97).

With this in mind, a symmetrized Hamiltonian for the s-G is [40,68]

$$H = -\tfrac{1}{4}\gamma_0^{-1}\int_{-\infty}^{\infty} dx\,[\cos 2\gamma_0 p + \cos\{-2\int_{-\infty}^{x} q(x',t)dx'\} - 2]$$

---

[6] For the Jost functions one finds that $\lambda(\zeta)$ is effectively defined by $A_{11}(\infty)$. Compare (117) below.

[7] Of course the operator $\partial/\partial x$ differentiates the functions as well as differentiating with respect to the variables x, e.g. $\partial f(q, q_x, \ldots, x)/\partial x = \partial f/\partial x + \partial f(q_x)/\partial q + \partial f(q_{xx})/\partial q_x + \ldots$.

with $p = -\frac{1}{2}\gamma_0^{-1}\sigma$, $q = -\frac{1}{2}\sigma_x$. From this

$$-\tfrac{1}{2}\sigma_{xt} = q_t = \frac{\delta H}{\delta p} = -\tfrac{1}{2}\sin -2\gamma_0\, p = -\tfrac{1}{2}\sin\sigma$$

$$\tfrac{1}{2}\gamma_0^{-1}\sigma_t = -p_t = \frac{\delta H}{\delta q} = -\tfrac{1}{2}\gamma_0^{-1}\int_x^\infty \sin\{-2\int_{-\infty}^{x'} q(x'')dx''\}dx'$$

or

$$\tfrac{1}{2}\gamma_0^{-1}\sigma_{tx} = \tfrac{1}{2}\gamma_0^{-1}\sin\sigma. \qquad (114)$$

The dimensionless number γ_0 is a coupling constant: note that it does not play a role in the equations of motion. It does play a role in the definition of the momentum conjugate to q. Thus, it plays a role in the Poisson brackets and is fundamental to the canonical quantization of the s-G equation.

The scattering problem (104) for the s-G is studied in terms of Jost functions. Consider the Jost function solutions for v, ϕ and $\bar{\phi}$, such that

$$\phi \sim \begin{bmatrix}1\\0\end{bmatrix} e^{-i\xi x}, \qquad \bar{\phi} \sim \begin{bmatrix}0\\1\end{bmatrix} e^{i\xi x}, \quad x \to -\infty;$$

$$\psi \sim \begin{bmatrix}0\\1\end{bmatrix} e^{i\xi x}, \qquad \bar{\psi} \sim \begin{bmatrix}1\\0\end{bmatrix} e^{-i\xi x}, \quad x \to +\infty. \qquad (115)$$

The eigenvalue ξ is real. Since there can be only two linearly independent functions belonging to ξ

$$\phi = a(\xi)\,\bar{\psi}(x,\xi) + b(\xi)\,\psi(x,\xi)$$

$$\bar{\phi} = \bar{b}(\xi)\,\bar{\psi}(x,\xi) + \bar{a}(\xi)\,\psi(x,\xi). \qquad (116)$$

The quantities $w \equiv b(\xi)/a(\xi)$ and $1/a(\xi)$ are the reflexion coefficient and transmission coefficient for functions

$$\bar{\psi} \sim \begin{bmatrix}1\\0\end{bmatrix} e^{-i\xi x}, \quad \text{as} \quad x \to \infty .$$

Both w and a^{-1} can be continued into the upper half ζ-plane: the zeros of $a(\zeta)$ there define the bound states of the scattering problem (104). The key result is that $\hat{A}[(1/a)\phi] = \partial/\partial t[(1/a)\phi] \Rightarrow a$ is a constant of the motion. We demonstrate this explicitly to show what is going on: we have for $x \to +\infty$ that

$$\frac{\partial}{\partial t}\frac{1}{a}\begin{bmatrix}1\\0\end{bmatrix} = (A_0 + f)\frac{1}{a}\begin{bmatrix}1\\0\end{bmatrix}$$

$$\frac{\partial}{\partial t}\frac{b}{a}\begin{bmatrix}0\\1\end{bmatrix} e^{i\xi x} = \begin{bmatrix}A_0 + f & 0\\ 0 & -A_0 + f\end{bmatrix}\left\{\begin{bmatrix}0\\1\end{bmatrix} e^{-i\xi x} + \frac{b}{a}\begin{bmatrix}1\\0\end{bmatrix} e^{+i\xi x}\right\} \qquad (117)$$

from which

$$A_0 = -f, \quad a(\xi,t) = a(\xi,0)$$
$$b(\xi,t) = b(\xi,0)\, e^{-2A_0 t} \qquad (118)$$

and these results can be continued for $\mathrm{Im}\zeta > 0$. Notice that $\hat{A}$ is replaced by $\hat{A} + f(\zeta,t)\hat{I}$, but $[f\,\hat{I}, \hat{L}] = 0$ in agreement with the analysis above (112). Further, only $\lim x \to +\infty$ of $\hat{A}$ is needed. Since the potential $\sigma_x/2$ vanishes for the s-G there, $\lim_{x\to\infty} \hat{A} = \hat{A}_0$ is essentially the linearized dispersion relation. Indeed from (111), since $\sigma \to 0 \pmod{2\pi}$, $|x| \to \infty$, $A_0 = (i/4\zeta)\ \mathrm{diag}\ [1,-1]$, whilst the linearised dispersion relation of the s-G (70) is $\omega = k^{-1} \to A_0 = i/4\zeta$ [see (74)]. The situation is quite general: for example, for the RMB equations (108), $B, C \to 0$ as $x \to \infty$ since $r, s \to 0$ as $x \to \infty$. Then $u \to -1$ and $A_0 = i\zeta\,[4\zeta^2 - \mu^2]^{-1} \equiv \Omega(\zeta)$. In general $\Omega(\zeta) = -\frac{1}{2}\, i\omega_q(-2\zeta) = \frac{1}{2}\, i\omega_r(2\zeta)$ in terms of the linear dispersion relations for q and r of (104) [68]. It is easily checked that $\omega_q(k) = \omega_r(k) = k[k^2-\mu^2]^{-1}$ for the RMB equations. It is clear that $\Omega(L^\dagger,y,t)$ is the natural extension of $\Omega(L^\dagger)$ and hence of the linearized dispersion relation in (106).

With $a(\zeta)$ a constant of the motion established by (118), it is expressed as the Wronskian $a(\zeta) = W(\phi,\psi) \equiv \phi_1\psi_2 - \phi_2\psi_1 \sim \phi_1\, e^{i\zeta x}$ for $x \to \infty$. In $W(\phi,\psi)$, ϕ_1 and ϕ_2 are the components of ϕ, i.e., $\tilde{\phi} = [\phi_1,\phi_2]$ and similarly for ψ. From $\hat{L}v = \zeta v$, the AKNS scattering problem (104) in terms of the potentials q,r, one finds that $\Phi \equiv \ln(\phi_1 \exp i\zeta x)$ satisfies

$$2\, i\zeta\, \Phi_{,x} = \Phi_{,x}{}^2 - q\, r + q\, \frac{\partial}{\partial x}\,(q^{-1}\, \Phi_{,x}). \tag{119}$$

As $|\zeta| \to \infty$, $\Phi \to 1$ so

$$\Phi_{,x} \sim \sum_{n=1}^{\infty} \mathcal{H}_n\, (2i\zeta)^{-n} \tag{120}$$

(say). Then from (119)

$$\mathcal{H}_{n+1} = q\, \frac{\partial}{\partial x}\,(q^{-1}\mathcal{H}_n) + \sum_{j+k=n} \mathcal{H}_j\, \mathcal{H}_k, \quad n = 1,2,\ldots.$$

$$\mathcal{H}_0 = -\, q\, r\, . \tag{121}$$

As $x \to \infty$, $\ln a(\zeta) \sim \Phi$ and since $a(\zeta) \sim 1$ as $|\zeta| \to \infty$

$$\ln a(\zeta) \sim \sum_{n=1}^{\infty} \zeta^{-n}\, (2i)^{-n} \int_{-\infty}^{\infty} \mathcal{H}_n\, dx = \sum_{n=1}^{\infty} \zeta^{-n}\, H_n. \tag{122}$$

Then $\ln a(\zeta)$ a constant of the motion yields $H_n = (2i)^{-n} \int_{-\infty}^{\infty} \mathcal{H}_n\, dx$ are constants of the motion and the $\mathcal{H}_n$ are conserved densities for the s-G $q = -r = -\sigma_x/2$ and these densities are T^2, T^4, etc., in agreement with (77).

By expansion about the ordinary point $\zeta = 0$ one finds a second infinite set of conserved densities. The point $\zeta = 0$ is chosen since it is a pole of the linearized dispersion relation for the s-G $\Omega(\zeta) = \zeta/4$. This choice makes $\Omega(\zeta)$ take its proper form from the evolution equation described by Hamilton's equations of motion. For the s-G this is just the s-G itself, Eq. (70), in light cone coordinates and evolution equation form. This analysis for the s-G is given in Refs. [40] and [68]. However, there is a trivial route to the first conserved density. Observe that $T^2 \equiv \sigma_x{}^2/2$ satisfies

$$\tfrac{1}{4}\gamma_0{}^{-1}\,(\sigma_x{}^2)_t + \tfrac{1}{2}\gamma_0{}^{-1}\,\{\cos\sigma - 1\}_x \;=\; 0 \tag{123}$$

because $\sigma_{xt} = \sin\sigma$. But this s-G equation is invariant under interchange of x and t (Lie invariance [74]) and therefore

$$\tfrac{1}{2}\gamma_0{}^{-1}\,\{\cos\sigma - 1\}_t + \tfrac{1}{4}\gamma_0{}^{-1}\,(\sigma_t{}^2)_x \;=\; 0. \tag{124}$$

The density $\frac{1}{2}\gamma_0^{-1}\{\cos\sigma - 1\}$ is the first of the sequence obtained by expansion of $\ell n\, a(\zeta)$ about $\zeta = 0$ [40,68], We now have in $\frac{1}{4}\gamma_0^{-1}\sigma_x^2$ a momentum density and in $\frac{1}{2}\gamma_0^{-1}\{\cos\sigma - 1\}$ a Hamiltonian density. We wish to express these in terms of the scattering data for the following reason: the scattering problem $\hat{L}v = \zeta v$ in (104) is again a canonical transformation; and the Hamiltonian in terms of scattering data again takes the simpler form of action-angle type depending only on the momenta. This is exactly the situation found by ZAKHAROV and FADEEV for the KdV [65] [compare (101)]. The expression of the Hamiltonian H and the momentum P in terms of scattering data is achieved by expressing $\ell n\, a$ in terms of the scattering data. This is done through CAUCHY's theorem [68]. The result, which is (129) below, shows that the s-G $\sigma_{xt} = \sin\sigma$ is a completely integrable infinite dimensional Hamiltonian system. Then by (130) the s-G (66) is also.

The symplectic form (97) in the case $u = q$, $p = \gamma_0\int_{-\infty}^{x} q\, dx'$ transforms into [40,68]

$$\bar{\bar{\omega}} = \sum dp_j \wedge dq_j + \int_{-\infty}^{\infty} d\xi [d\, P(\xi) \wedge dQ(\xi)]. \tag{125}$$

One set of complex canonical coordinates is defined in terms of the scattering data [40,68] by

$$\begin{aligned} &p_j = \gamma_0\, \ell n\, \zeta_j, \quad q_j = 2\ell n\, b_j, \quad j = 1,2,\dots,M. \\ &P(\xi) = (2\pi\, \xi\gamma_0)^{-1}\, \ell n\, \{a(\xi)\bar{a}(\xi)\}, \quad Q(\xi) = \arg b(\xi). \end{aligned} \tag{126}$$

The symmetrized Hamiltonian and momentum derived from the densities $\frac{1}{2}\gamma_0^{-1}$ $(\cos\sigma - 1)$ and $\frac{1}{2}\gamma_0^{-1}\sigma_x^2$, respectively, can be put in the forms [compare (114) and note we have now introduced a mass m in H]

$$\begin{aligned} &H(p,q) = -\frac{m^2}{4\gamma_0}\int_{-\infty}^{\infty} \{\cos(-2\int_{-\infty}^{x} q\, dx') + \cos(-2\gamma_0\, p) - 2\}\, dx \\ &P(p,q) = \frac{1}{2\gamma_0}\int_{-\infty}^{\infty} (q^2 + \gamma_0^2\, p_x^2)\, dx\, . \end{aligned} \tag{127}$$

By expressing these conserved quantities in terms of the scattering data [40,68] they can be expressed in the forms

$$\begin{aligned} &H = -\frac{m^2 i}{2\gamma_0}\left\{-2\sum_{j=1}^{K} (\zeta_j^{-1} - \xi_j^{*-1}) + 2i\sum_{j=1}^{L} \eta_j^{-1} + i\gamma_0 \int_{-\infty}^{\infty} d\xi\, \xi^{-1}\, P(\xi)\right\} \\ &P = +\frac{2i}{\gamma_0}\left\{-2\sum_{j=1} (\zeta_j - \zeta_j^*) - 2i\sum_{j=1}^{L} \eta_j - i\gamma_0 \int_{-\infty}^{\infty} d\xi\, \xi\, P(\xi)\right\} \end{aligned} \tag{128}$$

in which $2K + L = M$ [compare (126)]. The result for P suggests new momenta different from those quoted in (126) for which

$$\begin{aligned} &H = \sum_{j=1}^{K} \hat{h}_j + \sum_{j=1}^{L} h_j + \int_{-\infty}^{\infty} d\xi\, h(\xi)\, P(\xi) \\ &P = \sum_{j=1}^{K} \hat{p}_j + \sum_{j=1}^{L} p_j + \int_{-\infty}^{\infty} d\xi\, p(\xi)\, P(\xi) \end{aligned} \tag{129}$$

where $h_j p_j = 4m^2\gamma_0^{-2}$, $\hat{h}_j\hat{p}_j = 64m^2\gamma_0^{-2}\sin^2\theta$, $h(\xi)p(\xi) = m^2$ and $\theta_j = \arg\zeta_j$. These momenta with corresponding coordinates $\hat{q}_j, q_j$, etc., also close (125). Define now still new energies and momenta by $h(\xi) = \frac{1}{2}\{h'(\xi) + p'(\xi)\}$, $p(\xi) = \frac{1}{2}\{h'(\xi) - p'(\xi)\}$ and define similar quantities for h_j, p_j and $\hat{h}_j, \hat{p}_j$. Define a new mass $m' = 2m$ and drop the primed notation. Then

$$H = \sum_{j=1}^{K} [256m^2\gamma_o^{-2}\sin^2\theta_j + \hat{p}_j^2]^{\frac{1}{2}} + \sum_{j=1}^{L} [64m^2\gamma_o^{-2} + p_j^2]^{\frac{1}{2}}$$

$$+ \int_{-\infty}^{\infty} [m^2 + p^2(\xi)]^{\frac{1}{2}} P(\xi)\, d\xi$$

$$P = \sum_{j=1}^{K} \hat{p}_j + \sum_{j=1}^{L} p_j + \int_{-\infty}^{\infty} d\xi\, p(\xi)\ \ P(\xi). \tag{130}$$

The expression for H in this form was first achieved by TAKTADJAN and FADEEV [73] who used the LAX pair of first-order 4×4 matrix operators mentioned earlier. The route used here is via the 2×2 ZS-AKNS scattering problem (104). We can follow it because in light cone coordinates the s-G (66) is the evolution equation $u_t = \int_{-\infty}^{x} \sin u\, dx'$ in which the right side is a functional of u.

The expression (130) transforms the Hamiltonian of (127) to that for a collection of free relativistic particles of masses m, $8m\gamma_o^{-1}$ (the s-G kinks) and $16m\gamma_o^{-1}\sin\theta$ (the s-G breathers). The transformation is canonical and the result exact. It is an easy matter to canonically quantize it since the canonical coordinates q_j, $\hat{q}_j$, $q(\xi)$ can be found from the symplectic form. One finds the canonical pair $\gamma_o^{-1}\theta_j$, $4 \arg b_j$ for the internal coordinates of the breather. The phase space $0 \le \theta_j \le \pi/2$, $0 \le 4 \arg b_j \le 8\pi$ is compact with volume $4\pi^2\gamma_o^{-1}$. The corresponding breather spectrum in (129) is therefore discrete. One finds [40] a discrete set of breather masses for (130) given by

$$M_n = \frac{16m}{\gamma_o} \sin \frac{n\gamma_o}{16} \qquad n = 1,2,\dots,N \tag{131}$$

where N is the largest integer $\le 8\pi\gamma_o^{-1}$. This beautiful result has been reached by a number of authors by very different routes [75].

The s-G Hamiltonian in the form (130) provides an excellent model for the statistical mechanics of a soliton system. It is clear that the solitons and breathers are new elementary excitations; the continuum contribution is phonon-like. The complication in (130) is that although the Hamiltonian is separable, it depends on the initial data. In particular, the number of solitons L, of breathers K, and the occupation of phonon-like states $P(\xi)$ depends on the initial data. It is therefore necessary to trace over all sets of available initial data, namely over the set of C^∞ functions decaying exponentially at $x = \pm\infty$. BISHOP [76] takes the somewhat different view involving (x,t) space rather than momentum space where interactions between solitons, breathers and 'phonons' appear as phase shifts and emptied phonon states. It remains an interesting technical problem to ally the two different points of view.

Notice that, because the soliton and breather masses depend on γ_o^{-1}, perturbation theory in γ_o must be singular perturbation theory about sine-Gordon solitons and breathers (see [77] and [78]).

Finally, we note that we cannot hope to indicate all the ways in which the inverse spectral transform method is now being extended. We simply refer the reader to the paper [70] by NEWELL, recent work on the prolongation structure method of WAHLQUIST and ESTABROOK by DODD [79] and others, the work quoted by MANIN [60] and two remarkable papers by ZAKHAROV [80,81], especially.

This lecture II was published in part elsewhere [82]. An error of sign in the s-G Hamiltonian of (114) has been corrected and the sign of p is changed. This error occurs also in Pt.II of [40].

References

1. J.S. Russell, Report on Waves. British Association Reports (1844)
2. D.J. Korteweg and G. de Vries, Phil. Mag. 39, 422 (1895)
3. R.K. Bullough and P.J. Caudrey, 'History of the soliton' in "Solitons", Springer Topics in Modern Physics Series. R.K. Bullough and P.J. Caudrey Eds. (Springer-Verlag, Heidelberg) To be published 1978.
4. G.B. Whitham, Linear and Non-linear Waves (John Wiley & Sons, New York, 1974) pp.580-585
5. J.D. Gibbon and J.C. Eilbeck, J.Phys.A: Gen. Phys. 5, L132 (1972)
6. R.K. Dodd and R.K. Bullough, Proc. Roy. Soc. A 351, 499 (1976)
7. R.K. Bullough, Solitons in Interaction of Radiation and Condensed Matter Vol.I, IAEA-SMR-20/51 (International Atomic Energy Agency, Vienna, 1977) pp.381-469
8. J.M. Burgers, Adv. Appl. Mech. 1, 171 (1948)
9. S. Coleman, Classical lumps and their quantum descendants. Lectures at the 1975 International School of Subnuclear Physics "Ettore Majorana" (1975)
10. R. Jackiw, Rev. Mod. Phys. 49, 681 (1977)
11. A.A. Belavin and V.E. Zakharov, Pis'ma Zh. Eksp. Teor. Fiz. 25, 603 (1977) (JETP Lett. 25, 567 (1978))
12. H. Haken, in Synergetics in Cooperative Phenomena Edited by H. Haken (North Holland, Amsterdam, 1974)
13. J.A. Krumhansl and J.R. Schrieffer, Phys. Rev. B 11, 3535 (1975)
14. R.F. Dashen, B. Hasslacher and A. Neveu, Phys. Rev. D 10, 4130 (1974)
15. P.J. Caudrey, J.C. Eilbeck and J.D. Gibbon, Il Nuovo Cimento 25, 497 (1975)
16. J.D. Gibbon and J.C. Eilbeck, J. Phys.A: Gen. Phys. 5, L122 (1972)
17. P.J. Caudrey, J.C. Eilbeck and J.D. Gibbon, J. Inst. Math. Applics. 14, 375 (1975)
18. R. Hirota, Phys. Rev. Lett. 27, 1192 (1971)
19. M.J. Ablowitz, D.J. Kaup, A.C. Newell and H. Segur, Phys. Rev. Lett. 31, 125 (1973)
20. M.J. Ablowitz, D.J. Kaup, A.C. Newell and H. Segur, Studies in Appl. Math. 53, 249 (1974)
21. E. Fermi, J.R. Pasta and S.M. Ulam, Studies of nonlinear problems I, Los Alamos Rept. LA-1940 (May 1955) and Collected Works of E. Fermi Vol.II (Univ. of Chicago Press, 1965) pp.978-88
22. J. Gibbons, S.G. Thornhill, M.J. Wardrop, and D. ter Haar, On the theory of Langmuir Solitons, preprint Univ. of Oxford, Dept. of Theoretical Phys. Ref.36/76 (1976)
23. V.E. Zakharov, Zh. Eksp. Teor. Fiz. 62, 1745 (1972) (Soviet Phys. J.E.T.P. 35, 908 (1975)
24. G.L. Lamb, Rev. Mod. Phys. 43, 99 (1971)
25. J.C. Eilbeck, P.J. Caudrey, J.D. Gibbon and R.K. Bullough, J.Phys.A: Math. Nucl. & Gen. 6, 1337 (1973)

26. R.K. Bullough, P.M. Jack and P.W. Kitchenside, Physica Scripta (1978)

27. R.F. Dashen, B. Hasslacher and A. Neveu, Phys. Rev. D 11, 3424 (1975)

28. B.B. Kadomtsev and V.I. Petviashvili, Dokl. Akad. Nauk. SSR, 192, 753 (1970)

29. V.E. Zakharov and A.B. Shabat, Funkt. Anal. i Ego Prilozh. 8, 43 (1974)

30. S.V. Manakov, V.E. Zakharov, L.A. Bordag, A.R. Its and V.B. Matveev, Phys. Lett. 63A, 205 (1977)

31. H.C. Yuen and B.M. Lake, Phys. Fluids 18, 956 (1975)

32. V.E. Zakharov and A.B. Shabat, Zh. Eksp. Teor. Fiz. 64, 1627 (1973) (Sov. Phys. JETP 37, 823 (1973))

33. R.P. Feynman, R.B. Leighton and M. Sands, The Feynman Lectures on Physics Vol.III (Addison-Wesley, Reading, Mass. 1965) pp.21.14-21.18

34. H.M. Gibbs and R.E. Slusher, Phys. Rev. A 6, 2326 (1972)

35. T.A. Fulton and R.C. Dynes, Solid State Comm. 12, 57 (1973)

36. S.P. Novikov, Funkt Anal. i Ego Prilozh. 8, 54 (1974)

37. P.D. Lax, Comm. Pure and Appl. Maths. 28, 141 (1975).

38. R.K. Bullough, P.J. Caudrey, J.D. Gibbon, S. Duckworth, H.M. Gibbs, B. Bölger and L. Baede, Optics Communications 18, 200 (1976)

39. R.K. Bullough and P.J. Caudrey, Bumping spin waves in the B-phase of liquid ^{3}He. Preprint (April, 1977)

40. R.K. Bullough and R.K. Dodd, Solitons I Basic Concepts. II Mathematical Structures in Synergetics. A Workshop ed. by H. Haken (Springer-Verlag, Heidelberg, 1977) pp.92-103 and 104-119

41. N. Zabusky and M.D. Kruskal, Phys. Rev. Lett. 15, 240 (1965)

42. C.S. Gardner, J.M. Greene, M.D. Kruskal and R.M. Miura, Phys. Rev. Lett. 19, 1095 (1967)

43. J. Liouville, J. Mathematiques Pures et Appliquées (Paris) 18, (1) 71-72 (1853)

44. R.K. Dodd and R.K. Bullough, Proc. Roy. Soc. (London) A 351, 499 (1976)

45. R.K. Dodd and R.K. Bullough, "Integrability of nonlinear evolution equations: prolongations and solitons" To be published

46. Details communicated by W. Shadwick, Warsaw Meeting, September (1977)

47. F. Lund, Phys. Rev. Lett. 38, 1175 (1977); Proc. of NATO Advanced Study Institute on Nonlinear Problems in Physics and Mathematics (Istanbul, August 1977), ed. by A.O. Barut (D. Reidel Publishing Co., Dordrecht, Holland 1978)

48. V.E. Zakharov and A.B. Shabat, Zh. Eksp. Teor. Fis. (Soviet) 61, 118 (1971); JETP (Soviet) 34, 62 (1972)

49. K. Pohlmeyer, Comm. Math. Phys. 46, 207 (1976); New Developments in Quantum Field Theory and Statistical Mechanics, ed. by M. Levy and P. Nitter (Plenum Press, New York 1977) p.339

50. H.D. Wahlquist and F.B. Estabrook, J. Math. Phys. 16, 1 (1975); 17, 1293 (1976)

51. V.L. Ginzburg and L.P. Pitaevskii, JETP 34, 1240 (1958); L.P. Pitaevskii, JETP 35, 408 (1968); R.Y. Chiao, E. Garmire and C.H. Townes, Phys. Rev. Lett. 13, 479 (1964); P.L. Kelley, Phys. Rev. Lett. 15, 1005 (1965); T.B. Benjamin and J.E. Feir, J. Fluid. Mech. 27, 417 (1966); D.J. Benney and A.C. Newell, J. Math. Phys. 46, 133 (1967); V.I. Bespalov, A.G. Litvak and V.I. Tulanov, Nauka, 2nd All-Union Symposium on Nonlinear Optics, Collection of Papers, (Russian)(Moscow, 1968)

52. M.D. Kruskal, in Proceedings of Symposium on Nonlinear Evolution Equations Solvable by the Inverse Spectral Transform, Accademia dei Lincei, Rome, June 1977, ed. by F. Calogero (Pitman , London 1978); Proceedings of NATO Advanced Study Institute on Nonlinear Problems in Physics and Mathematics (Istanbul, August 1977) ed. by A.O. Barut (D. Reidel Publishing Co., Dordrecht, Holland 1978)

53. P.D. Lax, Comm. Pure Appl. Maths. 21, 467 (1968)

54. R.K. Bullough, in Proceedings of NATO Advanced Study Institute on Nonlinear Problems in Physics and Mathematics (Istanbul, August 1977), ed. by A.O. Barut (D. Reidel Publishing Co., Dordrecht, Holland 1978)

55. R.M. Miura, C.S. Gardner and M.D. Kruskal, J. Math. Phys. 9, 1204 (1968)

56. R.M. Miura, J. Math. Phys. 9, 1202 (1968)

57. F. Calogero, Nonlinear evolution equations solvable by the inverse spectral transform , invited lecture presented at International Conference on Mathematical Problems in Theoretical Physics, Rome University (June 6-15, 1977); A.C. Newell, The Inverse Scattering Transform, in Solitons, Springer Topics in Modern Physics Series, ed. by R.K. Bullough and P.J. Caudrey (Springer-Verlag, Berlin, Heidelberg, New York, 1978)

58. R. Hirota, in, for example, Direct Methods in Soliton Theory, in Solitons Springer Topics in Modern Physics Series, ed. R.K. Bullough and P.J. Caudrey (Springer-Verlag, Berlin, Heidelberg, New York, 1978); P.J. Caudrey, Proceedings of NATO Advanced Study Institute on Nonlinear Problems in Physics and Mathematics (Istanbul, August 1977), ed. by A.O. Barut (D. Reidel Publishing Co., Dordrecht, Holland, 1978); R. Hirota, Phys. Rev. Lett. 27, 1192 (1971); P.J. Caudrey, J.C. Eilbeck and J.D. Gibbon, J. Inst. Maths. Applics. 14, 375 (1974)

59. I.M. Gel'fand and L. Dikii, Uspeki mat. nauk. 30, 67 (1975); Russian Maths. Surveys 30, 77 (1975), Funkt. Anal. i Ego Prilog. 10, 18 (1976)

60. Yu. I. Manin , Itogi Nauki i Tekniki 11, 5 (1978)

61. S.P. Novikov, Funkt. Anal. i Ego Prilozh. 8, (3) 54 (1974); B.A. Dubrovin, I.M. Krichever and S.P. Novikov, Dokl. AN SSSR(1976)

62. C.S. Gardner, J. Math. Phys. 12, 1548 (1971)

63. H. Flanders: Differential Forms with Applications to the Physical Sciences (Academic Press, New York, 1963)

64. This argument and some earlier remarks were partly stimulated by access to notes by Prof. David Simms on Lectures by H.P. McKean at Calgary, 1978. We have not been able to check original sources and rely on our recollections

65. V.E. Zakharov and L.D. Fadeev, Funkt. Anal. i Ego Prilozh. 5, 18 (1971)

66. M. Toda, in Studies on a Nonlinear Lattice, Ark. for Der Fysiske Sem., Trondheim 2 (1974); On a Nonlinear Lattice - the Toda Lattice in Solitons, Springer Topics in Modern Physics Series, ed. by R.K. Bullough and P.J. Caudrey (Springer-Verlag, Berlin, Heidelberg, New York, 1978)

67. F. Calogero, A. Degasperis, in Solitons, Springer Topics in Modern Physics Series, ed. by R.K. Bullough and P.J. Caudrey (Springer-Verlag, Berlin, Heidelberg, New York, 1978); Proc. NATO Advanced Study Institute on Nonlinear Problems in Physics and Mathematics (Istanbul, August 1977), ed. by A.O. Barut (D. Reidel Publishing Co., Dordrecht, Holland 1978)

68. R.K. Dodd and R.K. Bullough, The Generalized Marchenko Equation and the Canonical Structure of the AKNS-ZS Inverse Method (to be published 1978)

69. R.K. Dodd and R.K. Bullough, Phys. Lett. 62A, 70 (1977)

70. A.C. Newell, "The general structure of integrable evolution equations" (Preprint 1977)

71. J.D. Gibbon, P.J. Caudrey, R.K. Bullough and J.C. Eilbeck, Lett. al Nuovo Cimento 8, 775 (1973)

72. M.J. Ablowitz, D.J. Kaup, A.C. Newell and H. Segur, Phys. Rev. Lett. 30, 1262 (1973)

73. L.A. Taktadjan and L.D. Fadeev, Teor. Mat. Fis. 21, 160 (1974)

74. R.K. Dodd and R.K. Bullough, Proc. Roy. Soc. (London) A 352, 481 (1977)

75. R.F. Dashen, B. Hasslacher and A. Neveu, Phys. Rev. D 11, 3424 (1975); V.E. Korepin and L.D. Fadeev, Teor. Mat. Fis. 25, 47 (1975); A Luther, Phys. Rev. B. 14, 2153 (1976)

76. A.R. Bishop, in Solitons and Condensed Matter Physics, ed. by A.R. Bishop and T. Schneider (Springer-Verlag, Heidelberg, 1978)

77. A.C. Newell, in Solitons and Condensed Matter Physics, ed. by A.R. Bishop and T. Schneider (Springer-Verlag, Heidelberg, 1978)

78. P.W. Kitchenside, A.L. Mason, R.K. Bullough and P.J. Caudrey, in Solitons and Condensed Matter Physics, ed. by A.R. Bishop and T. Schneider (Springer-Verlag, Heidelberg, 1978)

79. R.K. Dodd, Proc. NATO Advanced Study Institute on Nonlinear Problems in Physics and Mathematics (Istanbul, August 1977), ed. by A.O. Barut (D. Reidel Publishing Co., Dordrecht, Holland, 1978)

80. V.E. Zakharov and A.B. Shabat, Funkt. Analiz. i Ego Prilozh. 8, 43 (1974)

81. V.E. Zakharov, The inverse scattering method in Solitons, Springer Topics in Modern Physics Series, ed. by R.K. Bullough and P.J. Caudrey (Springer-Verlag, Berlin, Heidelberg, New York, 1978)

82. R.K. Bullough and R.K. Dodd, Solitons in Mathematics: Brief History in Solitons and Condensed Matter Physics, ed. by A.R. Bishop and T. Schneider (Springer-Verlag, Heidelberg, 1978).

Spin Wave Breathers in ^{3}He B

P.W. Kitchenside, R.K. Bullough, and P.J. Caudrey

Department of Mathematics, University of Manchester,
Institute of Science and Technology, P.O. Box 88
Manchester M60 1QD, U.K.

We have already reported elsewhere [1] a short study of the following problems: the solution of the double sine-Gordon equations [2,3,4]

$$u_{xx}-u_{tt} = \mp (\sin u+\tfrac{1}{2}\lambda\sin\tfrac{1}{2}u) \tag{1}$$

for boundary conditions $u \to c$, u_x, u_{xx}, etc $\to 0$, as $|x| \to \infty$, and initial data $u(x,0) = c$; $u_t(x,0) = a$, $|x| < \ell$, $u_t(x,0) = 0$, $|x| > \ell$. The number c is a particular chosen zero of $\sin u+\frac{1}{2}\lambda\sin\frac{1}{2}u$.[1] Most of the paper [1] was concerned with the case -ve sign and $\lambda=1$. We study this same problem again in the present paper, but we report additional results which provide a more comprehensive view of the behaviour of equations (1) under their chosen boundary conditions. Especially we are concerned with the creation and behaviour of breather solutions.

There are two cases of equations (1) which are directly relevant to the spin waves in the superfluid phases of ^{3}He below 2.6mK: these are the -ve sign and $\lambda=1$, and the +ve sign and $\lambda=0$. The latter is an initial value problem for the sine-Gordon equation (s-G) and can be solved by an inverse scattering method [1,5].[2] The double sine-Gordon equations which arise for $\lambda\neq 0$ are not solvable by any of the techniques presently available for solving nonlinear evolution equations [3,4,5,6]: there are, for example only three conservation laws [3] and only one of these is polynomial [3]; the systems are not completely integrable Hamiltonian systems [7]. Evidently only singular perturbation theory [8,9] and numerical integration are available to solve this problem. Despite success with perturbation theories for the case of the positive sign [8] we do not yet know how to handle similar perturbation theory for the negative sign: certainly $u_{xx}-u_{tt} = -\sin u$ for boundary conditions $u \to 0$ (mod 2π) is unstable and its multisoliton solutions are unstable, and we have not yet investigated how the perturbation is stabilised by the $\frac{1}{2}\lambda\sin\frac{1}{2}u$ term. This paper as did its predecessor [1] therefore confines its report to results of numerical work.

We mention very briefly the two cases of (1) with +ve sign and $\lambda=+1$. The boundary conditions and initial values have $c=0$ in the one case and $c = 2\pi$ in the other. The 'solitary wave' solutions in these two cases are

$$u = 4(\tan^{-1}e^{\Theta+\Delta_1} + \tan^{-1}e^{\Theta-\Delta_1}) \qquad (c=0) \tag{2a}$$

$$u = 2\pi+4(\tan^{-1}e^{\Theta+\Delta_2} - \tan^{-1}e^{\Theta-\Delta_2}) \qquad (c=2\pi) \tag{2b}$$

with

$$\Theta = (x-Vt)(1-V^2)^{-\frac{1}{2}} + \Theta_0, \qquad \Delta_1 = \ln(\sqrt{5}+2), \quad \Delta_2 = \ln(\sqrt{3}+2). \tag{2c}$$

---

[1] Because of a typing error c is given as zero (i.e. $c=0$) in all the problems posed in [1]. But in the examples studied c is taken as the appropriate zero of the right side.

[2] The most complete reference for a 2×2 inverse method for the s-G (1) in laboratory co-ordinates (Lorentz covariant form) is Kaup [17].

The essential point is that these solutions can be described in terms of the 2π-kinks and antikinks of the s-G and consist of a bound 2π-kink-kink pair (c=0 - evidently c=0 (mod 4π) only) and a bound 2π-kink-antikink pair ($c=2\pi$). We have noted elsewhere [2,4,6,7,8] that both bound pairs can wobble. Fig.1 shows the wobbling 4π-kink for c=0 [3]; Fig.2 shows the wobbling 0π-kink for $c=2\pi$. The latter is unstable to a sufficient separation of the 2π-kink and antikink in the initial data [8]. Both systems have been well described by singular perturbation theory [8,9,10]. The 4π wobbler (c=0) has been observed in the self-induced transparency (SIT) of degenerate D_1 transitions in sodium vapour [11]: in SIT the electric field envelope of an optical pulse is related to u_t and this is evidently double peaked [2]; a wobbling double peaked pulse has been observed [11]. The Fig.1 in reference [7] (these Proceedings) taken from [2] shows a numerical simulation of the break up of an 8π pulse into two such 4π wobblers.

We turn to the case -ve sign and $\lambda=1$. This case is strikingly different from the case of +ve sign. A Hamiltonian density is [12] $\mathcal{H} = \frac{1}{2}u_t^2+\frac{1}{2}u_x^2+2(\cos\frac{1}{2}u+\frac{1}{4})^2$. The potential energy minima are at either of the two roots δ and $4\pi-\delta$ of $u = 2\cos^{-1}(-1/4)$ and each is a zero of $-(\sin u+\frac{1}{2}\sin\frac{1}{2}u)$. There are two kink-like 'solitary wave' solutions which respectively take u from $u=\delta$ to $4\pi-\delta$ ("$4\pi-2\delta$" kink) and from $4\pi-\delta$ to $4\pi+\delta = \delta$ (mod 4π) ("2δ" kink). The energy minima at δ and $4\pi-\delta$ are <u>maxima</u> for the case of +ve sign and $\lambda=1$; the minima occur at 0, 2π (and 4π) in <u>this case</u> and for small λ at least one can expect that the problem can indeed be discussed in terms of the kinks of the s-G. Equally for the -ve sign and $\lambda=1$ the problem does not obviously easily relate to the s-G and one does not expect the kink solutions to be describable easily in terms of the 2π-kinks and antikinks of the s-G.

Explicitly the $4\pi-2\delta$ and 2δ kinks of the double s-G with -ve sign and $\lambda=1$ are

$$u = 2\pi+4\tan^{-1}(\sqrt{3/5}\ \tanh\tfrac{1}{2}\Theta),\quad u = 4\tan^{-1}(\sqrt{5/3}\ \tanh\tfrac{1}{2}\Theta) \qquad (3)$$

respectively: the arguments Θ are $\Theta = \kappa(x-Vt) + \Theta_0$ and $\kappa = \sqrt{15/16}(1-V^2)^{-\frac{1}{2}}$. In the rest frame these kinks have 'masses' 5.1097 and 11.3929 units. One easily realises (and see [12]) (i) that with boundary conditions $c=\delta$ i.e. $u \to \delta$ (mod 4π) and reading from left $(x \to -\infty)$ to right $(x \to +\infty)$ the $4\pi-2\delta$ kink must be followed by the 2δ kink, the order cannot be reversed and such kinks will bump in collision; (ii) with the same boundary conditions a $4\pi-2\delta$ kink-antikink pair can convert to a -2δ antikink-kink pair if the colliding $4\pi-2\delta$ pair has enough kinetic energy to produce the 2δ-pair rest mass. The threshold for this is V=0.8938. For a region of V below this threshold the $4\pi-2\delta$ kink-antikink pair bumps [4,6,12]. The behaviour just above threshold is particularly interesting [4,6,12] because of the loss of energy by radiation. The effective threshold is consequently about V=0.92 (see Fig.1b for V=0.925 in [6] or the Fig.11 of [4]).

The behaviour for much smaller V is also interesting. Kink-antikink pairs attract and could bind for small enough V. The Fig.3 shows a long lived breather-like state obtained by letting a $4\pi-2\delta$ kink and antikink with zero initial relative velocity (V=0) fall into each other. The state is long lived but the small amount of radiation shows that this bound state will have a finite lifetime. Evidently there is a threshold for the formation of this bound $4\pi-2\delta$ kink-antikink breather-like state. We find this to be close to V=0.35. The Fig.4 shows the behaviour of the $4\pi-2\delta$ kink antikink pair on this threshold of V=0.35. On the otherhand the 2δ-kink antikink pair (boundary condition $c=4\pi-\delta$ i.e. $u \to 4\pi-\delta$ $|x| \to \infty$) converts to a $-(4\pi-2\delta)$ antikink-kink pair with the emission of kinetic energy for all values of V. This kinetic energy is sufficient to cross any threshold for the formation of 2δ-kink-antikink breathers and such a breather is not formed.

We turn to the physics of spin waves in ^{3}He and study a possible means of creating them. A survey of the physics of spin waves in the A-phase (sine-Gordon case, +ve sign and $\lambda=0$) and the B-phase (double sine-Gordon case, -ve sign and $\lambda=1$)

---

[3] The Figures appear together at the end of the paper

must necessarily be brief. We have elsewhere [4,6,12] introduced an 'adiabatic' Hamiltonian density $\mathcal{H}$ which generalizes that proposed by Leggett [13] to describe the unusual NMR behaviours observed. We derive from this the equations of motion

$$\theta_t = \frac{\delta\mathcal{H}}{\delta r_3} = -\gamma B + \gamma^2\chi^{-1} r_3$$
$$r_{3,t} = -\frac{\delta\mathcal{H}}{\delta\theta} = \gamma^{-2}\chi\bar{c}^2\theta_{xx} - \frac{\delta H_D}{\delta\theta} \tag{4}$$

and therefore

$$\theta_{tt} - \bar{c}^2\theta_{xx} = -\gamma^2\chi^{-1}\delta H_D/\delta\theta \tag{5}$$

in one space dimension x (we confine discussion to this case). The parameters are: $\gamma = e/m_p c$ = gyromagnetic ratio (c=velocity of light), χ=static susceptibility, $\bar{c}$ a velocity $\sim v_F$; θ and r_3 are a pair of canonically conjugate variables such that γr_3 describes the magnetisation (r_3 is the expectation value of a spin one operator σ_W : σ_W has the eigenvalues $S_W^3 = \pm 1, 0$); B is a homogeneous external magnetic field imposed along x. Important results are [12,13] that the dipole interactions H_D are $H_D = -(3/5)g_D(T)\cos^2\theta$ (A-phase) and $H_D = +4/5g_D(T)(\cos\theta + \cos 2\theta)$ (B-phase). The first case yields the sine-Gordon equation in $u \equiv -2\theta$; the second case yields the double sine-Gordon equation, with negative sign and $\lambda=1$, in $u=-2\theta$.

In order to create spin waves we suppose that an additional inhomogeneous field $\Delta B_0(x)$ is removed at $t=0$. For $t>0$ the field B is a constant field B_0. It is easy to see from (4) that there is a jump induced in r_3, Δr_3, and a motion of θ which satisfy

$$\theta_t = \gamma^2\chi^{-1}\Delta r_3 - \gamma\Delta B_0(x)$$
$$\Delta r_{3,t} = -\delta H_D/\delta\theta + \gamma^{-2}\chi\bar{c}^2\theta_{xx} \tag{6}$$

with $\theta = \theta_0$ (a minimum of H_D), $\theta_t = -\gamma\Delta B_0(x)$, $\Delta r_3 = 0$, $\Delta r_{3,t} = 0$ at $t=0$ and $\theta \to \theta_0$, $\theta_x \to 0$, etc., as $|x| \to \infty$ for all $t \geq 0$. This is equivalent to $\theta_{tt} = -\gamma^2\chi^{-1}\delta H_D/\delta\theta + \bar{c}^2\theta_{xx}$ for $t \geq 0$, with $\theta = \theta_0$, $\theta_t = -\gamma\Delta B_0(x)$ at $t=0$, and $\theta \to \theta_0$, $\theta_x \to 0$, etc., as $|x| \to \infty$.

The A-phase (sine-Gordon) problem has a homogeneous case (recall $u \equiv -2\theta$ and note that we use $\Omega_{\ell A}$ for the longitudinal NMR frequency [12,13])

$$u_{tt} = -\Omega_{\ell A}{}^2\sin u;\ u = 0,\ u_t = 2\gamma\Delta B_0 \quad \text{at} \quad t=0. \tag{7}$$

However, for $u \to 0$, $u_x \to 0$ etc., as $|x| \to \infty$ for all $t \geq 0$, the solution does not remain homogeneous. It does so for the open-ended boundary condition $u_x = 0$, $|x| = \ell$ for all $t \geq 0$. The solutions in this case are $u = 2\cos^{-1}\mathrm{cn}(\omega t, k)$, $\omega = \gamma\Delta B_0$ and $0<k<1$; $u = \pi + 2\cos^{-1}\mathrm{dn}(\omega t, k^{-1})$, $\omega = \Omega_{\ell A}$ and $k>1$; $k \equiv \Omega_{\ell A}\gamma^{-1}\Delta B_0{}^{-1}$ is the modulus of the Jacobian elliptic function cn and k^{-1} of dn. When $k=1$, $u = 2\cos^{-1}\mathrm{sech}\,\omega t$. The magnetisation therefore rings with the frequencies of the Jacobian elliptic functions. These frequencies are $\pi\gamma\Delta B_0 K^{-1}$ for $\gamma B_0 > \Omega_{\ell A}$ and $\pi\Omega_{\ell A}(2K)^{-1}$ for $\gamma\Delta B_0 < \Omega_{\ell A}$ in which $K(k)$ is the complete elliptic integral of the first kind [$K(1) = \infty$, $K(0) = \pi/2$]. For small k, linearised theory, the longitudinal NMR has a steady oscillation at the frequency $\Omega_{\ell A}$ which is independent of B_0 and ΔB_0. This is in agreement with the unusual transverse and longitudinal NMR behaviour observed in the A-phase [13,14] (The transverse theory is not contained in the present analysis.)

Notice that when $k=1$ the trajectory $u = 2\cos^{-1}\mathrm{sech}\,\Omega_{\ell A}t$, $u_t = 2\Omega_{\ell A}\mathrm{sech}\,\Omega_{\ell A}t$, does not ring. This defines a critical field for no ringing $\gamma\Delta B_{0c} = \Omega_{\ell A}$. WHEATLEY [14] has measured ΔB_{0c} following an analysis similar to that given here: agreement between theory and experiment is not perfect. However, the assumption of homogeneity and the boundary condition $u_x = 0$, $|x| = \ell$ were not discussed. Physical boundary conditions certainly involve the effect of $\hat{\underset{\sim}{\ell}}$ vector textures [15] consistent with the condition $\hat{\underset{\sim}{\ell}}$ normal to the end surfaces. This problem is too hard to consider here. There is also the problem of forming composite 'solitons' [15,16] and the fact that the orbital vector $\underset{\sim}{\ell}$ does not necessarily provide a uniform background as a spin excitation passes (the length ℓ for e.g. $|x| = \ell$ has nothing to do with the orbital vector $\underset{\sim}{\ell}$ - see [12] for the role of $\underset{\sim}{\ell}$ in the A-phase). Here we assume $\underset{\sim}{\ell}$ does provide a uniform background and in order to determine one possible effect of boundary conditions we impose the <u>fixed ends</u> condition $u=0$ at $|x| = \ell$.

The sine-Gordon equation has been solved in a form readily applicable at present only for solutions $u(x,t)$ defined on the <u>real line</u> $-\infty<x<\infty$. We therefore approximate the problem of fixed ends by

$$u,u_x,\text{etc.}\to 0,|x|\to\infty;\ u(x,0)=0;\ u_t(x,0)=2\tau,|x|\lesssim\ell,u_t(x,0)=0,|x|>\ell.$$

KAUP [17] has used the inverse scattering method partially to solve this problem. The essential point is that for $\tau<1$ only stationary breathers are created. Each such breather is of the form

$$u = 4\tan^{-1}\{\cot\mu\sin(t\sin\mu)\operatorname{sech}(x\cos\mu)\}. \qquad (8)$$

The number of such breathers is the integral part of $(\ell\tau/\pi)+1/2$. In real units $\ell\tau \sim \gamma\Delta B_0 L\bar{c}^{-1}$ and $L\lesssim 1\text{cm}$, $\bar{c}\lesssim 3\times10^3\text{cm sec}^{-1}$ whilst $\gamma\Delta B_0 \sim \Omega_{\ell A} \sim 10^6$ rad sec^{-1} so $\ell\tau$ can be large, $\gtrsim 300$. The number of eigenvalues [17] $\gtrsim 800$. We are interested here however in the case $L\ll 1\text{cm}$ and $\ell\tau\lesssim 10$. The value $\tau=1$, where $\gamma\Delta B_0 = \gamma\Delta B_{0c} = \Omega_{\ell A}$, remains a critical field. For $\tau>1$ kink-antikink pairs begin to travel out in opposite directions. In terms of scattering data [5,17] pairs of eigenvalues ζ, $-\zeta^*$ move successively onto the imaginary axis and split up there. The maximum number of kink-antikink pairs which can be produced is therefore $\sim\ell\tau/\pi = \gamma\Delta B_0 L\bar{c}^{-1}\pi^{-1}$. Fig.5 shows one such radiating kink-antikink pair; four standing wave breathers remain in the central region $|x|\lesssim\ell$. The Figure 1 of [1] shows a corresponding figure for four outgoing kink-antikink pairs and two standing wave breathers. An important point is that kinks radiate to the left and antikinks to the right (for positive τ). Kinks and antikinks do not radiate together in the same direction.

Corresponding results for the B-phase double sine-Gordon case are more complicated. The homogeneous problem has phase plane trajectories

$$\tfrac{1}{2}u_t^2 = (\cos u+\cos\tfrac{1}{2}u)+C. \qquad (9)$$

There are two ringing free trajectories, namely when the constant $C=0$ and when $C=2$. The critical fields are $\gamma\Delta B_{0c_1} = \sqrt{3/5}\,\Omega_{\ell B}$ and $\gamma\Delta B_{0c_2} = \sqrt{5/3}\,\Omega_{\ell B}$. Further details are given in [12]. Preliminary experiments designed to observe the two critical fields are reported by WHEATLEY [14].

The arguments surrounding the boundary conditions for the A-phase problem apply in similar form to the B-phase problem; but the physical problem of textures is certainly different [15]. The initial value problem (1) (with negative sign, $\lambda=1$, and $c = \delta$ or $4\pi-\delta$) is not soluble by any known scattering method. Nevertheless, for fields $\Delta B_0<\Delta B_{0c_1}$ we can expect that standing wave breathers may be formed. The Fig.6a for $u_t(x,0) = 1$, $|x|<15$ ($u_t = 1$ is less than the critical value $u_t = 3/2$) is evidence supporting this expectation. However unlike the situation governing the comparable sine-Gordon problem the breathers do not necessarily stand. The Fig.6b shows a continuation of the Fig.6a. There appears to be a <u>radiating</u> breather system.

The results on threshold $u_t = 3/2$ are more complicated than those reported in [1]. The radiating kink-antikink pair for $u_t = 3/2$, $|x|<15$ (ref. [1] Fig.3) is confirmed. But a simple kink and antikink oppositely directed is also possible. Fig.7 shows results for $u_t = 3/2$, $|x|<14$. With this may be created a breather-like disturbance (a bound kink-antikink pair): Fig.8 shows results for $u_t = 3/2$,

$|x|<17$. Finally Fig.9 shows simple radiating breather-like excitations created from $u_t = 3/2$, $|x|<13$. Our conclusion from these results is that unlike the case of the sine-Gordon equation $\lambda=0$, $c=0$, the initial value problem (1) for the double sine-Gordon equation (negative sign, $\lambda=1$, $c=\delta$) is sensitive to ℓ as well as τ. There is a threshold at $u_t = 2\tau = 3/2$: below this threshold only breathers are formed but these may radiate out; on or above this threshold radiating kinks and radiating kink-antikink pairs are possible (a result critically different from the sine-Gordon case) and with these there may be radiating breather-like excitations.

We report again a result given already in [1] of crossing the second critical threshold at $u_t = 5/2$. The boundary condition on Fig.10 is $u = \delta$, $|x|\to\infty$, and from the right, $\delta \to 4\pi-\delta \to 4\pi+\delta \to 8\pi-\delta \to 4\pi+\delta \to 8\pi-\delta$. The sequence is therefore: (right to left) $4\pi-2\delta$ antikink, 2δ antikink, $4\pi-2\delta$ antikink, $4\pi-2\delta$ kink, $4\pi-2\delta$ antikink. This sequence confirms the conclusion [12] and noted earlier that (i) with boundary conditions $u\to\delta$, $|x|\to\infty$, a $4\pi-2\delta$ antikink (or a 2δ-kink) is first emitted; (ii) a $4\pi-2\delta$ antikink is followable by either a 2δ antikink or a $4\pi-2\delta$ kink. There is no evidence in this example of the formation of breather like excitations but it seems possible that by changing ℓ the final $4\pi-2\delta$ antikink-kink pair could bind and appear as an outgoing breather. We have no evidence yet of outgoing 2δ kink-antikink pairs. We have noted that if such pairs collide they convert to $4\pi-2\delta$ antikink-kink pairs and always provide enough rest mass energy as kinetic energy to overcome any tendency to breather formation. Thus an outgoing 2δ kink-antikink pair is possibly unstable to the formation of an outgoing $4\pi-2\delta$ antikink-kink pair.

As noted in [1] a conclusion important to the physics is that the critical fields for no ringing NMR remain critical for break up into solitons both in the A-phase and the B-phase. However, if break up into solitons occurs, as it surely must for most physically applicable boundary conditions, such break up would be accompanied by an apparent ringing signal at least at specific points $x>\ell$ outside the original region of inhomogeneous magnetic field. It is not easy to see that any averaged signal will necessarily be ringing free and this may have some bearing on the imperfect agreement between theory and experiment observed for the A-phase [14].

Despite the possibility of forming outgoing breathers above and below threshold in the B-phase, the point noted in [1], namely that it would seem to be possible to detect both kinks and antikinks above threshold at $u_t = 3/2$ in the B-phase, remains. Our results show that small changes ($\sim$ one or two correlation lengths $\xi \equiv \bar{c}\Omega_{\ell B}^{-1}$ in the initial distribution of u_t) change breather emission at threshold (Fig. 9, $|x|<13$) to single kink emission (Fig.7, $|x|<14$) to kink-antikink emission (Fig.5 of Ref.[1], $|x|<15$) to combined kink and breather emission (Fig.8, $|x|<17$). Evidently the emission depends fairly smoothly on increasing the energy in the initial data by increasing ℓ.

It is tempting to interpret these results in terms of the eigenvalues of an appropriate inverse scattering problem in a direct comparison with the sine-Gordon problem. It is clear that such eigenvalues are not constants of the motion and move in the complex plane. This is exactly the situation [8,9,10] in the case of (1) with positive sign and $\lambda=1$, but we cannot say if similar statements are meaningful for the negative sign. In these terms the pattern for the emitted breathers is not precisely clear: if the breathers emitted for $u_t = 3/2$, $|x|<13$ (Fig.9) represent in some way two eigenvalues, the single kinks emitted for $|x|<14$ (Fig.7) represent only one (although there may be standing wave breathers in Fig.7 also).

The pattern for the kinks is however plain: apparently below the threshold at $u_t = 5/2$ 2δ-kinks are not created. The sequence of kink emission for $3/2 < u_t(x,0) < 5/2$ <u>must</u> therefore be kink-antikink-kink etc (or antikink-kink-antikink etc) and for large enough ℓ we expect many such kink-antikink pairs. This contrasts with the A-phase where above the single threshold at $u_t = 2$ only a simple sequence consisting purely of kinks (or purely of antikinks) can be emitted. This fact may enable a magnetic detector to distinguish between the A- and B-phases and so provide further evidence on the correctness of the symmetries assigned to the order parameters in the two different phases.

Note: Time increases vertically i.e. runs backwards in all Figures except in Figs. 3 and 4.

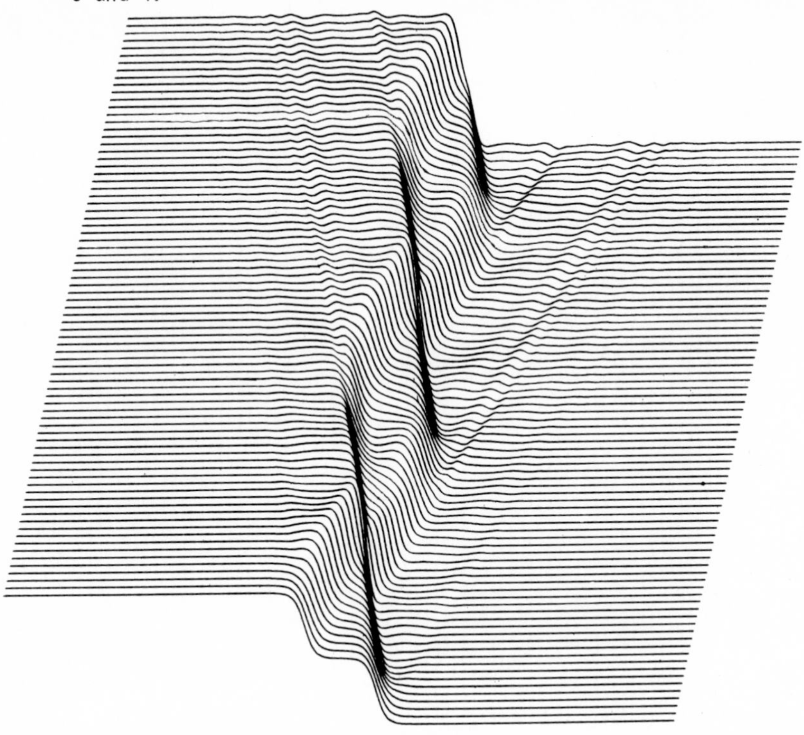

Fig. 1 The wobbling 4π-kink of the double sine-Gordon equation $u_{xx}-u_{tt}=\sin u+\frac{1}{2}\sin\frac{1}{2}u$ (case of positive sign); $u\to 0$, $|x|\to\infty$.

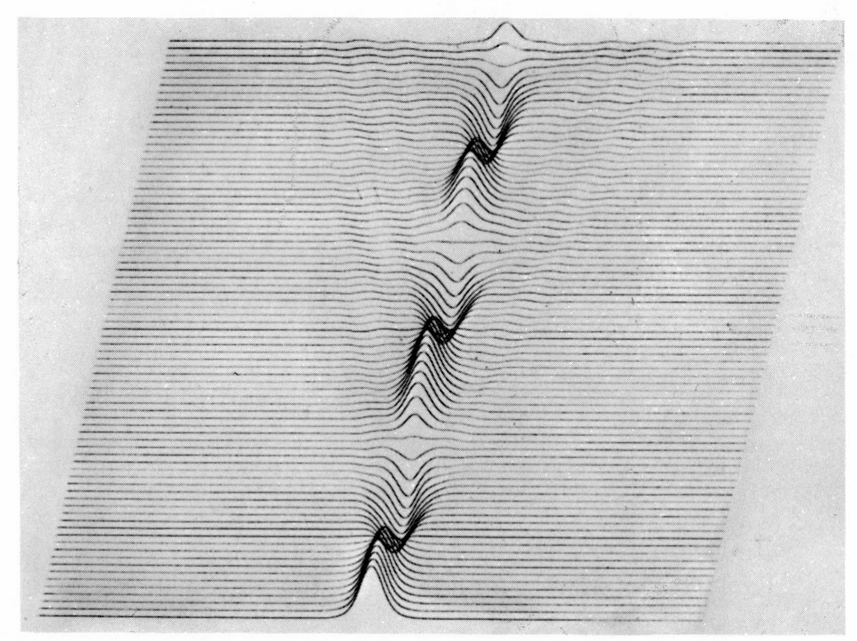

Fig. 2 The wobbling 0π-kink of the double sine-Gordon equation with positive sign and $u\to 2\pi$, $|x|\to\infty$.

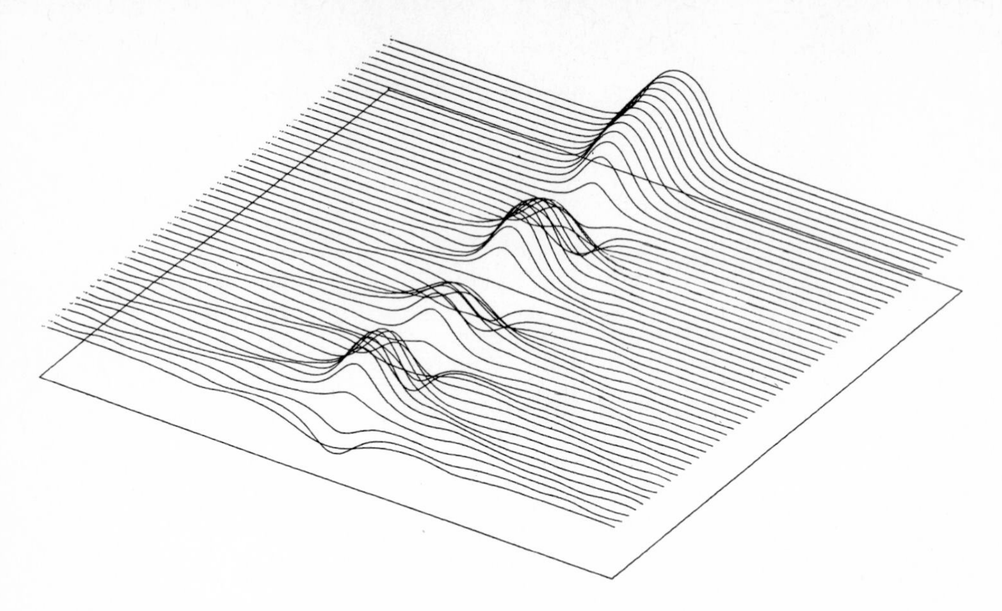

Fig. 3 The long lived breather-like state obtained by letting a $4\pi-2\delta$ kink and antikink fall into each other (V=0). Time runs forward in this figure.

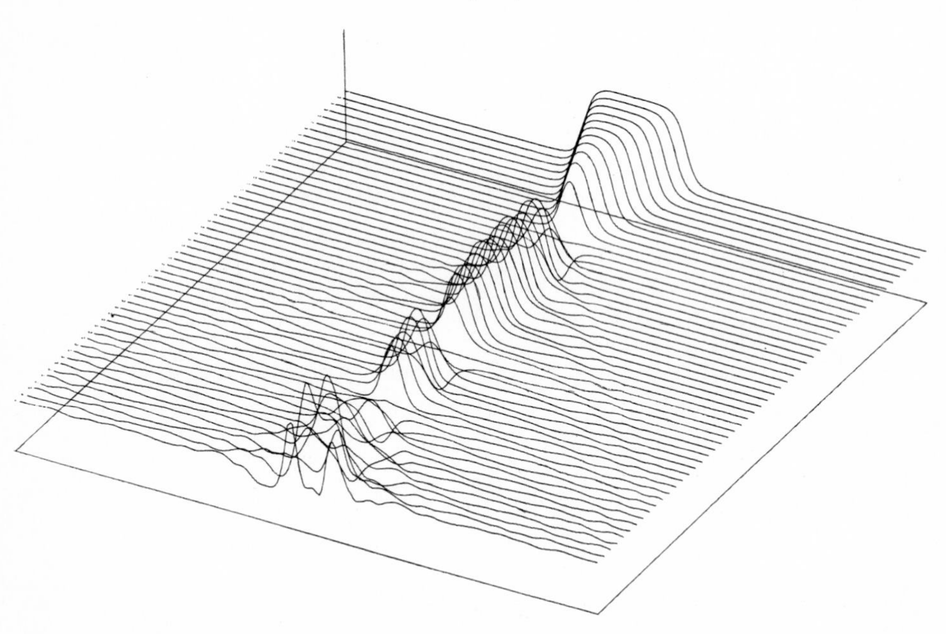

Fig. 4 The behaviour of the $4\pi-2\delta$ kink-antiking pair at the threshold (V = 0.35) for breather formation. Time runs forward in this figure.

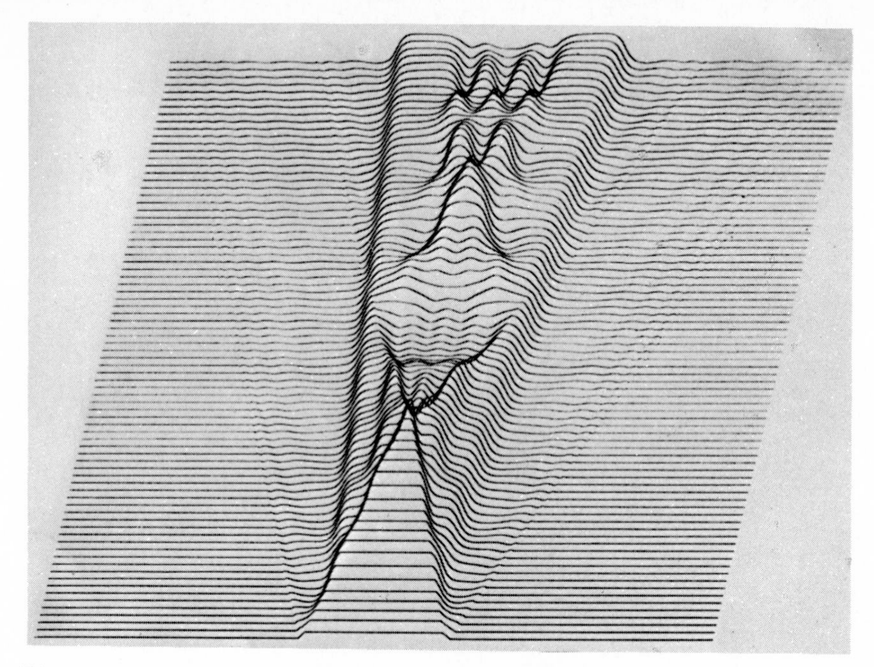

Fig. 5 Radiation of a single kink and a single antikink of the sine-Gordon equation; $u_t(x,0) = 2.1$, $|x|<15 = \ell$; $u_t = 2.1$ is above the threshold for kink formation at $u_t = 2$. Four standing wave kinks remain in the excitation region $|x|<15$. Note the initial homogeneous region for t=0.

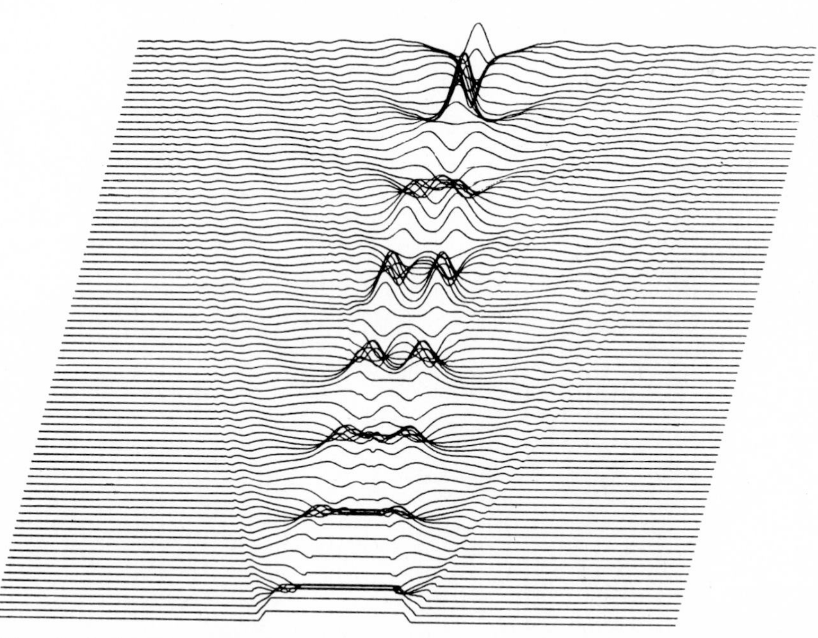

Fig. 6a Radiating breather-like excitations of the double sine-Gordon equation $u_{xx}-u_{tt}=-(\sin u+\frac{1}{2}\sin\frac{1}{2}u)$ (case of negative sign); $u_t(x,0) = 1$, $|x|<15$; $u_t = 1$ is less than the critical value $u_t = \frac{3}{2}$. In this Fig. the breathers appear to stand.

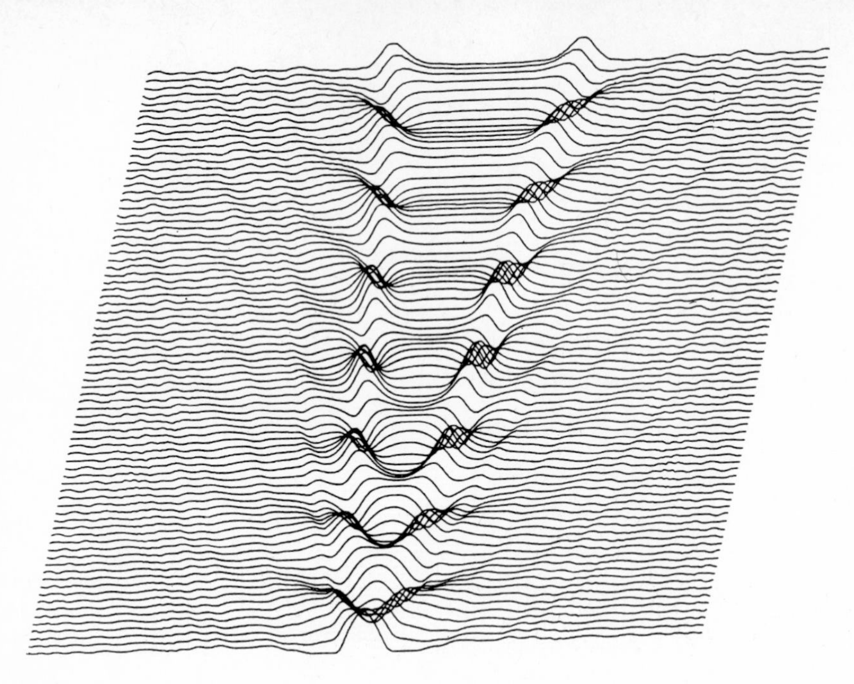

Fig. 6b Continuation of Fig. 6a showing that the breathers radiate.

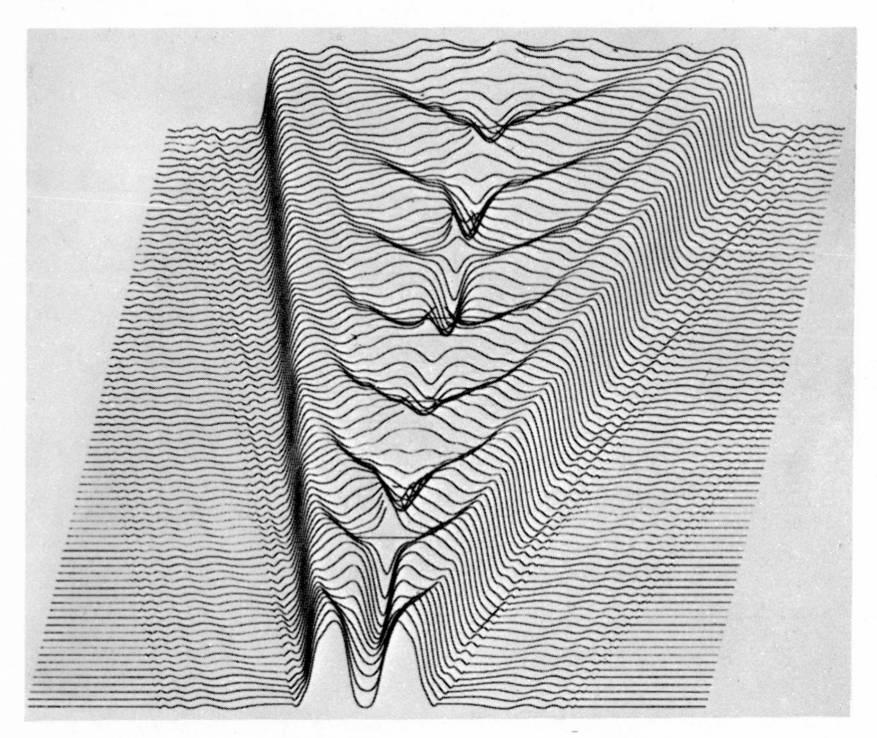

Fig. 7 Emission of a single 4π-2δ kink and antikink of the double sine-Gordon equation with negative sign: $u_t = \frac{3}{2}$, $|x|<14$; $u_t = \frac{3}{2}$ is the threshold condition for kinks.

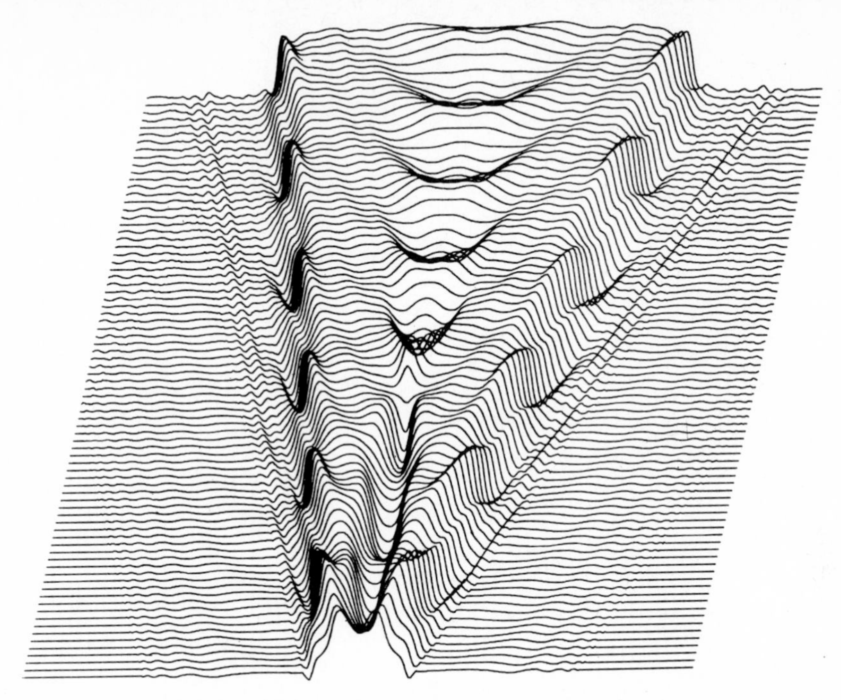

Fig. 8 Kink and breather-like emission for the double sine-Gordon equation with negative sign: $u_t = \frac{3}{2}$, $|x|<17$.

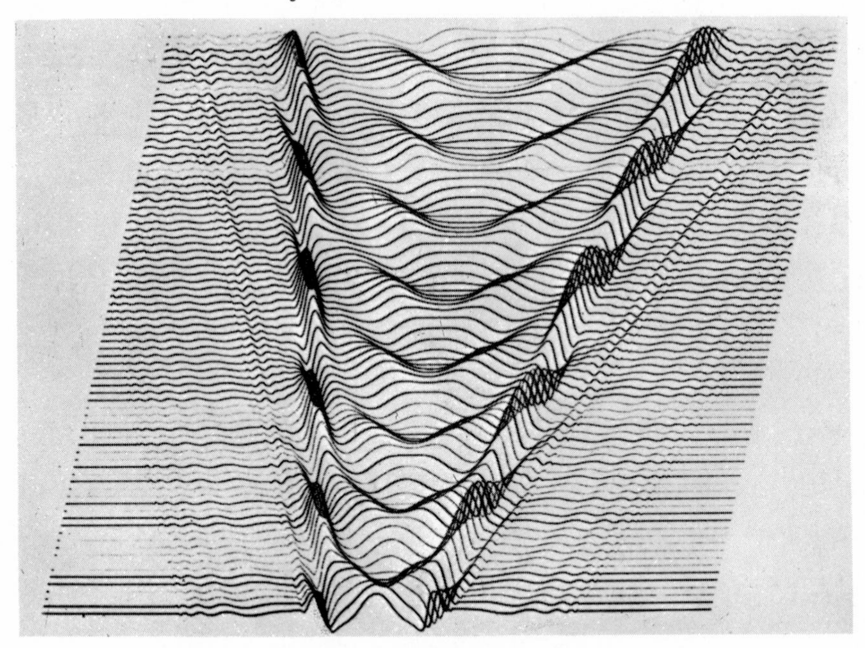

Fig. 9 Simple radiating breather-like excitations: $u_t = \frac{3}{2}$, $|x|<13$.

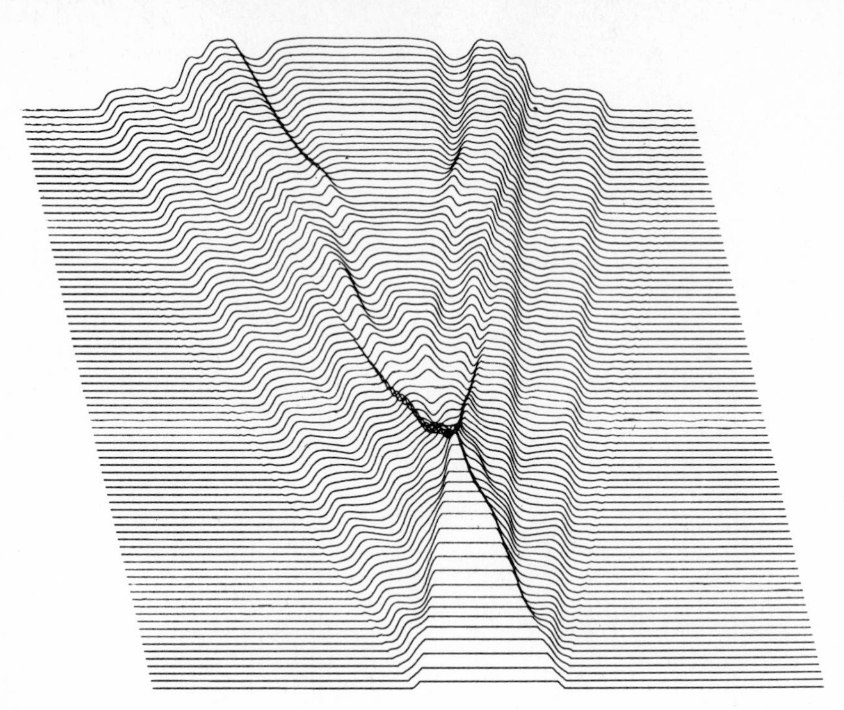

Fig. 10 Radiation of a 2δ-kink and a 2δ-antikink above the threshold at $u_t = 2.5$: $u_t(x,0) = 2.6$, $|x|<15$.

This paper was written when one of us (R.K.B) was working at NORDITA, Copenhagen, Denmark. He is grateful to NORDITA for the facilities which made the writing of this paper possible.

References

1. P.W. Kitchenside, R.K. Bullough and P.J. Caudrey, 'Creation of spin waves in ^{3}He B' in 'Solitons and condensed matter physics' A.R. Bishop and T. Schneider eds. (Springer-Verlag, Heidelberg, 1978) pp.331-336

2. S. Duckworth, R.K. Bullough, P.J. Caudrey and J.D. Gibbon, Phys. Lett. 57A, 19 (1976)

3. R.K. Dodd and R.K. Bullough, Proc. Roy. Soc. A351, 499 (1976); 352, 481 (1977)

4. R.K. Bullough and P.J. Caudrey, 'The multiple sine-Gordon equations in non-linear optics and in liquid ^{3}He' in 'Nonlinear evolution equations solvable by the spectral transform' F. Calogero ed. (Pitman, London, 1978) pp.180-224

5. R.K. Bullough and R.K. Dodd, 'Solitons I. Basic Concepts. II. Mathematical Structures" in 'Synergetics. A workshop' H. Haken ed. (Springer-Verlag, Heidelberg, 1977) pp.92-119

6. R.K. Bullough, 'Solitons in Physics' (Lectures given at the NATO Advanced Study Institute on 'Nonlinear equations in physics and mathematics', Istanbul, August 1977) Proceedings ed. by A.O. Barut (D. Reidel Publishing Co: Dordrecht, Holland, 1978)

7. R.K. Bullough and R.K. Dodd, 'Solitons in Physics and Mathematics' These Proceedings

8. P.W. Kitchenside, A.L. Mason, R.K. Bullough and P.J. Caudrey, 'Perturbation theory for the double sine-Gordon equation' in 'Solitons and condensed matter physics' A.R. Bishop and T. Schneider eds. (Springer-Verlag: Heidelberg, 1978) pp.54-57; A.C. Newell, 'Perturbed soliton systems' Ibid.

9. A.C. Newell, J. Math. Phys. 18, 922 (1977)

10. P.W. Kitchenside, Ph.D. Thesis, University of Manchester (1979); P.W. Kitchenside, A.L. Mason, P.J. Caudrey and R.K. Bullough. To be published.

11. R.K. Bullough, P.J. Caudrey, J.D. Gibbon, S. Duckworth, H.M. Gibbs, B. Bölger and L. Baede. Post deadline paper P7 IX International Quantum Electronics Conference, Amsterdam, June 1976 (Optics. Comm. 18, 200 (1976))

12. R.K. Bullough, P.J. Caudrey, P.W. Kitchenside, 'Bumping spin waves in the B-phase of liquid ^{3}He', to be published in J.Phys.C., Solid State Physics (1978); also see K. Maki, Phys. Rev. B 11, 4264 (1975); K. Maki and P. Kumar, Phys. Rev. B 14, 3920 (1976)

13. A.J. Leggett, Rev. Mod. Phys. 47, 331 (1975)

14. J. Wheatley, Rev. Mod. Phys. 47, 415 (1975)

15. K. Maki, 'Textures in ^{3}He' in 'Solitons and condensed matter physics' A.R. Bishop and T. Schneider eds. (Springer-Verlag, Heidelberg, 1978)

16. K. Maki and P. Kumar, Phys. Rev. Lett. 38, 558 (1977)

17. D.J. Kaup, 'Studies in Appl. Maths.' LIV, 165 (1975) Massachusetts Institute of Technology).

Part VII

Dynamical Systems

Catastrophic Effects in Pattern Recognition

R. Brause and M. Dal Cin

Institute for Information Sciences, University of Tübingen
D-7400 Tübingen, Fed. Rep. of Germany

1. Introduction

The purpose of this note is to report on an ongoing investigation into the steady state behavior of pattern recognition systems. The system considered in this paper generates the decision boundaries which separate pattern classes on the basis of a stochastic learning algorithm [1]. The inputs of this system, and hence, of the algorithm, are observed patterns drawn from a probability distribution $p(\underline{x})$, $\underline{x}$ a pattern vector. In the one dimensional case considered later on the decision boundaries are points in the pattern space. The decision of the system is based on a risk function R. That is, the system selects the boundaries that minimize R. As the distribution $p(\underline{x})$ of inputs is gradually changing the steady state boundaries will vary continuously most of the time. However, at certain instances we observe abrupt changes of the decision boundaries.

Abrupt changes in decision making were also investigated by E.G. Zeeman [2]. Our purpose is to derive the precise analytical condition for such catastrophic effects and to verify the results by simulation experiments.

2. A stochastic learning algorithm

When the number of pattern classes are given, the pattern recognition system tries to find an optimal separation of these classes. This search is controlled by a sequence of observed samples and is based on a risk function (even so the precise form of the risk function is not known to the system).

A typical configuration with two classes ω_1 and ω_2 and a two-dimensional distribution of patterns is shown in Fig. 1. (The boundary was obtained after 1000 iterations of the learning algorithm given in (3) below.)

Is there always one optimal boundary? Fig. 2 shows a situation with two optimal boundaries provided the boundary is a single straight line. It will be shown that in this case gradual changes of certain pattern distributions give rise to abrupt switches from one to two or more stationary states of the boundary.

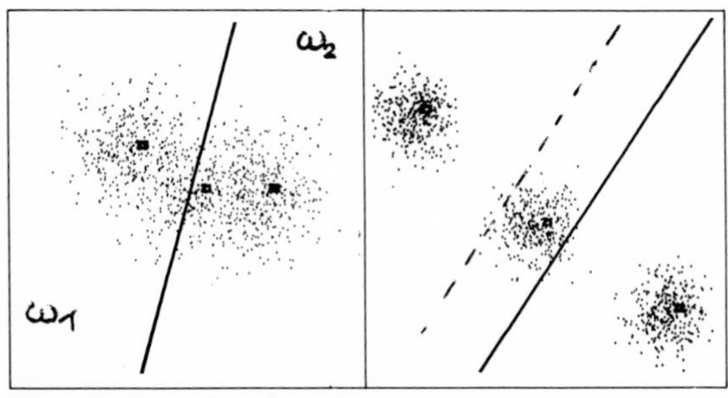

Fig. 1: Decision boundary

Fig. 2 Two optimal boundaries

Let a pattern be characterized by N features $x_1, x_2, \ldots, x_N$ where $x_i \in \mathbb{R}$. It is represented by a point $\underline{x} = (x_1, x_2, \ldots, x_N)$ of the set $X \subset \mathbb{R}^N$ of all possible patterns. The problem of classifying patterns into M classes $\omega_1, \omega_2, \ldots, \omega_M$ is equivalent to finding a partition of X into M corresponding disjoint subsets $X_1, X_2, \ldots, X_M$ which solve the problem. Let the i'th subset X_i be represented by a reference pattern $\underline{c}_i$ belonging to X_i.

Instead of a teacher the system is provided with M loss functions

$$L_1(\underline{x},\underline{c}_1),\ L_2(\underline{x},\underline{c}_2),\ \ldots,\ L_M(\underline{x},\underline{c}_M)$$

where $L_i(\underline{x},\underline{c}_i) \geqslant 0$ is the loss that will be incurred if the system classifies pattern $\underline{x}$ as belonging to X_i. Of course, the loss should be minimal if $\underline{x}$ actually belongs to ω_i. We assume that the system chooses $\underline{c}_i$ such that for patterns of class X_i the expectation value

$$R(\underline{c}_i) = \sum_{j=1}^{M} R_j(\underline{c}_i)\, P(\omega_i)$$

is minimal, where

$$R_j(\underline{c}_i) = \int_{X_i} L_i(\underline{x},\underline{c}_i)\, p(\underline{x}|\omega_j)\, d\underline{x} \tag{1}$$

is the risk due to a misclassification of patterns from ω_j into X_i $(i \neq j)$ and a bad choice of the reference pattern $\underline{c}_i$ $(i = j)$. $P(\omega_i)$ is the a priori probability of class ω_i. Hence, the system tries to minimize the risk function

$$R(\underline{c}_i) = \int_{X_i} L_i(\underline{x},\underline{c}_i)\, p(\underline{x})\, d\underline{x}, \quad i = 1, 2, \ldots, M \tag{2}$$

by choosing the optimal set $c = (\underline{c}_1, \underline{c}_2, \ldots, \underline{c}_M)$ of reference patterns. It can be shown that in this case, also, the total risk

$$\hat{R}(c) = \sum_{i=1}^{M} R(\underline{c}_i)$$

is minimal.

The method of stochastic gradient search proposed by Robbins and Monro [3] provides us with the following algorithm for finding the minimum of $\hat{R}(c)$.

Let $\underline{c}_i[n]$ be the reference pattern representing X_i after the n'th learning step and let $\underline{x}[n]$ be the next pattern shown to the system. Then a new set of reference patterns will be generated according to the following algorithm (or stability mechanisms).

$$\begin{aligned} \underline{c}_i[n+1] &= \underline{c}_i[n] - \gamma[n]\, \nabla_{\underline{c}_i} L(\underline{x}[n], \underline{c}_i)\Big|_{\underline{c}_i[n]} \\ \underline{c}_j[n+1] &= \underline{c}_j[n], \quad i \neq j, \end{aligned} \tag{3}$$

where index i is such that

$$L_i(\underline{x},[n], \underline{c}_i[n]) = \min_k \{L_k(\underline{x}[n], \underline{c}_k[n])\} \tag{4}$$

and $\gamma[n] \in \mathbb{R}$. Observe that knowledge of $p(\underline{x})$ is not necessary for this algorithm.

A steady state c^* of this algorithm is never reached by finitely many iteration steps but the convergence is guaranteed with

$$\lim_{n\to\infty} P(c[n] = c^*) = 1$$

and[+] $E(c^* - c[n]) = 0$,

+ E expectation cperator

if the following conditions for $\gamma[n]$ hold [3]:

(a) $\lim_{n\to\infty} \gamma[n] = 0$; (b) $\lim_{n\to\infty} \sum_{i=1}^{n} \gamma[i] = \infty$; and (c) $\lim_{n\to\infty} \sum_{i=1}^{n} \gamma[i]^2 = s < \infty$.

Next we derive a condition which tells us when switches of decision can occur.

3. A criterion for instabilities

A simple example of a loss function which will be used in the following is

$$L_i(\underline{x},\underline{c}_i) = \frac{1}{2}(\underline{x} - \underline{c}_i)^2. \tag{5}$$

Then, $\hat{R}(c^*) = \min_c \hat{R}(c)$ if

$$\nabla_{\underline{c}_i} R(\underline{c}_i)\Big|_{\underline{c}_i^*} = 0$$

$$= \int_{X_i} \nabla_{\underline{c}} L_i(\underline{x},\underline{c}_i^*)\, p(\underline{x})\, d\underline{x} \qquad \text{(The variation of } X_i \text{ vanishes.)}$$

$$= \int_{X_i} (\underline{x} - \underline{c}_i^*)\, p(\underline{x})\, d\underline{x}, \qquad i = 1, 2, \ldots, M.$$

Hence, the critical points are $c^* = (\underline{c}_1^*, \ldots, \underline{c}_M^*)$, where

$$\underline{c}_i^* = E(\underline{x}|X_i) = \int_{X_i} \underline{x}p(\underline{x})\, d\underline{x} \Big/ \int_{X_i} p(\underline{x})\, d\underline{x}. \tag{6}$$

Now, the boundary $\underline{d}_{ij}$ between two classes ω_i and ω_j is given by points $\underline{x}$ for which

$$L_i(\underline{x},\underline{c}_i) = L_j(\underline{x},\underline{c}_j).$$

Hence, the boundaries chosen by the system after the n'th learning step are determined by

$$\underline{d}_{ij}[n] = \frac{1}{2}(\underline{c}_i[n] + \underline{c}_j[n]) \tag{7}$$

with the steady states

$$\underline{d}_{ij}^* = \frac{1}{2}(\underline{c}_i^* + \underline{c}_j^*) = \frac{1}{2}(E(\underline{x}|X_i) + E(\underline{x}|X_j)).$$

In the case of $N = 1$, $M = 2$ this reduces to

$$d^* = \frac{1}{2}(E(x|x > d^*) + E(x|x \leq d^*)) \\ := \chi(d^*) \tag{8}$$

where now $X_1 = (-\infty,d)$ and $X_2 = [d,+\infty)$.

If $p(x)$ is symmetric (i.e. $p(x) = p(-x)$), the following relations hold (see Appendix):

(a) $\lim_{d\to\infty} [d/2 - \chi(d)] = 0$,

(b) $\chi(d) = -\chi(-d)$, hence, $\chi(0) = 0$, (9)

(c) $\frac{\partial}{\partial d}\chi(d)\Big|_{d=0} = 2p(0)\, E(x|x > 0)$.

That is, $\chi(d)$ approaches $d/2$ and $d = 0$ is a steady state of the algorithm.

Now, if $\chi(d)$ crosses the diagonal $f_1(d) = d$ below and above the d-axis, then the cross-points are also steady states, cf. Fig. 3. This certainly occurs if the derivative of $\chi(d)$ at $d = 0$ is greater than 1. Thus, if a symmetric probability distribution satisfies

$$m := 2p(0)\ E(x|x > 0) > 1, \tag{10}$$

then there are at least three possible steady states of the learning algorithm (3); $d^* = 0$ is unstable in this case.

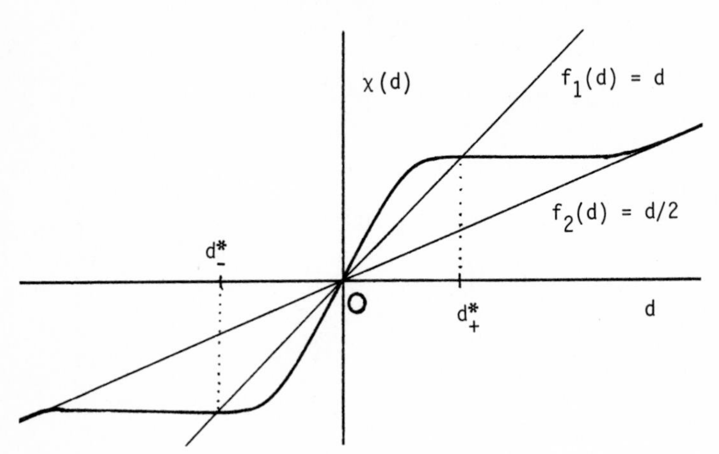

Fig. 3 Graph of χ

In the next section we show the performance of our learning algorithm. To this end, we choose the following family of pattern distributions:

$$P_{B,z}(x) = \frac{1-B}{2} N(-z,\sigma) + B\, N(0,\sigma) + \left(\frac{1-B}{2}\right) N(z,\sigma) \tag{11}$$

where

$$N(z,\sigma) = (\sqrt{2\pi\sigma})^{-1} \exp(-(x-z)^2/2\sigma)$$

and $0 \leq B \leq 1$. Thus, Eq. (10) is now

$$m(z,B) > 1. \tag{12}$$

We compare the performance of the stochastic learning algorithm with that of the following two learning algorithms

$$\bar{d}[n+1] = 1/2\ (1/n_1 \sum_{i=1}^{n_1} x_i + 1/n_2 \sum_{i=1}^{n_2} x_i),\ n = n_1 + n_2. \tag{13}$$

where the patterns are drawn sequentially from pattern distribution (11) and

$$\bar{d}[n+1] = \chi(\bar{d}[n]) = 1/2\ (E(x|x < \bar{d}[n]) + E(x|x \geq \bar{d}[n]), \tag{14}$$

The second algorithm utilizes the maximum amount of information available. Its steady states are the same as that for (3).

4. Simulation experiments

The computer simulations [4] of the stochastic learning algorithm (3) and its averaged versions (13) and (14) confirm the theoretical results. Diagram (4.1) shows the bifurcational splitting of the steady states d^* of the boundary $\bar{d}[n]$ for algorithm (14), when the control parameter $m \geq 1$ is linearly varied. (B=const =.5)

For every value of m(z), 20 iterations were initialized with 3 different starting values $\bar{d}[0] = -1, 0, +1$ and after a fixed number N_{max} of iterations the state $\bar{d}[N_{max}]$ of the boundary was recorded. State $d^* = 0$ is a stable solution for the determinate algorithm (14). Of course, this is no longer true for the stochastic algorithms (3) and (13), cf. Figs. 4.2 and 4.3, respectively.
After the N_{max}'th iteration the state of the boundary is stochastically distributed around the two stable states d^*_{+-}. (It can be shown analytically that $d^*(m)$ is linear in m if there are at least two patterns x_+, x_- between $\pm z$ and 0 where $p_{B,z}(x_\pm) = 0$.)

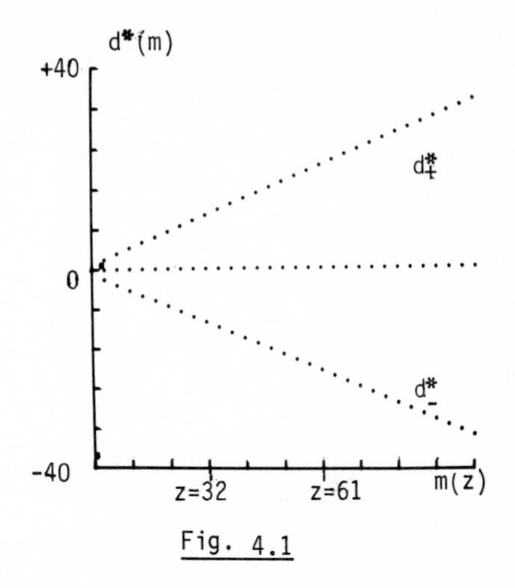

Fig. 4.1

The histograms of Figs. 5.1 - 5.3 show the abrupt switch of the stable solutions for the three algorithms when the control parameter m(z) is gradually changed from m(z) > 1 to m(z) < 1. Here too, the boundaries of the stochastic algorithms are distributed around the stable states. However, the variance is too great for us to see the switching.

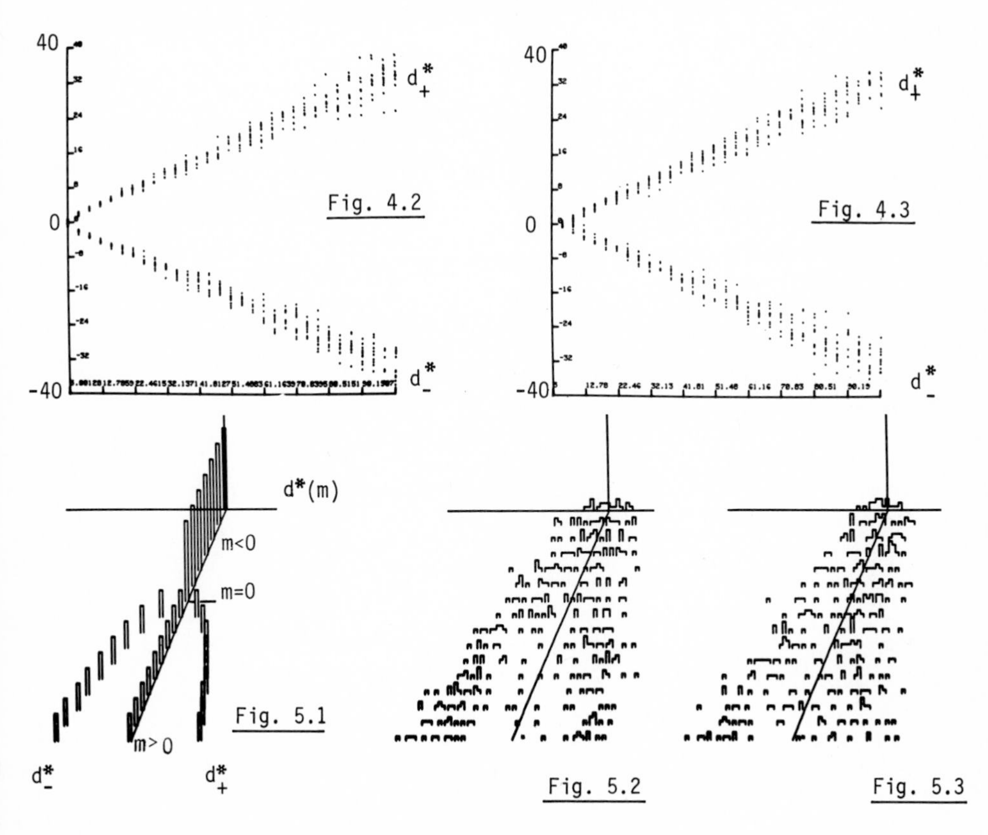

Fig. 4.2

Fig. 4.3

Fig. 5.1

Fig. 5.2

Fig. 5.3

Appendix: Proof of Eq. (9):

Let $I(a,b) = \int_a^b x\,p(x)\,dx \Big/ \int_a^b p(x)\,dx$, $I(-\infty,+\infty) = 0$, since $E(x) = 0$.

(a) $$\lim_{d\to\infty} (d/2 - \chi(d)) = \lim_{d\to\infty} \frac{1}{2} (d - I(d,\infty))$$

$$\underset{\text{l'Hopital}}{=} \lim_{d\to\infty} \frac{1}{2} [d - d\cdot p(d) / p(d)] = 0,$$

(b) $$2\chi(-d) = I(-\infty,-d) + I(-d,\infty) =$$

$$= -I(\infty,d) - I(d,-\infty) = -2\chi(d), \text{ since } p(x) = p(-x)$$

(c) $$2\chi'(d)\Big|_{d=0} = [\int_o^\infty p(x)\,dx]^{-2} [-dp(d)\cdot\int_d^{-\infty} p(x)\,dx +$$

$$+\int_d^{-\infty} x\,p(x)\,dx\,p(d) - dp(d)\cdot\int_d^\infty p(x)\,dx +$$

$$+ \int_d^\infty xp(x)\,dx\,p(d)]\Big|_{d=0} = 8p(0) \int_o^\infty xp(x)\,dx$$

$$= 4p(0)\,E(x|x \geqslant 0).$$

Acknowledgments

The authors are indebted to the late E. Pfaffelhuber, who helped provide the impetus for this investigation. The assistance of E. Dilger is also acknowledged.

References

1. Tou, J.T. and R.C. Gonzalez, Pattern Recognition Principles, Addison-Wesley Publishing Co., 1974
2. E.C. Zeeman, private communication (1978)
3. Robbins, H. and S. Monro, A stochastic approximation method, Ann. Math. Stat. 22, 400-407 (1951)
4. Brause, R., Mustererkennung mit stochastischem Lernalgorithmus, preprint, Institut für Informationsverarbeitung, Tübingen, 1978

Cusp Bifurcation in Pituitary Thyrotropin Secretion

Fritz J. Seif

Medizinische Poliklinik, Eberhard-Karls-Universität Tübingen
D-7400 Tübingen, Fed. Rep. of Germany

Summary

Patients with primary hyperthyroidism treated to attain normal serum concentrations of thyroxine (T_4) and triiodothyronine (T_3), show a bimodal distribution of pituitary thyrotropic responsiveness to exogenous protirelin (TRH). To the contrary, the hormonal constellation of primary hypothyroidism produces a unimodal distribution. THOM's catastrophe theory is applied to formulate a qualitative, macroscopic model of the thyrotropic responsiveness. The adapted cusp catastrophe demonstrates how the inhibition by T_4 and T_3 and the stimulation by TRH cooperate on the thyrotropic function. Moreover, a microscopic, stochastic model of a stimulus-secretion coupling system of pituitary thyrotropic cells is constructed to end up with a probability potential associated with a cusp catastrophe. Comparison of the macroscopic with the microscopic model shows that a reversible allosteric transition is an essential feature of the system.

1. Introduction

Adaptation of living systems to changes in the environment is generally achieved by regulatory processes of nervous and hormonal organs such that the system attains a stationary state. Diseases or major changes in the ambient world may cause large enough aberrations of the control parameters to bring about transitions to other states that are stable, functionally and structurally, and are separated from the former state by unstable loci. Such separated states can be observed in the thyrotropin secretion of the pituitary gland [18].

In a recent paper it has been shown that in man pituitary thyrotropic responseness to protirelin is highest in euthyroidism, reduced in primary hypothyroidism, and lowest in primary hyperthyroidism [17]. The change of responsiveness indicates that feedback inhibition of thyrotropin release by steadily increasing or decreasing concentrations of thyroid hormones does not act unidirectionally (monotonically) on the pituitary, rather another relationship must be employed. Further data accumulation demonstrates that in patients with hyperthyroidism treated in order to reduce thyroid hormone production to normal and to become euthyroid, pituitary thyrotropic responsiveness to a TRH stimulus does not turn normal in all cases, although euthyroidism is maintained for weeks [16]. On the contrary, the responsiveness of a population of treated patients shows a bimodal distribution, one peak of which is congruent with the normal responsiveness of euthyroidism, and the other one still coincides with the highly reduced responsiveness of hyperthyroidism (Fig.2). The bimodality of the distribution indicates that thyrotropic responsiveness can attain two different stable states with discontinuous transitions between them. Moreover, the hormonal constellation of primary hypothyroidism creates a unimodal distribution (Fig.2B).

Fruitful discussion with Prof.Dr. W. Güttinger and computational help of Dr. E. Dilger are gratefully acknowledged.

The data suggest the applicability of THOM's catastrophe theory [20,21] in order to formulate a mathematical model of the thyrotropic system. In the following this topological concept is used to explain the macroscopic observations and measurements. To arrive at a catastrophe-associated potential function, mathematical description is started from a microscopic subcellular model of pituitary thyrotropic function comprising formation of secretory granules, a stimulus-secretion coupling system, and hormone diffusion.

2. Physiological Preliminaries

The hormonal incretion of the thyroid gland is governed by specific functions of the anterior pituitary gland and the hypothalamus (Fig.1). The neuroendocrine tripeptide protirelin (TRH,thyrotropic releasing hormone) is transported by neurosecretion and the hypophyseal portal vessel system from neurosecretory cells in the hypothalamus to the thyrotropic cells of the anterior pituitary. Binding of protirelin to specific receptors at the cell membrane of the thyrotropic cells leads to secretory release of thyrotropin (TSH,thyroid stimulating hormone), which in turn stimulates synthesis and secretion by the thyroid gland of thyroid hormones, thyroxine and triiodothyronine. Besides the effects of ontogenetic differentiation and metabolic stimulation in target organs, these two hormones elicit a feedback inhibition of thyrotropin secretion in pituitary thyrotropic cells.

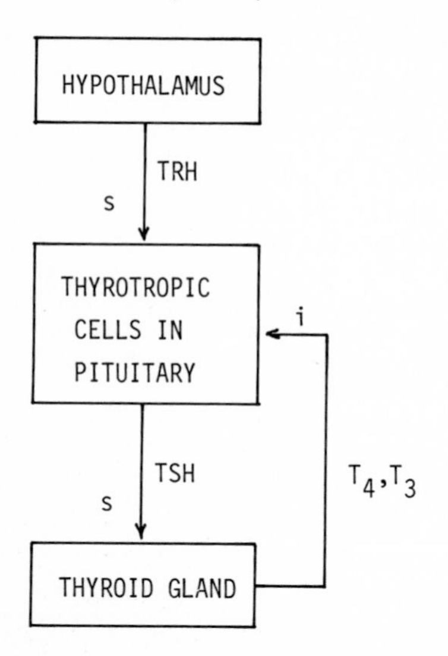

Fig.1 Regulatory principle of the thyroid gland with feedback inhibition at the pituitary level by thyroxine (T_4) and triiodothyronine (T_3).

TRH: protirelin
TSH: thyrotropin
s: stimulatory effect
i: inhibitory effect

Diseased thyroid glands can lead to a lack or an excess of thyroid hormones in the human organism, resulting in hypothyroid or hyperthyroid metabolism, respectively. Thyroid dependent hypometabolic states (primary hypothyroidism) thus produce an enhanced secretion of thyrotropin by the pituitary because of reduced inhibition, whereas thyroidal hypermetabolic states (primary hyperthyroidism) block thyrotropin secretion. The thyroid hormone dependent normal metabolic state of an organism is called euthyroidism.

3. Origin of Data

Healthy volunteers and patients with suspected thyroid disorders were screened by concentration measurements in serum of total thyroxine (T_4), total triiodothyronine (T_3), and basal thyrotropin (TSH(o)=H(o)). After withdrawal of blood samples for these measurements, at time t=o, 400 µg of TRH were injected intravenously as a bolus in order to stimulate TSH secretion (TRH test). In general the reactive peak level of serum TSH is reached between 20 and 35 min after TRH injection. Hence t=20 min was chosen as a second time point to evaluate TSH in serum (TSH(20)=H(t)). Measurements of T_4, T_3, and TSH were performed by radioimmune assays. In samples for T_4 and T_3 measurements the relative serum binding capacity r for radioactive triiodothyronine were estimated by equilibrium distribution in presence of resin as an indirect measure of thyroxine-binding globulin (T_3 resin uptake). A euthyroid standard serum yielding a serum-bound fraction of radioactive triiodothyronine to be p_o=o.7, served as normal reference for the calculation of $r = p/p_o$ with p as bound fraction of the serum under investigation.

Clinical signs and values of T_4, T_3, r, and TSH(o)=H(o) were used for the diagnostic evaluation of 314 untreated patients. 202 were euthyroid, 78 hyperthyroid, and 34 had primary hypothyroidism. In patients with hyperthyroidism the hormone measurements including TRH tests were repeated several times after initiation of antithyroid treatment. Thus 108 sets of hormone values were obtained of strictly euthyroid patients under treatment who formerly had been hyperthyroid (j=4 in Table 1).

Table 1 Hormone values (mean ± standard deviation) of diagnostically evaluated patients

Metabolic State	j	N_j	T_4 [μg/dl]	T_3 [ng/dl]	r	TSH(o) [μU/ml]	TSH(20) [μU/ml]	R
Euthyroid	1	202	7.3 ± 1.6	146 ± 37	o.97 ± o.09	1.2 ± o.6	11.8 ± 6.6	10.8 ± 6.4
Hypothyroid	2	34	4.6 ± 2.1	126 ± 63	1.02 ± o.10	17.7 ± 28.7	59.6 ± 64.4	4.3 ± 1.8
Hyperthyroid	3	78	13.6 ± 4.8	333 ± 141	o.87 ± o.11	o.8 ± o.3	o.8 ± o.3	1.0 ± o.1
Hyperthyroid treated	4	108	7.5 ± 1.5	144 ± 26	o.98 ± o.04	1.7 ± 2.1	§	§

R = TSH(20)/TSH(o); N_j: Number of value sets; § Since the values show a bimodal distribution (see also Fig.2), no mean and no standard deviation is calculated.

As shown in Table 1 the increase R = H(t)/H(o) of stimulated TSH value H(t) relative to the basal value H(o) is highest in euthyroidism and lowest in hyperthyroidism, whereas primary hypothyroidism yields an intermediate value of R [17]. In contrast, the group of hyperthyroid patients treated to euthyroidism does not have a unimodal histogram (frequency distribution) of R, but a bimodal one with one peak at R = 1 and the other at R = 7 (Fig.2D). They coincide fairly well with the hyperthyroid and euthyroid peak, respectively (Fig.2C and 2A).

4. Data Transformation

The hormonal data are obtained from peripheral blood and thus are only indirect measurements of the local concentrations in the pituitary gland. Therefore the peripheral data need transformation in order to arrive at an estimation of the local conditions in the pituitary thyrotropic system.

The feedback inhibition of TSH in the thyrotropic cells is brought about by a moiety of thyroxine and triiodothyronine, the intracellular concentration of which is not measurable and is assumed to be proportional to that of the free plasma or serum fraction of T_4 and T_3 not bound to plasma proteins. The respective serum

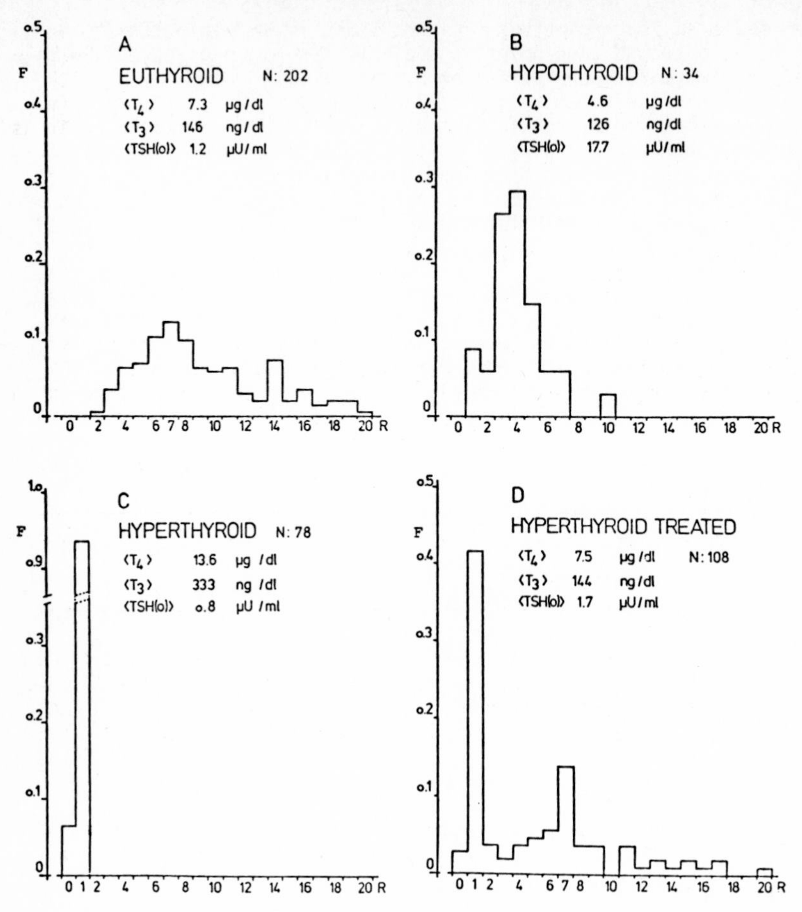

Fig.2 Histograms of the relative thyrotropin increase R 20 min after TRH stimulation of the evaluated groups of patients as in Table 1. F: relative frequency of R.

concentration of the free fractions we call ϑ_4 and ϑ_3. They only can be obtained by laborious methods not suited for routine measurements. Therefore we use indirect estimates by means of r.

For thyroxine the following binding kinetics to serum proteins {SP} are usually thought to be valid, although the serum proteins are not homogeneous and the assumption of uniform association and dissociation constants, k_1' and k_2', is a simplification. With T_4 as total serum concentration of thyroxine and the conservation equation $\vartheta_4 + \{SP\cdot\vartheta_4\} = T_4$, we can write

$$\vartheta_4 + \{SP\} \underset{k_2'}{\overset{k_1'}{\rightleftharpoons}} \{SP\cdot\vartheta_4\}.$$

Under physiological conditions equilibrium is quickly reached, i.e. $d\vartheta_4/dt = o$, which yields $\vartheta_4 = k_2'\{SP\cdot\vartheta_4\}/k_1'\{SP\}$. As more than 99.8% of T_4 is bound [19], we can set $\{SP\cdot\vartheta_4\} = T_4$. Moreover the concentration of unocupied, free serum proteins{SP} is proportional to $p_o r$. With c_1' as proportionality factor, we have $\{SP\} = c_1' p_o r$.

As estimate Θ_4 of the effective thyroxine concentration ϑ_4, we use the fraction

$$\Theta_4 = T_4/r = (k_1'c_1'p_o/k_2')\vartheta_4. \quad (1)$$

For triiodothyronine assumedly similar binding kinetics to serum proteins {SP} hold:

$$\vartheta_3 + \{SP\} \underset{k_4'}{\overset{k_3'}{\rightleftharpoons}} \{SP\cdot\vartheta_3\} .$$

For equilibrium distribution with $d\vartheta_3/dt = o$, and $k' = k_3'\{SP\}/k_4'$, the according differential equation reduces to

$$\vartheta_3 = \{SP\cdot\vartheta_3\}/k' \quad (2)$$

In presence of resin the serum protein bound fraction $p = p_o r$ of radioactive triiodothyronine is proportional to the bound fraction $\{SP\cdot\vartheta_3\}/T_3$ of the natural hormone in native serum. With the conservation $\vartheta_3 + \{SP\cdot\vartheta_3\} = T_3$ and the proportionality factor c_2', we find $p = p_o r = c_2'\{SP\cdot\vartheta_3\}/T_3 = c_2'(T_3 - \vartheta_3)/T_3$, or after rearranging

$$\vartheta_3 = (c_2' - p_o r)T_3/c_2' . \quad (3)$$

From (2) and (3) results $k' = p_o r/(c_2' - p_o r)$. In (3) we, arbitrarily, set $c_2' = 1$ and define Θ_3 as estimate of ϑ_3.

$$\Theta_3 = (1 - p_o r)T_3 \quad (4)$$

As Θ_3 cannot be negative, (4) is only applicable for $1-p_o r > o$, i.e. $r < 1/p_o = 1.42$.

It is further assumed that for thyroxine and triiodothyronine to be effective intracellularly in peripheral tissues and in pituitary thyrotropic cells, it needs them to be bound to an intracellular receptor with concentration η [24]. One receptor molecule binds up to $\alpha = \alpha_4 + \alpha_3$ individual thyroid hormone molecules to form a complex:

$$\alpha_4\vartheta_4 + \alpha_3\vartheta_3 + \eta \xrightarrow{k_5'} \{\alpha_4\vartheta_4\cdot\alpha_3\vartheta_3\cdot\eta\}.$$

It is immediately obvious that the concentration of the intracellular complex is proportional to ϑ_4 and ϑ_3, thus also to Θ_4 and Θ_3, besides to α_4, α_3, and η. This allows us to form the plausible product

$$P = \Theta_4\Theta_3 \quad . \quad (5)$$

Thus P represents the combined biological effectiveness of thyroxine and triiodothyronine. It further helps us to reduce the space $\mathcal{R}^4(\Theta_4,\Theta_3,H(o),R)$ by one dimension to $\mathcal{R}^3(P,H(o),R)$.

The TRH stimulated TSH secretion rate can be considered to be proportional to the pituitary TSH content $h(t)$. With k as secretory rate constant we can write, $dh(t)/dt = -kh(t)$. With the initial condition, $t=o$ and $h(o)$, we find the solution, $h(t) = h(o)\exp\{-kt\}$. After TRH stimulation the TSH increases accordingly. With v as TSH distribution volume and $H(t)$ as serum concentration of TSH, we formulate

$$v\,dH(t)/dt = -dh(t)/dt = kh(o)\exp\{-kt\} .$$

At $t=o$ we have the serum TSH concentration $H(o)$ and obtain the solution

$$v[H(t) - H(o)] = h(o)[1 - \exp\{-kt\}] \quad .$$

With $H(t) = RH(o)$ and $\exp\{-kt\} \simeq 1 - kt$ we obtain the following relation

$$R - 1 = kh(o)t/vH(o) . \quad (6)$$

This shows the proportional relationship of R-1 to the initial pituitary TSH content $h(o)$ and to the secretory rate constant k. Thus R-1 is proportional to the first approximation of the amount of thyrotropin secreted during t=20 min. From (6) it is further obvious, that R-1 is proportional to the initial thyrotropin secretion rate $kh(o)$.

5. The Cusp Catastrophe Metaphor

The creation of a bimodal distribution of pituitary thyrotropic responsiveness to protirelin after treatment of primary hyperthyroidism (Fig.2D) and the formation of a unimodal distribution of responsiveness in primary hypothyroidism (Fig.2B) by the same system suggest a bifurcative behavior and the applicability of a cusp catastrophe metaphor as a general model of the system.

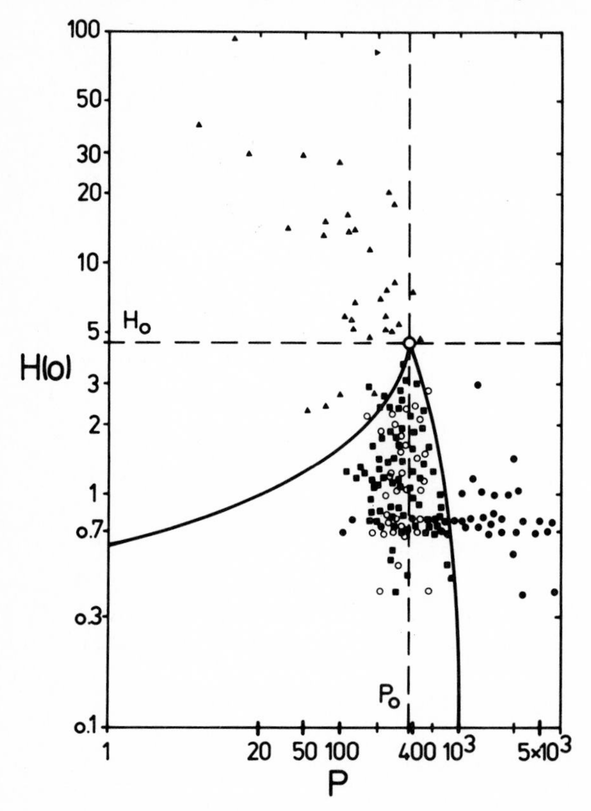

Fig.3 Distribution of data points and best fit to these data of the bifurcation set in the P-H(o)-plane (control plane). ■ Euthyroidism; ▲ primary hypothyroidism; ● hyperthyroidism; ○ treated hyperthyroidism.

Plotting H(o) versus P of the normal euthyroid population, $N_1=202$, we find the points (P,H(o)) spread over a triangular region with tip point $O(P_o,H_o)$, $P_o,H_o > o$, as shown in Fig.3. The points of hyper- and hypothyroidism lie mainly outside this region. But the triangular cuspidated area comprises, also, the data pairs $(P_i,H_i(o))$, $i=1,2,\ldots,N_4$, $N_4=108$ of hyperthyroid patients treated to normal euthyroid values of thyroxine and triiodothyronine and showing a bimodal histogram of R (Fig.2D). This bimodal distribution indicates that the pituitary thyrotropic system can attain 2 stable states expressed by R, as long as P and H(o) maps into the triangular region, the cusp. According to THOM's catastrophe theory [20,21,25,26] P and H(o) can be considered control parameters; they span the control plane of a simple cusp catastrophe with y as state parameter and the singularity $O(P_o,H_o,y_o)$ as organizing center of the associated universal unfolding of the potential

$$U(y) = y^4/4 + h'y^2/2 + p'y \tag{7}$$

with y_o=o. The potential U(y) describes the properties of the system in the vicinity of $O(P_o,H_o,y_o)$. In the region of the cusp this potential forms 2 minima corresponding to the 2 stable attractor states and one maximum being an unstable repellor state. These 3 extrema and other stationary values of (7) outside the cusp are found by setting $\partial U/\partial y = o$ and solving for y; thus we obtain the cubic equation

$$y^3 + h'y + p' = o \quad . \tag{8}$$

This polynomial of 3rd order together with $\partial^2 U/\partial y^2 = o$ and after eliminating y,gives

$$4h'^3 - 27p'^2 = o \quad . \tag{9}$$

For $h' < o$, (9) yields the bifurcation set (set of critical values) of (p',h') at which two of the stable states of (7) merge to one as two of the roots of (8) become complex. In other words, (9) describes the boundary of the cusp region as projection of the inflection lines of the S-shaped hypersurface (8) onto the control plane (p',h')(compare Fig.4).

6. Adaptation of Cusp to Data

The unfolding of the potential U(y) in the vicinity of $O(P_o,H_o,y_o)$ is considered to be a qualitative model of the secretory potential in the thyrotropic cells. But there exists a diffeomorphism $\phi : (p',h',y) \rightarrow (P,H(o),R)$, i.e. smooth transformation of variables that allow adaptation of (7),(8), and (9) to the data; it makes the model more quantitative. For the bifurcation set we assume to hold:

$$4c^3(H(o) - H_o)^3 - 27(P - P_o)^2 = o \quad . \tag{10}$$

Since $R = H(t)/H(o) > o$, and y in (8) will attain positive, negative, and zero values, we use the following transformation

$$\ln R = (y + y_o)/a \quad . \tag{11}$$

Substituting (10) and (11) into (8) we get

$$(a\ln R - y_o)^3 + c(H(o) - H_o)(a\ln R - y_o) + (P - P_o) = o \quad . \tag{12}$$

First estimates of the parameters a, c, y_o, H_o, and P_o can be obtained from fitting the curves of (10) and (11) to the data by eye. Thus by means of (12) a set of $R_i^{(c)}$, $i=1,2,3,\ldots N_1,N_1+1,\ldots,N_1+N_2,\ldots,N_1+N_2+N_3$; $N_1+N_2+N_3=314$, can be calculated by using P_i and $H_i(o)$, the calculated or measured individual data of P and H(o). Further $R_i^{(c)}$ can be compared with the measured individual data $R_i=H_i(t)/H_i(o)$. For fitting the parameters of (12) to the data by iterative computer calculations we define the optimality criterion

$$S = \sum_{i=1}^{N_1+N_2+N_3} (R_i - R_i^{(c)})^2 \quad ,$$

which has to become minimal.

Table 2 First parameter estimates of the cusp and their final values fitted to minimize S

Iteration	P_o	H_o	c	a	y_o	S
o	350.0	4.500	27.0	4.900	1.300	8399.2305
92	390.0	4.495	27.8	4.610	1.150	7620.4453

Starting from the first parameter estimates with S = 8399, after 92 iterations S settled around S = 7620 as shown in Table 2 .

7. A Secretory Model

The hormone measurements have shown that under certain conditions the stimulated thyrotropin serum concentrations distribute bimodally. It is inferred that pituitary thyrotropin secretion attains two different functional states. After data transformation the relationship of thyroid hormone concentration and basal thyrotropin secretion to the stimulated pituitary secretion of thyrotropin can be described macroscopically by a cusp catastrophe metaphor. In the following a stochastic microscopic model is constructed that will explain the different functional states of the thyrotropic cells by a catastrophe-associated potential. The microscopic model comprises the formation of secretory granules, a stimulus-secretion coupling system, and thyrotropin diffusion.

Pituitary thyrotropic cells (thyrotrophs) synthesize thyrotropin (TSH) and store it in subcellular secretory granules, which are membrane coated and of nearly equal size [8,10]. The amount or quantum of TSH packed into one granule we call Q. Thyrotropin is released from the thyrotrophs by exocytosis involving the fusion of the membrane of the secretory granule with the plasma membrane and the extrusion of the hormone quantum. The extruded quanta of thyrotropin form the extracellular hormone concentration X in the close vicinity of the thyrotrophs. From there thyrotropin diffuses into the blood stream with thyrotropin concentration X_o by rate $s(X - X_o)$. The functional coupling between the secretory granules and the TRH activated plasma membrane is brought about by a stimulus-secretion coupling system E_i [4], with n binding sites for granules; i = o,1,2,...,n indicates the number of sites occupied or the number of granules bound. It is assumed that E_o can experience a reversible conformational change to F by rate c_1 and c_2, thereby becoming independent of TRH. There are some indications that part of the stimulus-secretion coupling system is formed by contractile microfilaments [22]. Intracellular thyrotropin is synthesized and packed into granules by rate J_1 and eliminated by coalescence of granules with lysosomes with rate j_2. Association and dissociation of free granules Q with the coupling system E_i is governed by rates k_1 and k_2, respectively. The rate of exocytosis is k_3 and that of endocytosis k_4. The overall reaction system is formulated by reaction equations (13), (14), and (15) and shows some resemblance to models of allosteric transition [5,13].

$$\xrightarrow{J_1} Q \xrightarrow{j_2} \quad ; \quad E_o \underset{c_2}{\overset{c_1}{\rightleftarrows}} F \tag{13),(14}$$

$$\begin{array}{llllll} Q + E_o & \underset{k_2}{\overset{nk_1}{\rightleftarrows}} & E_1 & \underset{nk_4}{\overset{k_3}{\rightleftarrows}} & E_o + X & \\ & & \vdots & & & \\ Q + E_i & \underset{(i+1)k_2}{\overset{(n-i)k_1}{\rightleftarrows}} & E_{i+1} & \underset{(n-i)k_4}{\overset{(i+1)k_3}{\rightleftarrows}} & E_i + X & \overset{s}{\longrightarrow} X_o \\ & & \vdots & & & \\ Q + E_{n-1} & \underset{nk_2}{\overset{k_1}{\rightleftarrows}} & E_n & \underset{k_4}{\overset{nk_3}{\rightleftarrows}} & E_{n-1} + X & \end{array} \tag{15}$$

The coupling system is bounded by the conservation law

$$E = F + \sum_{i=o}^{n} E_i \quad . \tag{16}$$

We consider the coupling system (13) - (16) to be a Markovian process and use X, the amount of thyrotropin extruded into the extracellular space, as provisional state parameter. X-1 equals the X less the amount of the quantum of one granule and X+1 the amount plus one granule. The probability of the system to be in state X at time t is denoted by W(X,t). The transition probabilities are given by the reaction rates of the individual process [12,14]. Therefore the probability of the system to be in state X after time Δt has elapsed, can be formulated.

$$W(X,t+\Delta t) = W(X,t) + \Delta t W(X-1,t) k_3 \sum_{i=1}^{n} iE_i - \Delta t W(X,t) k_3 \sum_{i=1}^{n} iE_i$$

$$+ \Delta t W(X+1,t)\{[X+1] k_4 \sum_{i=0}^{n-1} (n-i)E_i + s[X+1-X_0]\}$$

$$- \Delta t W(X,t)\{X k_4 \sum_{i=0}^{n-1} (n-i)E_i + s[X - X_0]\} + \Omega(\Delta t) \qquad (17)$$

$\Omega(\Delta t)$ is a residual probability compensating for errors at large Δt. For $\Delta t \to o$, $\Omega(\Delta t)/\Delta t \to o$. In order to obtain the master equation (not shown) from both sides of (17) W(X,t) is subtracted and the results are divided by Δt. We introduce $\lambda = k_3 \sum_{i=1}^{n} iE_i$ and $\mu(X) = k_4 X \sum_{i=o}^{n-1} (n-i)E_i + s[X-X_0]$. With the limit $\Delta t \to o$ we can rewrite the master equation as partial differential equation

$$\partial W(X,t)/\partial t = \lambda W(X-1,t) - \lambda W(X,t) + \mu(X+1)W(X+1,t) - \mu(X)W(X,t) \quad . \qquad (18)$$

Further we introduce the notation $\Lambda(X)=\lambda W(X,t)$, $\Lambda(X-1)=\lambda W(X-1,t)$, $M(X)=\mu(X)W(X,t)$, $M(X+1)=\mu(X+1)W(X+1,t)$, and expand by Taylor series [11] ;

$$\Lambda(X-1) = \sum_{\nu=0}^{\infty} \{(-1)^{\nu}/\nu!\}(\partial/\partial X)^{\nu}\Lambda(X) \text{ and } M(X+1) = \sum_{\nu=0}^{\infty} \{(+1)^{\nu}/\nu!\}(\partial/\partial X)^{\nu} M(X).$$

Thus we can write (18) in the following way:

$$\partial W(X,t)/\partial t = \sum_{\nu=1}^{\infty} \{(+1)^{\nu}/\nu!\}(\partial/\partial X)^{\nu}\{[\lambda + (-1)^{\nu}\mu(X)]W(X,t)\} \quad . \qquad (19)$$

We simplify (19) by introducing

$$D_{\nu}(X) = \{1/\nu!\}[\lambda + (-1)^{\nu}\mu(X)] \quad , \qquad (20)$$

and obtain the generalized Fokker-Planck equation

$$\partial W(X,t)/\partial t = \sum_{\nu=1}^{\infty} (-1)^{\nu}(\partial/\partial X)^{\nu}\{D_{\nu}(X)W(X,t)\} \quad . \qquad (21)$$

We consider the Markovian process to be Gaussian, i.e. $D_{\nu}(X) = o$ for $\nu > 2$, and further assume $D_2(X)$ to be constant in the vicinity of the organizing center O, and thus to be independent of X, $D_2(X) = D_2$ = const.. Therefore we have the simple Fokker-Planck equation or Kolmogorov's first equation

$$\partial W(X,t)/\partial t = -(\partial/\partial X)\{D_1(X)W(X,t) - D_2(\partial/\partial X)W(X,t)\} \quad . \qquad (22)$$

$J = D_1(X)W(X,t)-D_2(\partial/\partial X)W(X,t)$ is called probability current [7,11].

In case of the stimulus-secretion coupling system being at rest, λ=o and $\mu(X)$=o, and then stimulated by an episodic release from the hypothalamus of TRH, or, as performed in our studies, stimulated by a bolus injection of TRH, we postulate the transition to a new steady state of the system or another detailed balance of the Markovian process is reached nearly instantaneously after the initiation of the TRH stimulus. The new steady state is maintained at least for the following 20 min of observation, i.e. the system undergoes an activation in the form of a Heaviside function. Therefore we can assume that $\partial W(x,t)/\partial t = o$, J = o, and that W(X,t) is independent of time for t>o. In consequence (22) reduces to a first order differential equation in X with the initial conditions $X(t=o) = X_0$ and $W(X_0) = W_0$; $D_2(d/dx)W(X) - D_1(X)W(X) = o$. Its solution is easily found to be

$$W(X) = W_o \exp\{1/D_2 \int_{X_o}^{X} D_1(X)dX\} \quad . \tag{23}$$

This equation describes the probability density $W(X)$ as function of X and other parameters introduced in (13) - (16). W_o is to be chosen such that $\int_o^{+\infty} W(X)dX = 1$. In analogy to thermodynamic potentials we call, in the following, $V(X)$ probability potential

$$V(X) = \int_{X_o}^{X} D_1(X)dX \ . \tag{24}$$

According to (20) we find for $\nu = 1$

$$D_1(X) = k_3 \sum_{i=1}^{n} iE_i - k_4 X \sum_{i=o}^{n-1} (n-i)E_i - s(X - X_o) \quad . \tag{25}$$

$D_1(X)$ we call drift coefficient and D_2 diffusion coefficient of the Markovian process.

The amount Q of unbound secretory granules can be considered as another state parameter of the system, and by using (17) - (22) in the way as before we arrive at a probability density $W(Q)$ similar to (23). Equivalent to (25) we can write for $\nu=1$

$$D_1(Q) = k_2 \sum_{i=1}^{n} iE_i - k_1 Q \sum_{i=o}^{n-1} (n-i)E_i + J_1 - j_2 Q \quad . \tag{26}$$

From (13) - (16) we derive other drift coefficients

$$D_1(F) = c_1 E_o - c_2 F \ , \tag{27}$$

$$D_1(E_o) = c_2 F - c_1 E_o - nk_1 Q E_o + k_2 E_1 + k_3 E_1 - nk_4 X E_o \quad , \tag{28}$$

$$D_1(E_i) = -(n-i)k_1 Q E_i + (i+1)k_2 E_{i+1} + (i+1)k_3 E_{i+1} - (n-i)k_4 X E_i$$
$$+(n+1-i)k_1 Q E_{i-1} - ik_2 E_i - ik_3 E_i + (n+1-i)k_4 X E_{i-1} \quad , \tag{29}$$

for $i = 1,2,\ldots, n-1$,

$$D_1(E_n) = k_1 Q E_{n-1} - nk_2 E_n - nk_3 E_n + k_4 X E_{n-1} \quad . \tag{30}$$

Since we assume the system to be in a steady state during TRH stimulated secretion, only the most probable values of the probability potential $V(X)$ are interesting. These values of X producing extrema of $V(X)$, are obtained by setting $\partial V(X)/\partial X = o$, i.e. the drift coefficient vanishes, $D_1(X) = o$. Equivalently we must set $D1(Q), D_1(F), D_1(E_i) = o$ for all i. As shown by OPPENHEIM et al. [14] and KURTZ [9] the macroscopic equations of chemical reaction systems are equal to the formulation of equivalent Markov chain models in the limit of an infinite volume. Equally the drift coefficients $D_1(.)$ in our notation are identical with the macroscopic equations.

We introduce $\gamma=c_1/c_2$, $x=k_4X/(k_2+k_3)$, $x_o=k_4X_o/(k_2+k_3)$, $q=k_1Q/(k_2+k_3)$, and $j_1 = k_1J_1/(k_2+k_3)$, and find from (27) and (28) - (3o) that $F = \gamma E_o$, and $E_i=\binom{n}{i}(q+x)^i E_o$. With $\sum_{i=o}^{n} E_i = E_o(1+q+x)^n$ we can rewrite (16) as $E=E_o\{\gamma+(1+q+x)^n\}$. Immediately we find

$$\sum_{i=1}^{n} iE_i = n(q+x)E_o(1+q+x)^{n-1} \text{ and } \sum_{i=o}^{n-1} (n-i)E_i = nE_o(1+q+x)^{n-1} \quad . \tag{31),(32}$$

Thus (25) and (26) are transformed to

$$D_1(X) = \frac{k_2+k_3}{k_4} D_1(x) = \frac{+nE(1+q+x)^{n-1}(k_3 q - k_2 x)}{\gamma + (1+q+x)^n} - \frac{k_2+k_3}{k_4} s(x-x_o), \tag{33}$$

$$D_1(Q) = \frac{k_2+k_3}{k_1} D_1(q) = \frac{-nE(1+q+x)^{n-1}(k_3 q - k_2 x)}{\gamma + (1+q+x)^n} + \frac{k_2+k_3}{k_1}(j_1 - j_2 q). \tag{34}$$

The common term of $D_1(x)$ and $D_1(q)$ we set equal to u,

$$u(q,x) = nE(1+q+x)^{n-1}(k_3q-k_2x)/(k_2+k_3)\{\gamma+(1+q+x)^n\} \quad , \tag{35}$$

and define $u(q,x)$ as new state parameter. It describes the stimulated transmembrane flow of thyrotropin accomplished by the stimulus-secretion coupling system E_i. Eqs. (33) and (34) are rewritten

$$D_1(X) = (k_2+k_3)\{k_4u-s(x-x_o)\}/k_4 \; , \; D_1(Q) = (k_2+k_3)\{-k_1u+j_1-j_2q\}/k_1 \; . \quad (36),(37)$$

For steady state solutions with $D_1(X)=o$, $D_1(Q)=o$, and $k_1,k_2,k_3,k_4>o$ we find from (36) and (37)

$$q = j - \kappa u \; , \qquad x = x_o + Bu \; , \quad (38),(39)$$

with $j=j_1/j_2$, $\kappa=k_1/j_2$, and $B=k_4/s$. Setting $E/(k_2+k_3)=\rho$ and replacing q and x in (35) by (38) and (39), we get a polynomial in u

$$\{\gamma+[(1+j+x_o)+(B-\kappa)u]^n\}u = n\rho[(k_3j-k_2x_o)-(k_3\kappa+k_2B)u][(1+j+x_o)+(B-\kappa)u]^{n-1}. \tag{40}$$

This polynomial describes the steady state solutions of the transmembrane thyrotropin flow after TRH stimulation. As pointed out before it suffices to obtain a polynomial of 3rd order to describe the experimental data satisfactorily, i.e. in (40) we can choose n=2. By abbreviating with $(1+q+x_o)=\beta$, $B-\kappa=\delta$, $k_2B-k_3\kappa=\zeta$, and $k_3j-k_2x_o=\varepsilon$, we get from (40)

$$\delta^2u^3 + 2\delta(\beta+\rho\zeta)u^2 + (\gamma+\beta^2+2\rho\beta\zeta-2\rho\varepsilon\delta)u - 2\rho\beta\varepsilon = o \quad . \tag{41}$$

In (12), for convenience of writing we set $a\ln R = Z$, and after rearranging we obtain the expression

$$Z^3 - 3y_oZ^2 + \{3y_o^2+c(H(o)-H_o)\}Z - y_o^3-c(H(o)-H_o)y_o + (P-P_o) = o \quad . \tag{42}$$

As shown in (6) R is closely related to thyrotropin secretion rate kh(o). Therefore R is also conceptually akin to the transmembrane thyrotropin flow u . Because of these relationships it is presumed that the following hold: $u \equiv Z = a\ln R$, and $x_o \equiv H_o$. By comparison of coefficients of (41) with (42), we can derive some conditions that must be fullfilled so that system (13) - (16) describes the experimental data. We find $\delta = -1$, from which follows $k_4/s < k_1/j_2$. Further we get $y_o = (2/3)(\beta+\rho\zeta)$ and $y_o > o$.

From (13) to (16) it is obvious that $\sum_{i=o}^{n} E_i = E - F$ contributes an important part to the secretion rate of thyrotropin. The association of the E_i's to the TRH stimulated cell membrane is governed by $\gamma = c_1/c_2$. Under the above introduced conditions, the explicit solution for γ is found to be

$$\gamma = \{[(1+j+H(o))-y_o]^3 + c(H(o)-H_o)[(1+j+H(o))-y_o] + (P-P_o)\}/(1+j+H(o)) \; . \tag{43}$$

Since c_1 and c_2 are always positive, i.e. $\gamma > o$. This inequality implies, according to (43), as necessary condition, that $y_o < 1+j+H(o)$.

Eq.(43) can be rearranged to form

$$(1+j+H(o)-y_o)^3 + c(H(o)-H_o-\gamma)(1+j+H(o)-y_o) + (P-P_o-\gamma y_o) = o \quad , \tag{44}$$

which represents the manifold of a cusp catastrophe with the organizing center $O'(P_o+\gamma y_o, H_o+\gamma, y_o)$. For $\gamma \to o$, its bifurcation set becomes identical with (10), and $1+j+H(o)-y_o$ attains two stable solutions inside the region of the cusp. As the measurements have shown that the basal thyrotropin concentrations H(o) behave monotonically smooth and y_o=const., the discontinuous behavior of the thyrotropic responsiveness R can no longer be explained by the stimulus-secretion coupling system (13) - (16), with $\gamma = o$, but then must reside in $j = f(J_1) = J_1k_1/j_2(k_2+k_3)$, which reflects hormone synthesis and granula formation. Moreover, for stimulus-secretion coupling system to be valid and in order to explain the discontinuous behavior of responsiveness, the reversible allosteric transition $E_o \rightleftarrows F$, i.e. $\gamma >> o$, is an essential feature of the thyrotropic system.

8. Discussion

THOM's catastrophe theory [20,21] was applied to construct biological models, especially to describe behavior of the heart beat and nerve action potential [26], ontogenetic morphogenesis [3], ciliary movement [23], and denaturation of proteins [1,2]. The theory's context, relevant to physics and biology, was recently reviewed [6].

In this paper catastrophe theory is used to formulate a coherent concept of individual macroscopic data of pituitary thyrotropic function in man. Thyrotropin release from the pituitary is a function of thyroid hormone feedback inhibition and hypothalamic TRH stimulation. The hypothalamic controlability of thyrotropin release is expressed by the measure R of thyrotropin responsiveness to a constant TRH stimulus. Moreover, the splitting factor, the basal thyrotropin H(o) is inversely related to thyroid hormone concentration P, thus inhibiting at the pituitary level. By means of the resultant cusp catastrophe (Fig.4), three stable operational regions of physiological significance can be distinguished:

1. Euthyroidism (region A) is characterized by low endogenous basal thyrotropin secretion reined by thyroid hormone feedback inhibition and shows concomitantly high responsiveness of thyrotropin secretion to a TRH stimulus. This region represents the functional states predominantly controlled by the hypothalamus and the thyroid gland.

2. Hyperthyroidism (region C) shows inhibition of endogenous basal thyrotropin secretion and also inhibition of thyrotropin release induced by TRH. It is the state of complete feedback inhibition by thyroid hormones with concomitant abolition of hypothalamic control. The same is true to region D.

3. Hypothyroidism (region B) is featured by enhanced basal thyrotropin because of lack of thyroid hormone feedback inhibition and shows reduced thyrotropin responsiveness to TRH stimulation.

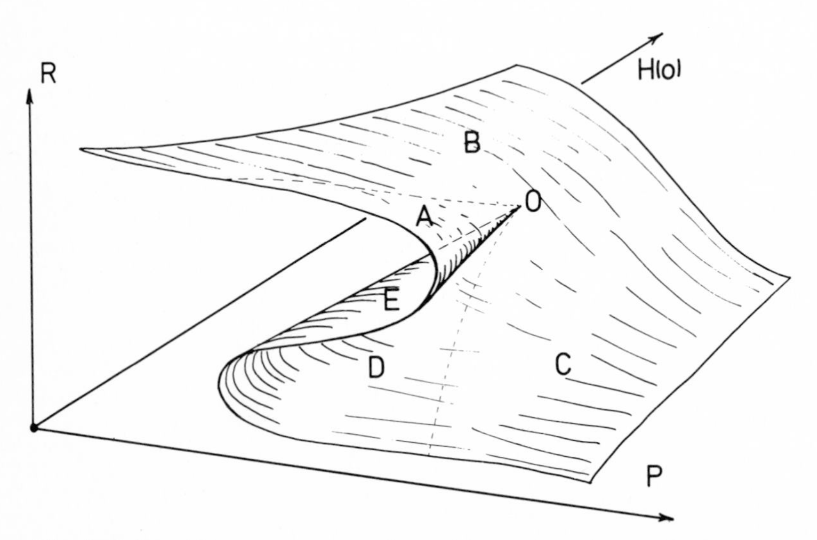

Fig.4 Schematic cusp catastrophe representing the stationary states of pituitary thyrotropic responsiveness R as function of P and H(o). A: Region of euthyroidism; B: region of primary hypothyroidism; C: region of hyperthyroidism; D: region of hyperthyroidism treated to euthyroid values of P with still highly reduced R; E: region of unstable stationary states; O: organizing center of the catastrophe.

For the cusp model to be compatible with physiological reality, it must entail the spontaneous return to the normal euthyroid behavior, if once the pituitary had undergone the transition to the hyperthyroid mode by temporary increase of thyroid hormone concentration P. The hyperthyroid mode, even after reduction of P to euthyroid values, means no thyrotropin secretion and consequently no stimulation of the thyroid gland. The resultant lack of thyroid hormone secretion leads finally to a reduction of thyroid hormone concentration and of metabolic effects in tissues, and also in the pituitary, until P decreases so far to the hypothyroid range that the inverse transition to higher responsiveness R occurs. Only after this return transition will the pituitary show euthyroid behavior, when the reinstituted thyrotropin secretion has elicited thyroid stimulation and in turn an increase to euthyroid values of thyroid hormone concentration P. For this to happen, a stimulatable, normally functioning thyroid gland is necessary. This hysteresis behavior of the pituitary for a certain range of basal thyrotropin secretion is teleonomically meaningful, since the time constant of decrease of thyroid hormone dependent metabolism is much larger than the time constant of thyrotropin and thyroid hormone increase. This behavior further explains the observation that a patient must be guided to a hypothyroid state for a short time, if euthyroid responsiveness shall be reached from hyperthyroidism.

In generalizing our results the cusp catastrophe can be considered the topological functional structure of the most probable, thus stable stationary state brought forth by a "thyrotropic secretory potential". This model visualizes discontinuous transitions from the operational region of euthyroidism to hyperthyroidism by enhancing continuously thyroid hormone concentration, whereas the opposite transition is brought about by reducing smoothly thyroid hormones to end up in the boundary region between hypothyroidism and euthyroidism. Although these discontinuous transitions are solutions derived from equation (12), the real system may show somehow different but still precipitous transitions in the close vicinity of the organizing center 0 because of microscopic fluctuations in the system. The available data do support discontinuities but cannot prove certain individual transitions because ergodicity is implied in the mathematical treatment of the data, i.e. $R_t(i) = R_i(t)$.

The used ultrastructural microscopic system of secretory granules and stimulus-secretion coupling is *one* possible model that can explain the measured macroscopic data. Besides, other plausible functional relations can be invented to meet the data equally well, e.g. an autocatalytic chemical process [18]. Whatever model is used, two features are essential: 1. an overall process of 3rd order, e.g. reactions with E_i, i = o,1,2 , and 2. a process that increases order of the system by maintaining its operational point away from thermodynamic equilibrium, e.g. conformational change of proteins, as $F \not\approx E_0$. In the system under consideration higher order is produced not by an independent parameter - biological systems never allow independence - but rather by the effects of the thyroid hormones, mediated by the basal thyrotropin secretion, the splitting factor, i.e. H(o) is "also" a function of P.

In its application to our data catastrophe theory provides a general qualitative model or metaphor [15,20,21] and moreover a conceptual link between the local stochastic and ultrastructural transportation process of secretory granules and the global macroscopic behavior of the thyrotropin secretion system. The cusp model shows how an inhibitory and a stimulatory effect cooperate on a single macroscopic function and visualizes once again that extrapolations of data beyond certain critical parameter values (bifurcation set) are not valid, which is true *par excellence* of biosystems.

References

1. Benham C.J.& J.J.Kozak: Denaturation: An Example of a Catastrophe. II. Two-state Transitions. J. theor. Biol. 63(1976), 125 - 149.

2. Benham C.J.& J.J.Kozac: An Example of a Catastrophe. III. Phase Diagrams for Multistate Transformations. J. Theor. Biol. 66(1977), 679 - 693.

3. Cooke J. & E.C. Zeeman: A Clock and Wavefront Model for Control of the Number of Repeated Structures during Animal Morphogenesis. J. Theor. Biol. 58(1976), 455 - 476.

4. Douglas W.W.: Stimulus-Secretion Coupling: The Concept and Clues from Chromaffin and other Cells. Br. J. Pharmacol. 34(1968), 451 - 474.

5. Go N.& Y. Anan: Regulatory Functions of Allosteric Enzymes in Far-from-Equilibrium Systems. J. theor. Biol. 66(1977), 475 - 483.

6. Güttinger W.: Catastrophe Geometry in Physics and Biology, p. 2 -3o. In: M. Conrad, W. Güttinger, and M. DalCin, ed., "Physics and Mathematics of the Nervous System". Springer Verlag, Berlin-Heidelberg-New York, 1974.

7. Haken H.: Syergetics, an Introduction: Nonequilibrium Phase Transitions and Self-Organization in Physics, Chemistry, and Biology. Springer Verlag, Berlin-Heidelberg-New York, 1977.

8. Kurosumi K. & H. Fujita: Functional Morphology of Endocrine Glands. Georg Thieme Publ. Stuttgart and Igaku Shoin Ltd. Tokyo, 1974.

9. Kurtz T.G.: The Relationship between Stochastic and Deterministic Models for Chemical Reactions. J. Chem. Phys. 57(1972), 2976 - 2978.

10. Landolt A.M.: Ultrastructure of Human Sella Tumors. Acta Neurochirurgica, Suppl. 22, Springer Verlag, Wien-New York, 1975.

11. Lax M.: Fluctuation and Coherence Phenomena in Classical and Quantum Physics. p. 271 - 478. In: M. Chrêtien, E.P. Gross, and S. Deser, ed., "Statistical Physics, Phase Transitions and Superfluidity". Gordon and Breach Science Publ., New York-London-Paris, 1968.

12. McQuarrie D.A.: Stochastic Approach to Chemical Kinetics. J. Appl. Prob. 4(1967), 413 - 478.

13. Monod J., J. Wyman & J.P. Changeux: On the Nature of Allosteric transitions: A Plausible Model. J. Mol. Biol. 12(1965), 88 - 118.

14. Oppenheim I., K.E. Shuler & G.H. Weiss: Stochastic and Deterministic Formulation of Chemical Rate Equations. J. Chem. Phys. 5o(1969), 460 - 466.

15. Rosen R.: The Generation and Recognition of Patterns in Biological Systems. p. 222 - 341. In: D.E. Matthews, ed., "Mathematics and the Life Sciences". Springer Verlag, Berlin-Heidelberg-New York, 1977.

16. Sanchez-Franco F., M.D. Garcia, L. Cacicedo, A. Martin-Zurro, F. Escobar del Rey & G.Morreale de Escobar: Transient Lack of Thyrotropin (TSH) Response to Thyrotropin-Releasing Hormone (TRH) in Treated Hyperthyroid Patients with Normal or Low Serum Thyroxine (T_4) and Triiodothyronine (T_3). J. Clin. Endocrinol. &Metab. 38(1974), 1o98 - 1102.

17. Seif F.J.: Mathematical Model of Pituitary Thyrotropic Function. Experientia 33(1977), 1243 -1244.

18. Seif F.J.: Cusp Catastrophe Model of Pituitary Thyrotropic Function Stimulatable by Protirelin. Biomedizinische Technik 23(1978), Suppl.p.131.

19. Sterling K. & M.A. Brenner: Free Thyroxine in Human Serum: Simplified Measurement with the Aid of Magnesium Precipitation. J. Clin. Invest. 45(1966),153-163.

2o. Thom R.: Topological Models in Biology. p. 89 - 116. In: C.H. Waddington, ed., "Towards a Theoretical Biology", Vol.3,Edinburgh Univ.Press, Edinburgh,1970.

21. Thom R.: Stabilité structurelle et morphogénèse. Essai d'une théorie générale des modèles. W.A. Benjamin Inc., Reading, Mass., 1972.

22. Trifaro J.M.: Common Mechanismes of Hormone Secretion. Ann. Rev. Pharmacol. Toxicol. 17(1977), 27 - 47.

23. Varela F.J., J.C. Rowley III. & D.T. Moran: The Control of Ciliary Movements: An Application of the Cusp Catastrophe. J. theor. Biol. 65(1977), 531 - 553.

24. Wahl R., D. Geiseler & E. Kallee: Adsorption Equilibria of Thyroid Hormones in the Liver Cell. Eur. J. Biochem. 80(1977), 25 - 33.

25. Woodcock A.E.R.: Catastrophe Theory and the Modelling of Biological Systems, p. 342 - 385. In: D.E. Matthews, ed., "Mathematics and the Life Sciences". Springer Verlag, Berlin-Heidelberg-New York, 1977.

26. Zeeman E.C.: Differential Equations for the Heartbeat and Nerve Impulse, p. 8 - 67. In: C.H. Waddington, ed., "Towards a Theoretical Biology", Vol 4, Edinburgh Univ. Press, Edinburgh, 1972.

Chaos

Otto E. Rössler

Institute for Physical and Theoretical Chemistry, University of Tübingen
D-7400 Tübingen

and

Institute for Theoretical Physics, University of Stuttgart
D-7000 Stuttgart, Fed. Rep. of Germany

Contents

1. Introduction

'Chaos' is the canonic translation into Greek of the Hebrew term 'tohu-wa-bohu' found in the first chapter of the Bible. 'Wa' means 'and,' and 'bohu' certainly means the same thing as 'tohu.' However, since the word appears only once in the Bible and there is no continuous oral tradition, one can only guess what 'tohu' means. A possible English analogue is 'topsy(-and-)turvy.' A mathematical redefinition thus seems admissible.

This has been done by Yorke [1] (see May [2]), who proposed the word 'chaos' as a label for a kind of dynamical behavior characterized by the triad: infinite number of periodic trajectories; uncountable number of nonperiodic trajectories; hyperbolicity (instability) of all (or the overwhelming majority of; as is proposed here) trajectories in the regime. A favorite example of Yorke and May was the logistic difference equation

$$x_{t+1} = \alpha\, x_t(1 - x_t). \qquad (1)$$

If, for example, $\alpha = 4$, the equation fulfills the three constraints. This is easily seen by looking at iterates: The equation as it is produces an inverted parabola that is non-negative between $x = 0$ and $x = 1$ and has two fixed points (intersections with the line $x_{t+1} = x_t$), one at zero and one at the right-hand shoulder. The second iterate is double-peaked in the same interval, having four fixed points, the third has four peaks with eight fixed points, and the nth has 2^n fixed points, all of them unstable (numerical value of the slope of the tangent > 1). So there is a countable number of periodic solutions (fixed points) of repelling type. In between, there are uncountably many nonperiodic (wandering) solutions.

Julia, a pupil of Poincaré's, in 1918 first considered iterates of a quadratic polynomial and found, for example, that for most parameter values in the region of interest (α between 3.57 [3] and 4 in the case of Eq. (1)), there is exactly one periodic attractor contained in the infinite set of periodic solutions [4].

It is that periodic solution which (roughly speaking) comes closest to the maximum of the map. For any solution running precisely through the maximum is an arbitrarily strong attractor when considering closer and closer neighbors [5]; only infinite periodicity of this solution (as when $\alpha = 4$) cancels the effect. Many more subsequent authors have to be mentioned (cf. [6 - 8]). Sharkovsky [9] indicated a canonical bifurcation sequence for smooth single-maximum maps.

A constitutive single property of chaotic systems is hyperbolicity. In one-dimensional examples like Eq. (1) hyperbolicity means that lateral distances between adjacent points on the original interval are (on the average) expanded by a factor of more than one (say 2). This has a curious consequence [10]: digital computers cease to be of avail in calculating orbits beyond about one hundred iterates. For, even if a computer has a hundred (binary) digits, the missing 101st digit will be blown up to cover the whole interval at the 100th iteration. 'Embedded' periodic attractors with a period larger than about one hundred (or some finite multiple) thus are undetectable numerically. They also presumably play no role physically.

Chaotic behavior in hand-held calculators (or one-dimensional functions/single-variable difference equations/one-dimensional endomorphisms, respectively) has attracted much attention because of the apparent simplicity of the class of examples. As it turns out, they form a singular special case in a wider universe.

2. Continuous Chaos

The basic ingredients were discovered by Poincaré (although he did not use the term 'chaos'). Pondering Hill's reduced model of the three-body problem of celestial mechanics [11], Poincaré detected the possibility of not looking at the trajectories themselves, but rather at a 2-dimensional 'surface of section' (cross-section) through the hairdo of threads [12,13]. Accustomed to thinking of dynamical processes in terms of stationary stream-lines of a fluid in laminar motion (see [14]), Poincaré realized that when such a flow is 3-dimensional, there are not just invariant one-dimensional threads (trajectories) and invariant 3-dimensional domains (from which there is no escape), but also invariant 2-dimensional sheets never left by trajectories. On a cross-section, invariant sheets should impose as lines [15]. The upshot was that such lines may intersect in a new way.

When trajectories intersect in 2-dimensional dynamical systems, they do so in invariant singular points with characteristic asymptotic properties. These points had been classified by Poincaré before (for example, the 'saddle' and the 'node,' to mention only the non-rotatory ones). When the 'same' thing now happens in a 2-dimensional cross-section--where the invariant lines correspond to sheets--then the intersection point may indeed again be a critical point (saddle, node, etc.); in its neighborhood, each line will be punctuated, by every trajectory given on it, the more densely the closer the distance to the invariant point is. But there is a second possibility now: intersection in a non-singular point. The trajectory through this point is, like any other non-singular trajectory, bound to hit the cross-section at some other point on its own sheet next time. Only--there are two sheets now. Both invariant sheets therefore have to cross each other again and again. If the two manifolds (sheets) originate from the same singularity, for example, a saddle point (as considered by Poincaré), then that saddle will be both the source and the sink of the trajectory running through the intersection point. Because of this doubly asymptotic relationship (homoclinicity) to the saddle, Poincaré [13] called the intersection point a 'homoclinic point.' For positive t, the unstable manifold of the saddle will 'double up' on the stable manifold (in order to make possible all of the intersections), and vice versa for negative t. So a 'grid' of infinitely many intersection points is formed in the neighborhood of the saddle. In other words, there are infinitely many, infinitely long sequences of homoclinic points implied by a single one.

All this is a bit complicated and counterintuitive, as Poincaré [13] noted. See Fig. 1 for a picture. Birkhoff [16] later added Fig. 2. It shows that in

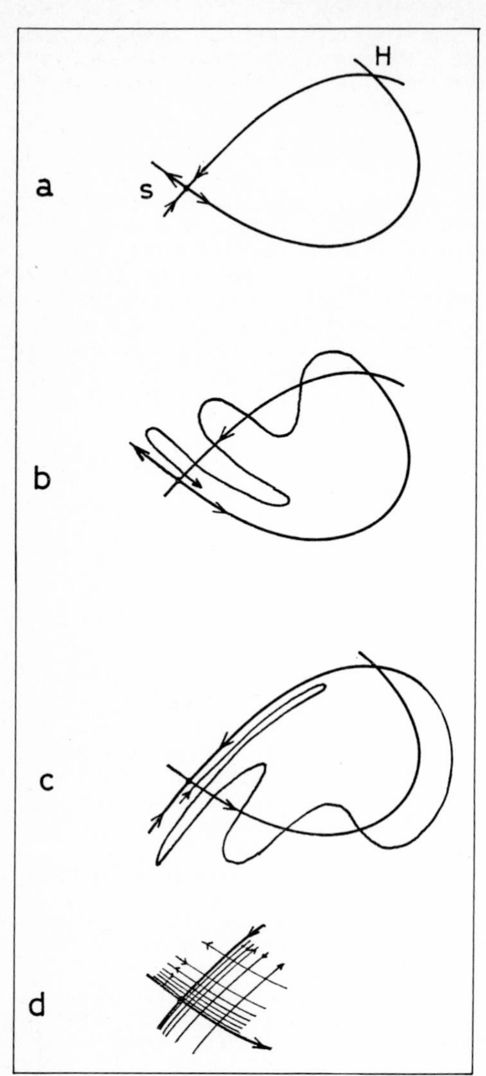

Fig. 1 Poincaré's homoclinic point H in a two-dimensional cross-section through a three-dimensional flow
s = saddle point in the cross-section
a: Basic situation
b: Further crossings of the saddle's stable and unstable manifold, at positive t
c: The same, at negative t
d: Poincaré grid (magnified)

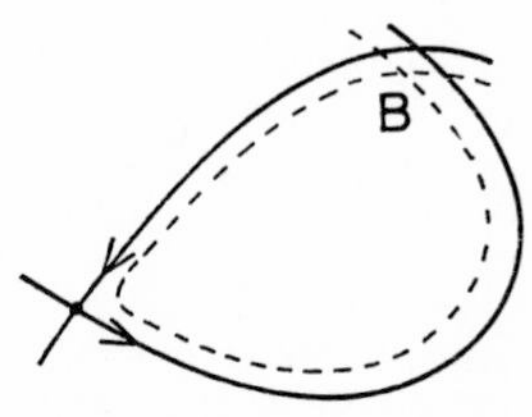

Fig. 2 Birkhoff's point (B) near H

the neighborhood of every point homoclinic to a saddle, there is a whole family of self-intersecting invariant sheets (only one is depicted). Since they all are traversed at differing rates (those lines coming closer to the saddle being punctuated more densely there), the family is bound to comprise an infinite number of cases in which the self-intersection occurs in a periodic point. Therefore, each (transversal) homoclinic point implies an infinite number of periodic trajectories of different periodicities in its neighborhood. Smale [17] complemented the picture by drawing it in the form of Fig. 3. Assuming linearity of the horizontal parts, he showed that, in addition, there is also an uncountable number of nonperiodic trajectories present, namely a Cantor set. This was done by looking at all those points to the right of the saddle which, under iteration, never return to the right side. In this way, more and more 'stripes' are lost to the left side, with a Cantor set of stripes (in fact, stars) remaining.

Beyond the 'mess' of the three-body problem as envisioned by Poincaré, further examples of (as it turned out) the same phenomenon were found soon. Hadamard's example of periodic and nonperiodic geodesics on a surface of negative curvature [18] came second. It also was a Hamiltonian example. A subclass of it found later are Anosov flows [19,17]. For books summarizing the state of knowledge on 'recurrent' and 'random' motions in Hamiltonian systems, see [20 - 22].

Smale [17] stressed that his result did not depend on the Hamiltonian (in a cross-section area-conserving) nature of the flow. The same applies to Poincaré's point H and Birkhoff's point B. (See Figs. 1 - 3.) Smale's construction was motivated [17] by his trying to get a geometric picture of the behavior of the periodically forced van der Pol oscillator [23]. This system is an ordinary (non-Hamiltonian) dynamical system, but nonetheless shows--as discovered by Cartwright and Littlewood in 1945 [24]--at certain parameter values an infinite number of periodic trajectories ('subharmonics') and an uncountable number of nonperiodic ones. It thus provides the first example of chaos in a continuous dynamical system of ordinary (non-Hamiltonian) type.

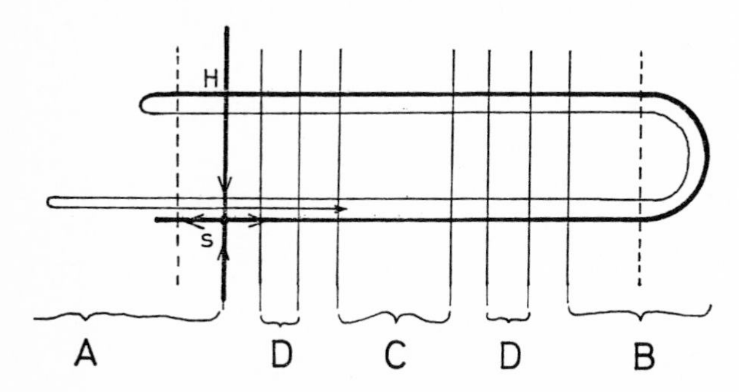

Fig. 3 Smale's Cantor set (S) near H. Compare Fig. 1a. The region between the two vertical dashed lines is assumed to be a linear map.
A: Region of points which remain to the left side of the saddle's stable manifold
B: Region mapped into A after one time step
C: Region mapped into B after one time step
D: Region mapped into C after one time step (and into A after three) and so forth
S: Gap set (set of points staying to the right for $t \to \infty$)

Any periodically forced 2-variable system creates a flow on a three-manifold. Therefore--even though a cross-section is 2-dimensional--the whole flow cannot easily be comprehended. Fortunately, however, there exists an invertible transformation which renders the flow 3-dimensional in any finite region [25,26].

Knowing now that 3-dimensional examples are possible, it is reasonable to ask for pertinent cases in the literature. In 1963, E.N. Lorenz described a 3-variable autonomous nonlinear differential equation with two quadratic terms, producing what he called a 'deterministic nonperiodic flow' [27]. Unexpectedly, a cross-section through the flow (in the regime considered by Lorenz [27]) was not invertible, that is, was not a diffeomorphism (as all previous surface-of-section studies in the spirit of Poincaré had been concerned with [17]), but belonged to a somewhat more complicated kind of map [28,29]. It is an endomorphism (that is, a non-invertible map) because there is one line in the map which has two images (two points)--or, more precisely, no image at all. The reason is, simply, that the recurrent portion of the flow (the 'chaotic convolute') contains a saddle-point in 3-space. So there is one sheet, the 2-dimensional stable manifold of that saddle in 3-space, which after hitting the cross-section once, never returns.

The behavior of such abstract maps was investigated by Guckenheimer [28] and Williams [30]. Most invariant lines in the map are (multiply) cut. However, by artificially transplanting a continuous bridge between the two nonunique points (in fact, after having excised them along with an appropriate neighborhood, so that the bridge can be diffeomorphic), again a homoclinic point between (non-cut) invariant manifolds can be found.

The Lorenz equation was arrived at empirically through approximating a turbulent Navier-Stokes equation by its Fourier modes and then considering only the first three of the infinite set of ordinary differential equations obtained [27]. A rest of the 'weather-like' behavior of the original equation turned out to be preserved, much to the astonishment of the author. Similarly behaving, but mostly somewhat more complicated, equations were later found independently in different contexts (a hydromechanical gadget [31] and a two-dynamo model [32]).

The Cartwright-Littlewood [24] and the Lorenz [27] example together suggest that many more 3-variable differential equations might exist which also produce chaos ('continuous chaos'). Therefore, an independent, synthetic approach toward building a 'zoo' of such systems appears possible. A zoo bears the promise that for some of its members a simple proof of their chaotic nature may be possible. Moreover, emergence of a comprehensive view (classification) may be facilitated.

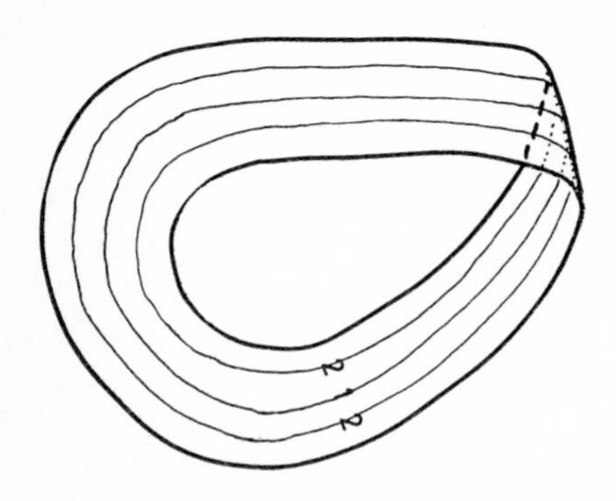

Fig. 4 Two-dimensional Möbius strip with lines. 'Nondivergent case' (see text). 1,2 = internal lines.

3. Some Origami

Origami is the Japanese art of paper folding, yielding paper airplanes, cats' heads, etc. A simple origami device of the last century is the well-known Möbius strip. It is tempting to consider similar paper games in which not just surfaces, but also lines play a role. Lines can be drawn on a surface by using a pencil. (In the preceding section, invariant surfaces were introduced into a universe of hair lines. Here now, the opposite takes place.) An example is depicted in Fig. 4: A Möbius strip with parallel lines on it generates a family of figure 8-shaped closed lines all surrounding one which is 0-shaped.

If we do the same thing with a Möbius strip on which the lines diverge between one end and the other, before the two ends are glued together with a twist, only the 0-shaped trajectory remains closed while the 'eights' are all open, one running into the next. That is, we have an ordinary unstable limit cycle which is asymptotic to transients of double periodicity.

The game is not finished. Fig. 5 show two further strips, one orientable, the other non-orientable again. They have the same behavior as the corresponding non-wrapped analogues (Fig. 4 and the same strip non-twisted, respectively). If now again an amplification of lateral distances along the strip is introduced, the systems of Fig. 6 are obtained. They still look deceptively simple, but there is a new feature now: the gluing is not just end-to-end, but rather involves a partial overlapping. (Running the machines backwards is therefore no longer possible in

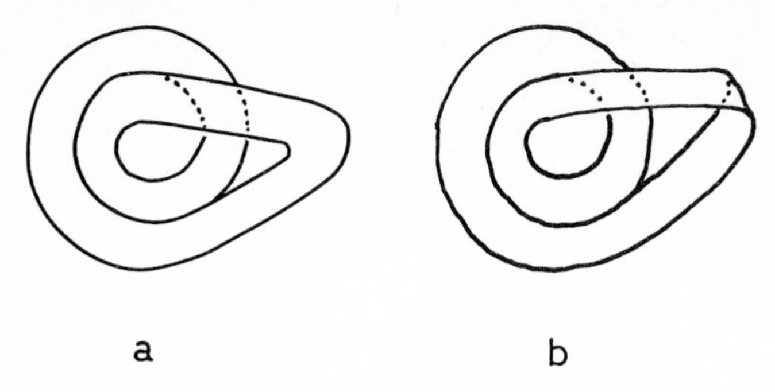

Fig. 5 Once wrapped-up normal strip (left) and once wrapped-up Möbius strip (right). Nondivergent cases. Compare with Fig. 4

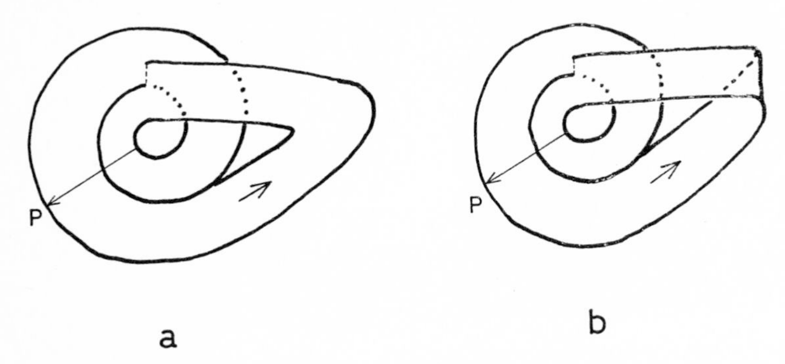

Fig. 6 'Divergent' analogues to the paper flows of Fig. 5. P = Poincaré cross-section

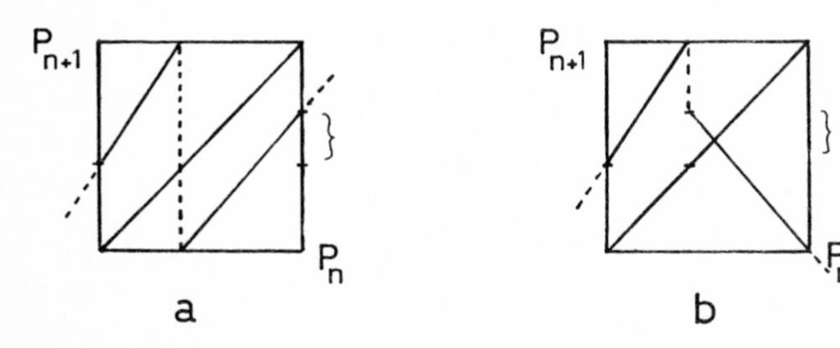

Fig. 7 Poincaré maps applying to the two flows of Fig. 6. The brackets mark the sizes of the glued-together zones. The dashed elongations of the maps obtain when the paper strips in Fig. 6 are broadened (see text).

a unique way.) This minor difference is, apparently, the reason why now we no longer find only one closed curve of repelling type, as in the end-to-end cases above, but rather an infinite number. Also, internal lines no longer spill over the edges eventually. On the contrary: lines lying originally outside the regions depicted in the Fig. (think of the strips in Fig. 6 as being broadened by some added lateral material) are sucked into them--and this despite the amplification of all lateral distances (hyperbolicity) inside.

Thus we have an attracting chaotic regime in either case. This is confirmed by Fig. 7. Here the trick is, once more, to look at a cross-section. The two quadratic boxes correspond to the paper flows drawn in Fig. 6. The dotted elongations outside correspond to the added lateral width; they carry into the quadratic boxes under iteration. Also, there is again a countable number of unstable periodic orbits (fixed points in higher and higher--for example, graphically obtained--iterates). Thus, the technique sketched in the Introduction can be applied again.

Evidently, there is a one-to-one relationship between the class of one-dimensional endomorphisms and the class of 'origamis with lines on them.' Eq. (1), for example, is realized by the flow of Fig. 8. A first example of a non-trivial flow on a 'branched manifold' was considered by Williams [33]; it is related to the system of Fig. 6b.

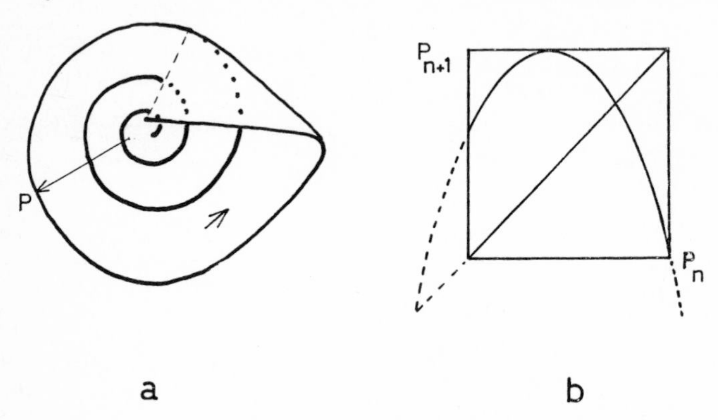

Fig. 8 Paper flow (left) possessing a single-maximum smooth map (right) as a cross-section. The map depicted in b corresponds to Eq. (1) with $\alpha \simeq 3.6$

At this point, two avenues of possible further investigation open up. One is conceiving not only of transparent strips but also of transparent 'jelly' (as a kind of 3-dimensional paper). By drawing lines on it (or rather, into it) and then gluing ends together with partial submersion (that is, overlap), 'higher' forms of chaotic behavior may be found. Such 'paper' is available in the form of computers with 3-dimensional graphic display (see below, Fig. 17). The other avenue consists in taking advantage of the paper flows directly--by 'translating' them into differential equations.

4. The Reinjection Principle

In Fig. 9a, a simple 3-dimensional flow is depicted. There is an unstable spiral 'downstairs' on a Z-shaped slow manifold. The same spiral, only somewhat displaced in a direction parallel to the edges of the slow manifold, applies 'upstairs.' As a result, a 'reinjection loop' is formed which, depending on the geometry of the arrangement, may have the form shown in the figure. In Fig. 9b, we see a paper model (of the kind considered in the preceding section) that fits right into the flow of Fig. 9a.

The flow of Fig. 9a and the paper generalization (Fig. 9b) were described in [34] as an implementation of a 'soft watch.' A simple differential equation of the system of Fig. 9a is

$$\begin{aligned} \dot{x} &= -y-z \\ \dot{y} &= x+0.15\,y \\ \varepsilon\dot{z} &= (1-z^2)(x-1+z)-\delta z. \end{aligned} \tag{2}$$

In the limit of $\varepsilon \to 0$, the equation realizes the corresponding paper flow strictly: the trajectories then are non-invertible. (If in addition $\delta \to 0$ also, the limiting 2-variable flow becomes 'piecewise linear with hysteresis' [35], so that its

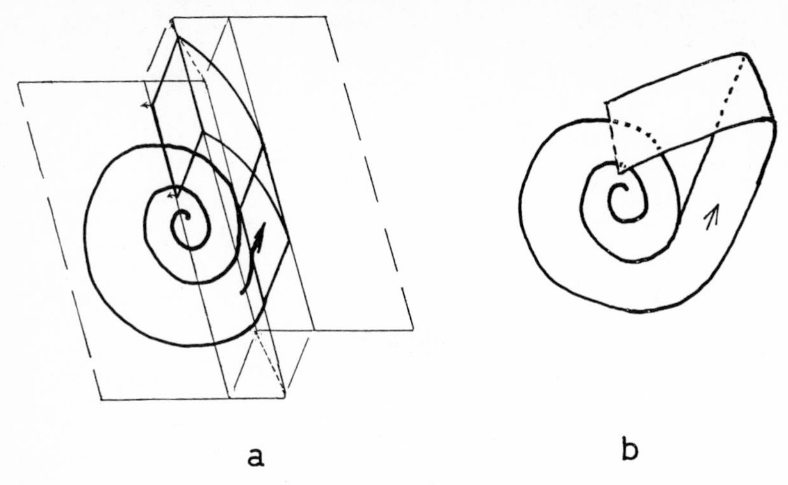

Fig. 9 An abstract dynamical system of relaxation type (a) and corresponding paper flow (b)

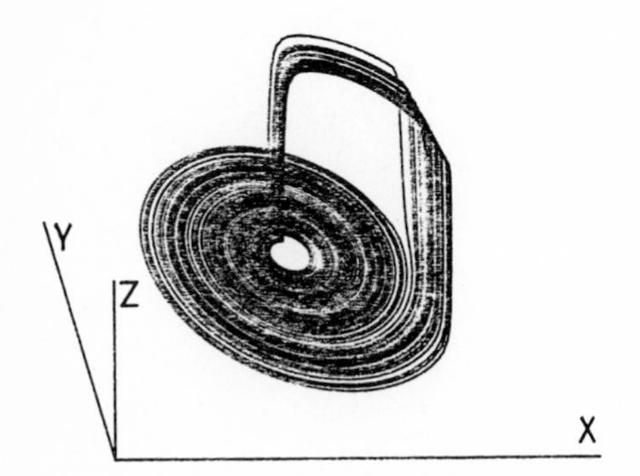

Fig. 10 Continuous chaos in Eq. (2) Compare Fig. 9a.
Numerical siumlation on an HP 9845A desk computer with peripherals, using a standard Runge-Kutta-Merson integration routine. Parameters: $\varepsilon = \delta = 0.03$. Initial conditions: $x(0) = 1, y(0) = 0, z(0) = -1$; $t_{end} = 700$. Axes: -4 . . . 4 for x, -3 . . . 4 for y, -1 . . . 1 for z

cross-section can in principle be calculated analytically.) A computer simulation is shown in Fig. 10: the equation functions as expected. However, as the caption shows, ε is not zero but 0.03, that is, quite finite. A cross-section through the simulated flow therefore cannot have the form of a one-dimensional non-invertible map (endomorphism), but must have the form of a 2-dimensional invertible map (diffeomorphism). There is no reason to assume it non-diffeomorphic, for there is no singularity involved in the recurrent portion of the flow.

Whereas the ideal paper model (Fig. 9b or 9a) has a 'hole' between the non-reinjected and the reinjected portions of the flow, this hole is covered in the simulated 'non-ideal' equation (Fig. 10). This behavior follows from the 'continuous dependence on initial conditions' theorem on ordinary differential equations [15], or, more simply, from the fact that in invertible flows there are no gaps between hairs.

Having thus slipped into the class of non-degenerate differential equations, a simulation of a further simplification of Eq. (2) may be added (Fig. 11). On the left, the simulation in 3-space is provided, on the right, a corresponding paper model, but with 'finite width' (a 'flint-stone model,' so to speak). A cross-section presumably looks like a blown-up 2-dimensional version of the one-dimensional 'hairpin map' of Fig. 8b: Fig. 12.

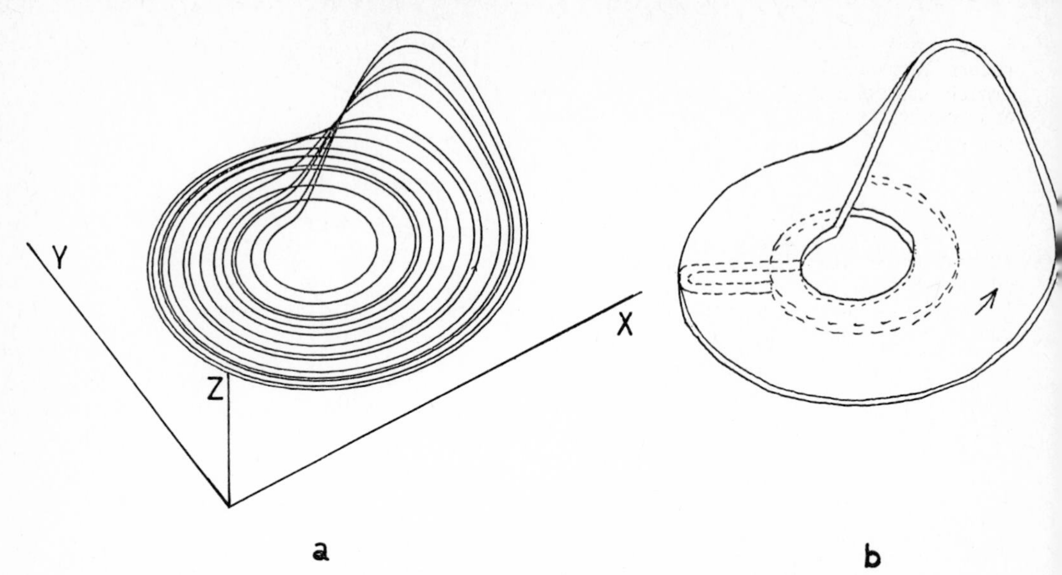

Fig. 11 a: Numerical simulation of an equation analogous to Eq. (2) but 'reduced' (replace the third line in Eq. (2) by $\dot{z} = 0.2 + z(x - 10)$[63]). Initial conditions: $x(0) = 10$, $y(0) = 1$, $z(0) = 0$; $t_{end} = 116$. Axes: -20 . . . 20 for x, -20 . . . 15 for y, 0 . . . 15 for z
b: Corresponding 'thick paper' model (exaggerated)

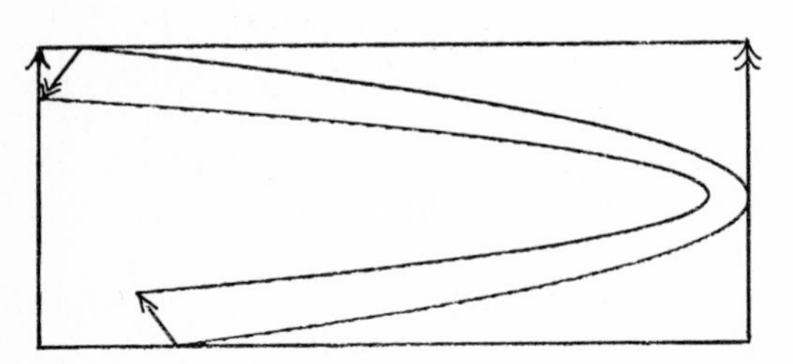

Fig. 12 Walking-stick map. Compare Fig. 11b. The arrows serve to facilitate identification of original (rectangular box) and image (inset), respectively.

Eq. (2) has been complemented recently by a somewhat more complicated analogue in which the pertinent single-variable difference equation applying in the limit $\varepsilon \to 0$ can be calculated more easily [36]. Moreover, the resulting one-dimensional map is differentiable. Thus the transition proposed above (hairpin → walking-stick [29]) should proceed smoothly in this example.

5. More Complicated Kinds of Three-Variable Continuous Chaos

Further paper flows and their corresponding differential equations describing other types of 3-dimensional chaos (screw type, Lorenz type, anti-Lorenz type) are possible [29,35,37,38]. The possibility of realizing a Lorenz type flow in a differential equation with algebraic constraints ('implicit differential equation') was also seen by Takens [39]. In each case, chaos provably exists in the limit $\varepsilon \to 0$.

The difference to the above example is never too great: instead of having a 'direct' return loop downstairs, the loop first goes to the right and then makes a U-turn (which inverts ends); or, instead of using the same 2-dimensional system upstairs and downstairs, two different dynamics are assumed; or, there is more than one singularity (like a focus plus a saddle) involved on one side; etc.

Interestingly, many of these flows (as well as the underlying paper models) can be shown to be equivalent to a second set which derives from a (partly folded over and then glued together) torus in the first place. For example, there is a direct connection between Eq. (2) and the 3-dimensional autonomous equivalent of a periodically forced van der Pol oscillator [26]. The main difference is that the 'cliffs' are rounded rather than straight, corresponding to the presence of a few additional (unnecessary) non-linear terms in the equation.

Some of the more exotic 'animals' in the zoo certainly are still missing. For a preliminary classification, see [38]. When attention is confined to the one-dimensional limiting cross-sections, only 'folded' and 'cut' maps remain.

All of the above-mentioned systems tested so far have non-singular analogues (compare Eq. (2)). The related 2-dimensional cross-sections which presumably apply in a neighborhood of the one-dimensional ones, at nonsingular values of the stiffnes parameter ε, have accordingly been classified into 'walking-stick' (that is, folded) and 'sandwich' (that is, cut) maps [29]. Empirically, these maps apply not only in an asymptotic, but in a rather large neighborhood.

Furthermore, the same maps apparently apply to chaotic systems which do not have a readily formulated singular analogue. The flow of Fig. 11 is obviously closely related to that of Fig. 10. But the case of chaos in a simple 3-variable quadratic mass action system [40], for example, is not so easily linked to a singular prototype. The same holds true for the simplest equation producing 'toroidal chaos' [38,35]; or for chaos in the periodically forced FitzHugh equation [41]; or for the different types of Lorenzian [27] and non-Lorenzian chaos (O.E.Landford, cited in [42]; [43,44]) in the Lorenz equation and an analogous simpler equation [29,38]; etc. Nonetheless, a 'relatedness' to one of the singular equations can be construed in all of these cases [35].

A first exception to this rule is chaos in a simple 3-variable control system [45]. This flow requires '3-dimensional paper' for its explanation, with the gluing-together being end-to-end and smooth. The example thus corresponds to a class of paper flows standing in between the preceding class and that opened up by 3-dimensional paper with reinjection (see below). Nonetheless, the cross-section is of a familiar (walking-stick) type.

Proof techniques for the presence of a walking-stick or sandwich map in an arbitrary (far from singular) 3-variable differential equation producing chaos have yet to be developed. The situation is not hopeless because not all details of a (say) folded-over cross-section have to be shown to be present. A few estimates on portions of the map may be sufficient [46].

In order to understand better what happens in a neighborhood of the non-invertible singular case, and perhaps also farther away from it, it is desirable to have on hand an explicit example of a map which belongs to one of the abstract classes postulated.

6. An Explicit Walking-Stick Map

Fig. 12 of the preceding section was actually generated by computer. It corresponds to the following 2-variable difference equation:

$$\begin{aligned} x_{t+1} &= 3.8\, x_t(1-x_t) - 0.1\, y_t \\ y_{t+1} &= \varepsilon\,(y_t - 1.2)(1 - 1.9\, x_t). \end{aligned} \qquad (3)$$

For $\varepsilon \to 0$, Eq. (3) becomes identical to Eq. (1) after one time-step. Yet for any nonzero value of ε, however small, all of the qualitative features of Figs. 1 - 3 are present.

The map of Fig. 12 is a diffeomorphism (in the initial frame shown in the figure), as is verified by the non-vanishing of the Jacobian determinant in that region (and by the fact that other constraints implying non-uniqueness [22] are absent). A simulation (first iterate of the frame--as in Fig. 12--plus 1st - 4000th iterate of an arbitrarily picked initial point) is shown in Fig. 13a; Fig. 13b shows the second iterate of the original domain. The only critical point of the map as drawn (at $x = 0.73369803377 \ldots$, $y = 0.087658364728 \ldots$) is also displayed. In Fig. 13c, the stable and unstable manifolds of this point, a saddle point, have been entered together with two points H, homoclinic points.

Eq. (3) thus provides an example to the abstract Figs. 1 - 3 of the first section. A similarly simple diffeomorphism (with only one quadratic term) producing chaos was recently given by Hénon [47], cf. [48,49]. Hénon's map can be interpreted as a single-variable difference equation involving two time-steps. It is in fact contained in the present map as a singular special case (replace the parameter +1 in front of the y_t term in the second line of Eq. (3) by zero). The present map has the asset that the folding involved is of the same (paste-layering) type as postulated for the simple continuous system of the preceding section. Eq. (3) thus provides a model of what actually happens in Eq. (2) and related differential equations in their own singular limits.

As a side-remark, the 'logistic difference equation' contained in the first line of Eq. (3) can be replaced by any other smooth single-maximum function. Eq. (3) therefore stands for a whole class of walking-stick maps. For example, a function which is actually found in an equation analogous to Eq. (2) in its own singular limit may be entered. Under certain favorable conditions concerning the nature of the overlap in the differential equation (the outer layer may not be folded-over much more strongly than the inner one), the resulting 2-dimensional map may even become a quantitative model over a certain finite range of the stiffness parameter ε. Another interesting property of Eq. (3) is that it possesses not one but two ordinary horseshoe maps in the sense of Smale [17] in its second iterate, [44,38]. Each uses the other for its sink--Smale [17] assumed a point attractor instead--so that together they form a ('non-minimal' [35]) attractor. This composed attractor may be called the 'dollar attractor' because of the characteristic intersection pattern of the two stable and the two unstable manifolds involved:

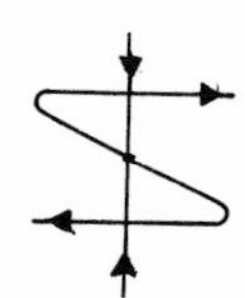

The 'dollar' structure just mentioned at the same time constitutes the (about) simplest example of a 'cycle' in the sense of Smale [17] and Newhouse [50]. Providing an explicit example, Eq. (3) thus illustrates in a nutshell the modern problem of Ω-instability in diffeomorphisms and flows (see [51 - 53] for monographs on the subject). By playing around with the equation's parameters (reducing the constant 1.2), the 'gap' between the two layers of the image can be markedly reduced, thus probably allowing formation of a 'thick' Cantor set in the sense of Newhouse [50], or rather, of two of them. If so, the equation would show 'structural instability in a structurally stable way' (namely, in an open dense set in parameter space [50]) and also an infinite number of attracting periodic solutions [54].

The cycle itself (without thickness assumption) already implies the existence of an uncountable number of structurally different flows close to the given one [55]. A bit of the complexity of the situation is illustrated in Fig. 14: a kind of 'supergrid' has now formed in the neighborhood of the saddle. All of the 'long' windings

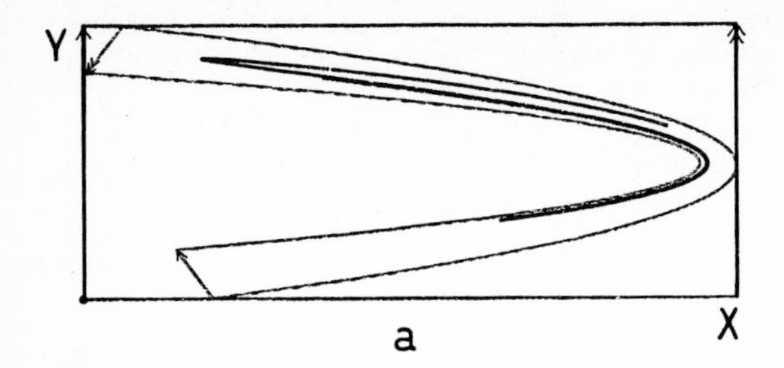

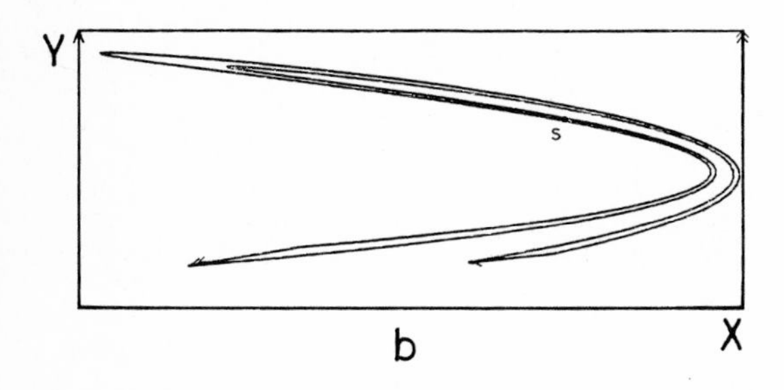

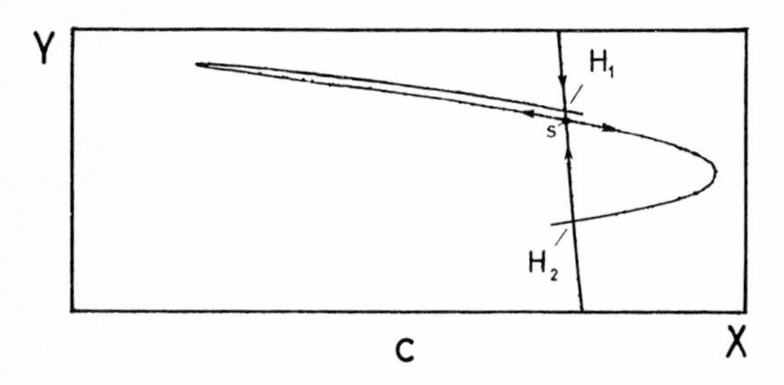

Fig. 13 Numerical computation of Eq. (3) at $\varepsilon = 0.2$.
Axes: 0.06 . . . 0.976 for x and -0.26 . . . 0.25 for y

a: Chaotic attractor inside the walking-stick map: 4000 iterates of the lower left corner point (11-digit accuracy)

b: Second iterate of the box. s = fixed point of the box

c: Numerically computed stable and unstable manifolds of the saddle point s in the map. H_1, H_2 = homoclinic points. Compare Fig. 1a

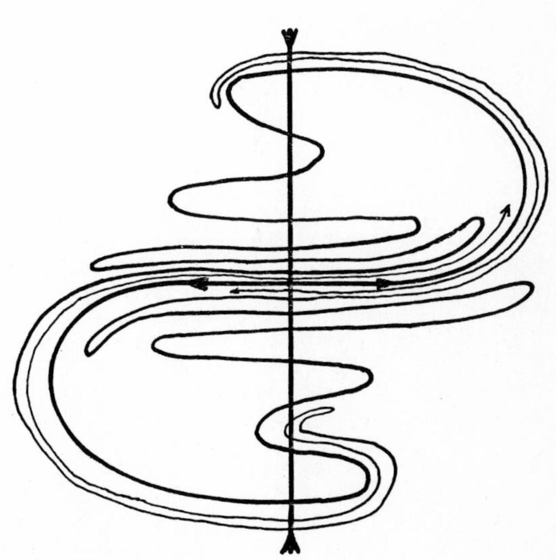

Fig. 14 Cycle formed in Fig. 13c. Non-quantitative picture. Compare Fig. 1b

of one unstable manifold (as it 'doubles up' on the saddle's two stable manifolds) are wrapped around the other (and in between its layers), in order to return then along with the latter's long windings, and so on _ad infinitum_.

If ε in Eq. (2) is reduced, this structure is maintained. Only at the singular limit ($\varepsilon = 0$) does it disappear. Thus, the simple 'gluing together' mechanism of Figs. 6 and 8 is not invalidated. It merely acquires a new dimension.

7. Hyperchaos

Concerning flows based on '3-dimensional paper with reinjection,' one straightforward possibility is illustrated in Fig. 15. There is an expanding 3-dimensional flow in the form of a solid screw. The flow is, upon reaching a certain plane threshold, reinjected toward another plane inside the screw. Note that on the way back, two folding-overs occur, one along the axis of the screw (between windings), the other in a perpendicular direction (within windings). A cross-section therefore will look like a doubly-folded towel in its first iterate [35]; compare Fig. 16 for an explicit non-degenerate example [56].

In analogy with the construction of Eq. (2), perhaps a (this time 3-variable) linear system plus switching variable can be found as a straightforward realization of the paper flow [57]. The following equation [56] is more simple (although its cross-section involves a few additional foldings):

$$\begin{aligned} \dot{x} &= -y - z \\ \dot{y} &= x + 0.25\,y + w \\ \dot{z} &= 2.2 + xz \\ \dot{w} &= -0.5\,z + 0.05\,w. \end{aligned} \tag{4}$$

Eq. (4) contains only one nonlinear (quadratic) term. A simulation is presented in Fig. 17. The bottom line shows the excitable ('switching') variable z as a function of t. It is activated only from time to time, namely, whenever the linear 'main flow' (in x, y, w-space, presented stereoscopically on top) exceeds z's threshold (at $x \approx 0$) from below. During the brief periods of activation of z, a 'reinjection event' takes place in the main flow. The corresponding trajectorial segments (mediating between more or less linear segments) have been dashed. The bottom part of the figure has been simulated for a much longer time. Note the 'two time scales of chaos' visible in the bottom line.

Because Eq. (4) is so simple, analogous higher-dimensional systems, also comprising only one nonlinear term but producing chaos with even more independent directions of hyperbolic instability, may be possible as well. Turbulence in partial differential equations (cf. [58]) probably involves not only the lowest level of the present hierarchy (ordinary chaos), but also its higher forms. This feature may be constitutive for 'turbulence.' When analyzing the power spectra of systems like those presented in Figs. 16 and 17, but with more and more folded-over dimensions implied, a tendency toward a 1/f structure--as characteristic of natural turbulence (see [59,60])--may be found. In chaotic 3-dimensional systems (and one-dimensional endomorphisms, respectively), such a spectrum arises only as an exception (see [8]). Another possible means of distinguishing experimentally between lower and higher forms of chaos is the 'sonogram' (intensity plotted as a function of frequency (ordinate) and time (abscissa)) or its generalization, the 'sonochrome' (which in addition displays phase [61]). A tape with the sound of hyperchaos--z(t) in Eq. (4)--has been prepared. The hoarse, 'snoring' tone of chaos [29] has given way to an even more husky (and unsettling) tonal picture.

Abstractly speaking, 'hyperchaos' as described here is related to the concept of a 'basic set with 2-dimensional unstable manifold' [62]. These sets are a generalization of the sets formed by Poincaré's intersection between one-dimensional manifolds [17]. When looking at the second iterate in the map of Fig. 16, a 'hyper-

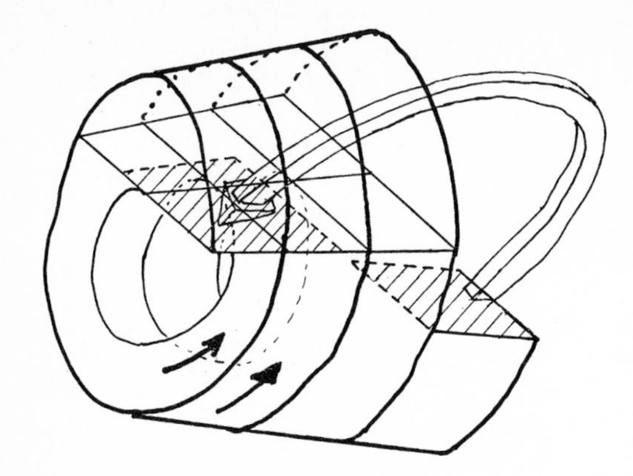

Fig. 15 Nontrivial reinjection between different portions of a 3-dimensional flow. Example of a generalized ('three-dimensional') paper model.

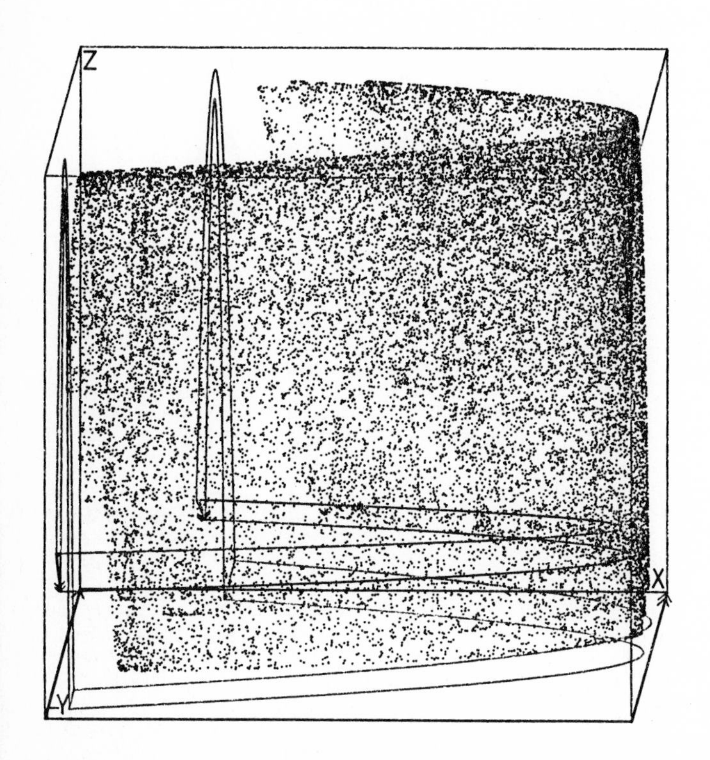

Fig. 16 Folded towel map. Numerical computation of a three-dimensional analogue of the walking-stick diffeomorphism (Eq. (3)): $x_{t+1} = 3.8x_t(1-x_t) - 0.05(y_t + 0.35)\cdot(1-2z_t)$, $y_{t+1} = 0.1((y_t + 0.35)(1-2z_t) - 1)(1-1.9x_t)$, $z_{t+1} = 3.78\,z_t(1-z_t) + 0.2y_t$. The first iterate of the brick-shaped original domain is shown together with 25 000 iterates of the front lower left corner point (hyperchaotic attractor). Note that the image of the lower front rectangle (bolder, with arrows) is lying innermost within the 'folded towel.' Axes: 0.085 . . . 0.971 for x, -0.121 . . . 0.121 for y, 0.075 . . . 0.973 for z.

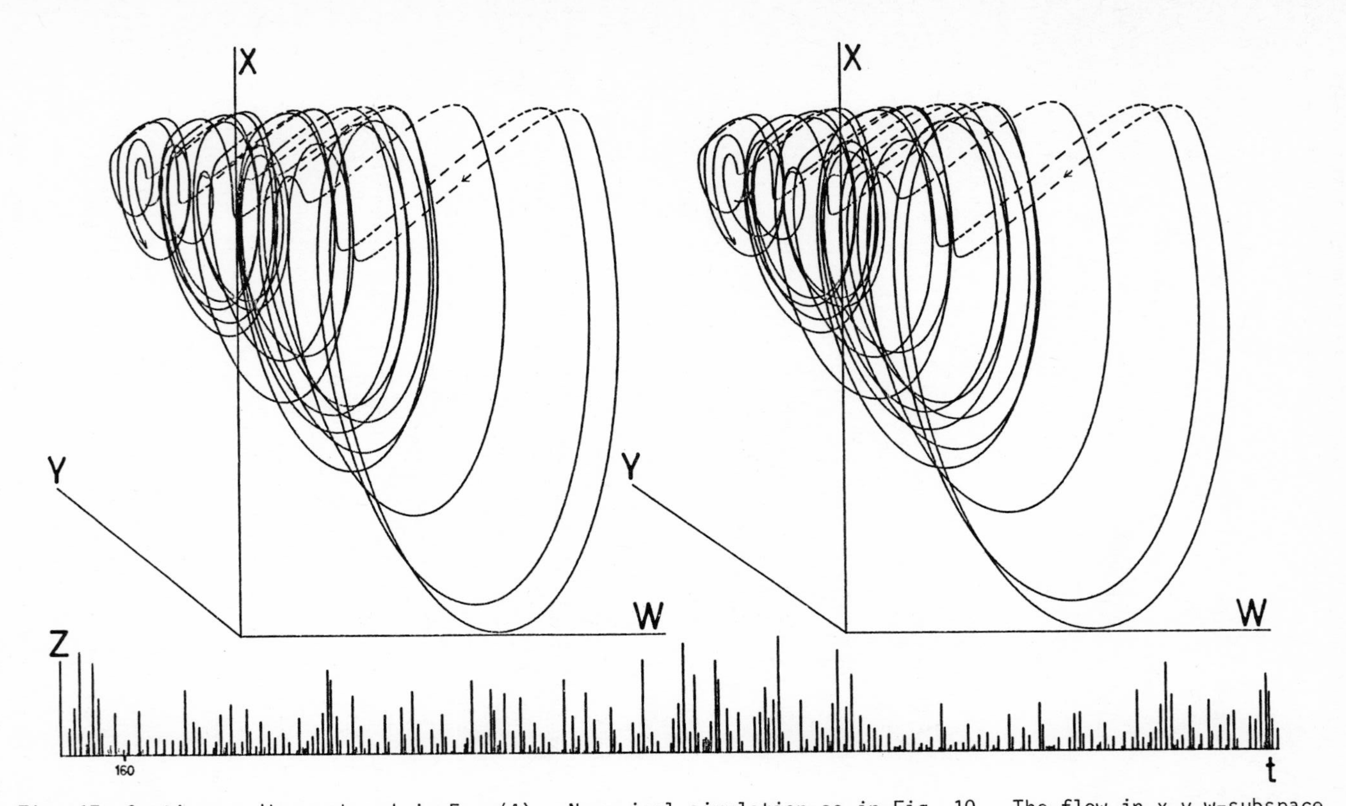

Fig. 17 Continuous 'hyperchaos' in Eq. (4). Numerical simulation as in Fig. 10. The flow in x,y,w-subspace is shown in two different parallel projections (stereoplot). Compare Fig. 15. The ('spiking') time behavior of the fourth variable (z) is shown underneath. Dashing (in the upper picture) occurs when z exceeds a value of 20 (visibility threshold in the lower picture). Initial conditions: $x(0) = -19$, $y(0) = z(0) = 0$, $w(0) = 15$; $t_{end} = 160$ (upper picture) and 3020 (lower picture). Axes: -110 . . . 40 for x, -60 . . . 60 for y, 0 . . . 400 for z, 15 . . . 70 for w. A stereoscopic impression can be obtained by fixing the tip of a pencil held about 20 cm before one's eyes in front of the picture, so that the two pictures behind are aligned. The distance can then be varied till of the 4 blurred pictures behind, the 2 innermost ones merge. They will be in focus soon.

horseshoe'(as sketched in [26, Fig. 2]) is found. Its qualitative implications have yet to be derived.

Since playing with 3-dimensional paper strips is only at its beginning, it was perhaps premature to attach the label 'hyperchaos' to the first phenomenon encountered.

8. Threads vs. Domains

Why do 3-dimensional flows, with precisely one hair-line running through every regular point in 3-space, produce a type of behavior so different from the familiar pictures in two dimensions?

One possible clue lies in an emancipation from the concept that trajectories are the most important natural entities in flows. In one dimension, trajectories are all there is. In two dimensions, they play the role of a powerful indicator of everything that goes on. In three dimensions, they apparently lose their grip.

In two dimensions as well as higher dimensions, flows can be said to consist of a streaming of whole domains, with suspended 'particles' (and hence trajectories) only playing the role of markers. In this modest role, particles nonetheless do remarkably well in two dimensions--because they enclose domains. This 'dual' relationship between domains and threads breaks down in three dimensions.

One may think of saving the familiar picture by instating invariant 2-dimensional manifolds into the domain-bounder's role in three dimensions. Poincaré attempted doing this and failed, that is, he discovered 'homoclinicity to' (implying an infinite splintering of invariant domains) instead. One reason is that nothing prevents a 'criss-crossing' in three dimensions: a portion of flow that was horizontally expanding throughout at first may come into a region where there is vertical expansion throughout, and so on. What actually is going around in loops, to be twisted and whirled, is not threads but whole domains (or areas, respectively, when speaking in terms of a cross-section).

This view certainly is one-sided. It may nonetheless be used to explain why area-oriented concepts--like the 'paste-bound' interpretations of the maps of Figs. 12 and 16 advocated in [37,35]--demonstrate a highly robust (though infinitely extended and infinitely fine-layered) attractor. This stability (and the whole attractor) disappears when substructures (manifolds and threads) are made the objects of one's focus. But the stability of a forest also belongs to a different level than that of its trees.

9. Epilogue

There is one more aspect to a large 'zoo': its inhabitants tend to have many relatives in the outside world.

An irregularly dripping faucet, a nonperiodic pulsar, the measured trajectories of a well-stirred (or non-stirred) reaction system, the spike trains of a nerve cell (or netlet), the ripples in a brook, most natural sounds (not to mention the weather, the brain, the economy, or the trajectory of collective opinion in a mass of scientists): all these phenomena may, under careful scrutiny, turn out to be governed by equations which have relatives in the zoo described above.

In other words, chaos is in danger of becoming fashionable in the applied sciences --as a so-called paradigm. Most natural systems have more than two variables and are not quite linear. And chaos has beautiful pictures. But the rigorous standards of playfulness should be preserved.

Chaos is a reflection of the built-in tendency of matter to get entangled. Hylogenesis, morphogenesis, biogenesis, noogenesis all reflect the same nontrivial synergistic tendency. Chaos is its crystal.

Summary

Poincaré's postulate of a 'homoclinic point' in the three-body problem of celestial mechanics opened up the possibility of arbitrarily complicated motions in simple deterministic systems. Julia, and later Cartwright and Littlewood and Lorenz, provided the first down-to-earth examples. A synthetic approach is possible using 'paper models with lines on them.' The underlying 'reinjection principle' leads to two main classes in three dimensions, 'folded' and 'cut' chaotic flows. There exists a whole zoo of simple 3-variable systems. In four dimensions, scarcely the first steps have been made. Several explicit maps are available to check qualitative concepts. 'Higher' types of chaos may play a role in turbulence.

References

1. Li, T.Y. and J. Yorke (1975). Period three implies chaos. Amer. Math. Monthly 82, 985-992

2. May, R.M. (1974). Biological populations with nonoverlapping generations: Stable points, limit cycles, and chaos. Science 186, 645-647

3. Hoppensteadt, F.C. and J.M. Hyman (1977). Periodic solutions of a logistic difference equation. SIAM J. Appl. Math. 32, 73-81

4. Julia, G. (1918). Mémoire sur l'itération des fonctions rationelles. J. Math. Pur. Appl., Série 8, 1, 1-18

5. Smale, S. and R.F. Williams (1976). The qualitative analysis of a difference equation of population growth. J. Math. Biol. 3, 1-5

6. Myrberg, P.J. (1963). Iteration of the real polynomials of second degree III (in German). Ann. Acad. Sci. Fenn. Ser. A, 336/3, 1-18

7. Gumowski, I. and C. Mira (1969). Sensitivity problems related to certain bifurcations in the nonlinear recurrence relations. Automatica, the Journal of IFAC 5, 303-317

8. Grossmann, S. and S. Thomae (1977). Invariant distribution and stationary correlation functions of one-dimensional discrete processes. Z. Naturforsch. 32a, 1353-1363

9. Sharkovsky, A.N. (1964). Coexistence of the cycles of a continuous mapping of the line into itself. Ukrain. Mat. Z. 16, 61-71

10. Guckenheimer, J. (1979). Dynamical systems. In: Lectures in Applied Mathematics, Vol. 17 (F.C. Hoppensteadt, ed.), Providence, R.I.: Amer. Math. Soc.

11. Hill, G.W. (1878). Researches in the lunar theory. Amer. J. Math. 1, 5-26

12. Poincaré, H. (1890). Sur le problème des trois corps et les équations de la dynamique. Acta Math. 13, 1-271

13. Poincaré, H. (1899). Les Méthodes Nouvelles de la Mécanique Céleste, Vol. III, Chapter 27; p. 387. Paris: Gauthier-Villars

14. Birkhoff, G.D. (1920). Recent advances in dynamics. Science 51, 51-55

15. Hartman, P. (1964). Ordinary Differential Equations. New York: Wiley

16. Birkhoff, G.D. (1927). On the periodic motions of dynamical systems. Acta Math. 50, 359-379

17. Smale, S. (1967). Differentiable dynamical systems. Bull. Amer. Math. Soc. 73, 747-817

18. Hadamard, J. (1898). Les surfaces à curbures opposés et leurs lignes géodésiques. Journ. de Math. (5)4, 27-73

19. Anosov, D.V. and J.G. Sinai (1967). Some smooth ergodic systems. Uspekhi Mat. Nauk 22, 107-172

20. Arnold, V.I. and A. Avez (1967). Théorie Ergodique des Systèmes Dynamiques. Paris: Gauthier-Villars

21. Moser, J. (1973). Stable and Random Motions in Dynamical Systems. Princeton, N.J.: Princeton Univ. Press

22. Gumowski, I. (1979). Monograph on Point Recurrences (forthcoming)

23. Levinson, N. (1949). A second order differential equation with singular solutions. Ann. Math. 50, 127-153

24. Cartwright, M.L. and J.E. Littlewood (1945). On nonlinear differential equations of the second order: I. The equation $\ddot{y} - k(1-y^2)y + y = b\lambda k \cos(\lambda t + \alpha)$, k large. J. Lond. Math. Soc. 20, 180-189

25. Rössler, O.E. (1977). Quasiperiodic oscillation in an abstract reaction system. Biophysical J. 17, 281a (abstract)

26. Rössler, O.E. (1979). Chaotic oscillations: An example of hyperchaos. In: Lectures in Applied Mathematics, Vol. 17 (F.C. Hoppensteadt, ed.), Providence, R.I.: Amer. Math. Soc.

27. Lorenz, E.N. (1963). Deterministic nonperiodic flow. J. Atmos. Sci. 20, 130-141

28. Guckenheimer, J. (1976). A strange, strange attractor. In: The Hopf Bifurcation and Its Applications (J.E. Marsden and M. McCracken, eds.), pp. 368-381, New York: Springer-Verlag

29. Rössler, O.E. (1976). Different types of chaos in two simple differential equations. Z. Naturforsch. 31a, 1664-1670

30. Williams, R.F. (1978). The bifurcation space of the Lorenz attractor. In: Bifurcation Theory and Applications in Scientific Disciplines (O. Gurel and O.E. Rössler, eds.), Proc. N.Y. Acad. Sci. 316

31. Moore, D.W. and E.A. Spiegel (1966). A thermally excited nonlinear oscillator. Astrophys. J. 143, 871-887

32. Cook, A.F. and P.H. Roberts (1970). The Rikitake two-disc dynamo system. Proc. Camb. Phil. Soc. 68, 547-569

33. Williams, R.F. (1974). Expanding attractors. Public. Math. de l'Institut des Hautes Etudes Scientifiques 43, 169-203

34. Rössler. O.E. (1976). Chaotic behavior in simple reaction systems. Z. Naturforsch. 31a, 259-264

35. Rössler, O.E. (1978). Continuous chaos--four prototype equations. In: Bifurcation Theory and Applications in Scientific Disciplines (O. Gurel and O.E. Rössler, eds.), Proc. N.Y. Acad. Sci. 316, 376-394

36. Mira, C. (1978). Dynamique complexe engendrée par une équation différentielle d'ordre 3. Proceedings "Equadiff 78" (R. Conti, G. Sestini, and G. Villari, eds.), Florence, May 24 - 30, 1978, pp. 25-37

37. Rössler, O.E. (1977). Chaos in abstract kinetics: Two prototypes. Bull. Math. Biol. 39, 275-289

38. Rössler, O.E. (1977). Continuous chaos. In: Synergetics--A Workshop (H. Haken, ed.), pp. 184-197. Heidelberg/New York: Springer-Verlag

39. Takens, F. (1976). Implicit differential equations: Some open problems. Springer Lecture Notes in Math. 535, 237-253

40. Rössler, O.E. (1978). Chaotic oscillations in a 3-variable quadratic mass action system. In: Proc. Int'l Symp. Math. Topics in Biology, Kyoto, Sept. 1978, pp. 131-135, Publ. Kyoto Research Institute for Math. Sci.

41. Rössler, O.E., R. Rössler, and H.D. Landahl (1978). Arrhythmia in a periodically forced excitable system. Sixth Int'l Biophysics Congress, Kyoto, Japan. Abstracts Vol. p. 296

42. Marsden, J.E. (1977). Attempts to relate the Navier-Stokes equations to turbulence. In: Turbulence Seminar (A. Chorin, J. Marsden, and S. Smale, orgs.), Springer Lecture Notes in Math. 615, 1-22.

43. Hénon, M. and Y. Pomeau (1976). Two strange attractors with a simple structure. In: Springer Lecture Notes in Math. 565, 29-68

44. Rössler, O.E. (1977). Horseshoe-map chaos in the Lorenz equation. Phys. Lett. 60A, 392-394

45. Rössler, R., F. Götz, and O.E. Rössler (1979). Chaos in endocrinology. Biophys. J. 25(2), 216a

46. Rössler, O.E. (1978). Chaos and strange attractors in chemical kinetics. In: Synergetics--Far from Equilibrium (A. Pacault and C. Vidal, eds.), pp. 107-113, Springer-Verlag

47. Hénon, M. (1976). A two-dimensional mapping with a strange attractor. Comm. Math. Phys. 50, 69-78

48. Curry, J.H. (1979). A homoclinic point in the Hénon map. In: Lectures in Applied Mathematics, Vol. 17 (F.C. Hoppensteadt, ed.), Providence, R.I.: Amer. Math. Soc.

49. Mira, C. and I. Gumowski (1979). In preparation

50. Newhouse, S.E. (1970). Nondensity of Axiom A(a) on S^2. In: Global Analysis, Summer Institute Berkeley 1968 (S. Chern and S. Smale, eds.), Providence, R.I., Amer. Math. Soc: Proc. Symp. Pure Math. 14, 191-202

51. Nitecki, Z. (1971). Differentiable Dynamics. Cambridge, Mass.: M.I.T. Press

52. Chillingworth, D.R.J. (1976). Differential Topology with a View to Applications. Research Notes in Mathematics 9, London: Pitman

53. Abraham, R. and J.E. Marsden (1978). Foundations of Mechanics. 2nd enlarged ed. Reading, Mass.: Benjamin/Cummings

54. Newhouse, S.E. (1974). Diffeomorphisms with infinitely many sinks. Topology 13, 9-18

55. Newhouse, S. and J. Palis (1976). Cycles and bifurcation theory. Astérisque 31, 44-140

56. Rössler, O.E. (1979). An equation for hyperchaos. Submitted to Phys. Lett. A.

57. Rössler, O.E., I. Gumowski and C. Mira (1979). In preparation

58. Kuramoto, Y. (1978). Diffusion-induced chaos in reaction systems. Progr. Theor. Phys. 64, Suppl.

59. Voss, R.F. and J. Clark (1975). "1/F" noise in music and speech. Nature 258, 317-318

60. Thomas, H. (1978). Instabilities and fluctuations in systems far from equilibrium. In: Proc. 5th Int'l Conf. Noise in Physical Systems, Bad Nauheim, March 1978

61. Johannesma, P.I.M. (1979). In preparation

62. Franks, J.M. (1977). The dimensions of basic sets. J. Differential Geometry 12, 435-441

63. Rössler, O.E. (1976). An equation for continuous chaos. Phys. Lett. 57A, 397-398

Index of Contributors

Numbers refer to the first page of each contribution; authors addresses appear on these pages.

D. D. Joseph

Stability of Fluid Motions I

(in 2 parts)

1976. 57 figures. XIII, 282 pages
(Springer Tracts in Natural Philosophy, Volume 27)
ISBN 3-540-07514-3

Contents: Global Stability and Uniqueness. – Instability and Bifurcation. – Poiseuille Flow: The Form of the Disturbance whose Energy Increases Initially at the Largest Value of v. – Friction Factor Response Curves for Flow through Annular Ducts. – Global Stability of Couette Flow Between Rotating Cylinders. – Global Stability of Spiral Couette-Poiseuille Flows. – Global Stability of the Flow Between Concentric Rotating Spheres. – Appendices: Elementary Properties of almost Periodic Functions. – Variational Problems for the Decay Constants and the Stability Limit. – Some Inequalities. – Oscillation Kernels. – Some Aspects of the Theory of Stability of Nearly Parallel Flow.

D. D. Joseph

Stability of Fluid Motions II

(in 2 parts)

1976. 39 figures. XIV, 274 pages
(Springer Tracts in Natural Philosphy, Volume 28)
ISBN 3-540-07516-X

Contents: The Oberbeck-Boussinesq Equations. The Stability of Constant Gradient Solutions of the Oberbeck-Boussinesq Equations. – Global Stability of Constant Temperature-Gradient and Concentration Gradient States of a Motionless Heterogeneous Fluid. – Two-Sided Bifurcation into Convection. – Stability of Supercritical Convection-Wave Number Selection Through Stability. – The Variational Theory of Turbulence Applied to Convection in Porous Materials Heated from below. Stability – Problems for Viscoelastic Fluids. – Interfacial Stability.

"...well written and amply illustrated, with particular attention given to the presentation of good pictures of flow phenomena. ...will constitute a standard work in theoretical fluid dynamics..."
Contemporary Physics

Physics and Mathematics of the Nervous Systems

Proceedings of a Summer School organized by the International Centre for Theoretical Physics, Trieste, and the Institute for Information Sciences, University of Tübingen, held at Trieste, August 21–31, 1973
Editors: M. Conrad, W. Güttinger, M. Dal Cin

1974. 159 figures. XI, 584 pages
(Lecture Notes in Biomathematics, Volume 4)
ISBN 3-540-07014-1

With contributions by H. B. Barlwo, V. Braitenberg, H. Bremermann, M. Conrad, M. Dal Cin, E. Dilger, P. Fatt, G. Falk, J. George, W. Güttinger, H. Hahn, R. Heim, B. D. Josephson, C. R. Legéndy, R. B. Livingston, W. Merzenich, E. G. Meyer, E. Neumann, H. H. Pattee, E. Pfaffelhuber, W. Precht, J. G. Roederer, O. E. Rössler, J. G. Taylor, B. T. Ulrich, R. Vollmar

This important volume represents the record and product of a summer school devoted to exploring major theoretical ideas and methodologies applicable to brain function and related problems. Topics covered include fundamental ideas, methods, and applications of dynamical systems, automata, and information theory; molecular basis of nerve impulse and brain function; biophysics of nerve cells and sensory perception; neural networks, cerebellum and cerebral cortex; artificial intelligence. It is hoped that the coverage, which is unique, will serve a much needed catalytic function in an area which is felt by many workers to be ready for both theoretical and experimental break-throughs.

Springer-Verlag
Berlin
Heidelberg
New York

J. Schnakenberg

Thermodynamic Network Analysis of Biological Systems

Universitext
1977. 13 figures. VIII, 143 pages
ISBN 3-540-08122-4

Contents: Introduction. – Models. – Thermodynamics. – Networks. – Networks for Transport Across Membranes. – Feedback Networks. – Stability.

This book is devoted to the question: what can physics contribute to the analysis of complex systems like those in biology and ecology? It adresses itself not only to physicists but also to biologists, physiologists and engineering scientists. An introduction into thermodynamics particularly of non-equilibrium situations is given in order to provide a suitable basis for a model description of biological and ecological systems. As a comprehensive and elucidating model language bondgraph networks are introduced and applied to quite a lot of examples including membrane transport phenomena, membrane excitation, autocatalytic reaction systems and population interactions. Particular attention is focussed upon stability criteria by which models are categorized with respect to their principle qualitative behaviour. The book intends to serve as a guide for understanding and developing physical models in biology.

Turbulence

Editor: P. Bradshaw
2nd corrected and updated edition. 1978. 47 figures, 4 tables. XI, 339 pages
(Topics in Applied Physics, Volume 12)
ISBN 3-540-08864-4

Contents: *P. Bradshaw:* Introduction. – *H.-H. Fernholz:* External Flows. – J. P. Johnston: Internal Flows. – *P. Bradshaw, J. D. Woods:* Geophysical Turbulence and Buoyant Flows. – *W. C. Reynolds, T. Cebeci:* Calculation of Turbulent Flows. – *B. E. Launder:* Heat and Mass Transport. – *J. L. Lumley:* Two-Phase and Non-Newtonian Flows

There are several books which survey turbulence in depth, but none which adequately treats it in depth as the most important fluid-dynamic phenomenon in engineering and the earth sciences. This book is a unified treatment of most of the turbulence problems of aeronautical, mechanical, and chemical engineering, meteorology and oceanography. Each chapter is written by an expert in one of these disciplines, but emphasizes phenomena rather than hardware details so as to make the material accessible to non-specialists. As well as a descriptions of phenomena, the book contains detailed discussions of methods for calculating turbulent flow fields and heat transfer.

Solitons and Condensed Matter Physics

Proceedings of the Symposium on Nonlinear (Solitons) Structure and Dynamics in Condensed Matter Oxford, England, June 27–29, 1978
Editors: A. R. Bishop, T. Schneider
1978. 120 figures, 4 tables. XI, 341 pages
(Springer Series in Solid-State Sciences, Volume 8)
ISBN 3-540-09138-6

Contents: Introduction. – Mathematical Aspects. Statistical Mechanics and Solid-State Physics. – Summary.

The papers in this volume survey the applications of nonlinear (soliton) mathematics to condensed matter physics. They exhibit the common mathematical structure underlying applications with different physical manifestations and highlight some of the more pressing and universal mathematical problems now facing the nonlinear physicist. The conference was attended by mathematicians and physicists, but the primary emphasis is on physics contexts rather than on mathematical details. Topics considered include: completely integrable systems; topology; singular perturbation theory; molecular dynamics simulations; statistical mechanics and lattice dynamics; nonlinear transport; low-dimensional systems; epitacial registry; Josephson junctions; superfluid ^{3}He; dislocations. This emphysis on applied aspects and its rapid publication will make the coherent review of the present state-of-the art a valuable aid to researchers and graduate students in condensed matter physics and applied mathematics.

Springer-Verlag
Berlin
Heidelberg
New York